OIL CRISIS

by

C. J. CAMPBELL

Multi-Science Publishing Company Ltd

In the litigious society in which we now live, I have to include the following:

The views, descriptions, methods, analysis and comments in this book are presented solely for the purposes of study and review. Any references to persons, corporations or other organizations are fair comments made in good faith carrying no imputations of whatsoever nature. The author assumes no liability whatsoever for any loss or damage arising directly or indirectly from the material contained in this book. By notice hereby conveyed, persons purchasing, reading or using the book, or the material contained therein, are to understand that they do so solely at their own risk.

Reprinted October 2005

ISBN 0906522 39 0
Multi-Science Publishing Co. Ltd
5 Wates Way, Brentwood, Essex, CM15 9TB, UK
2005

Contents

LIST OF ILLUSTRATIONS

Chapter 4

PREFACE

T HE PRECURSOR of this book, *The Coming Oil Crisis*, was written in 1996 at a time when few people outside the oil industry had an inkling that there were any threats to oil supply other perhaps than those arising from political events in the Middle East. It was accordingly ahead of its time. Now, eight years on, the World has witnessed amazing changes, not least those stemming from the events of September 11th 2001, which almost put the World on a war-footing, and had tangible expression with the invasion of Iraq. We now find ourselves with virtually no spare oil capacity, and prices have been soaring in recent months. It seems therefore a good moment to update the arguments and data. In general, the conclusions of *The Coming Oil Crisis* have proved to be not too far wide of the mark, but many points of detail and emphasis have changed, and the general understanding of the subject has improved.

In short, we can say that it is now clear that the First Half of the Age of Oil draws to a close. It has lasted 150 years since the pioneering wells were drilled in Pennsylvania and on the shores of the Caspian. The growing supply of cheap oil-based energy that has marked this epoch led to a rapid expansion of industry, transport, trade and agriculture, allowing the population to expand six-fold, exactly in parallel with oil. It also gave rise to an explosion of financial capital, which led in turn to the subject of Economics, by which to understand and manage money.

Now, the Second Half of the Age of Oil dawns, meaning that the supply of this priceless commodity, which underpins our entire way of life, will decline towards eventual exhaustion, along with all the many other things that depend upon it. It probably heralds the End of Economics as current understood. Some see this as a doomsday message, but others have confidence that human ingenuity will be able to adapt and even find better solutions. The challenge comes in the transition, which promises to be a time of great international tension. Time is short to make the critical changes and adopt new mind-sets to match the new situation. It is hoped that the pages of this new study will provide basic insights that will help people understand their predicament in order to plan and prepare.

The original idea of merely revising *The Coming Oil Crisis* proved to be more difficult than at first anticipated. In particular, Chapter 8, which gives a forecast of future production, had to be rewritten in its entirety to take into account the new geopolitical circumstances following the declaration of the so-called War on Terror by the United States. These matters lie far from the geological controls affecting oil production, on which professional qualifications can be claimed, but sensitive as they are, they have to be addressed in forecasting the future. The official account of events lacks credibility in many important regards forcing the reader to consider what are dismissively termed Conspiracy Theories, which cannot be ignored. On balance, it seemed better to present what appears to be a logical interpretation, even if its inclusion may be taken by detractors as ammunition with which to reject the other findings that rest on more solid scientific foundations.

The preface to *The Coming Oil Crisis* is reproduced below.

Ballydehob, Ireland
1st February 2005

Preface to *The Coming Oil Crisis*

It turns out that writing a book isn't easy. You set forth and try to nurse the thing along, not knowing quite what you are going to say until you say it. No sooner have you said it than you regret what you have said and wonder if you have got it right. Why did I try?

Chance has given me certain experiences and insights which provide compelling evidence that conventional oil production is about to peak. As it becomes scarcer and controlled by fewer hands, it will inevitably become much more expensive. That in turn may herald a major discontinuity in the way the world lives.

This is not a new subject. It was much debated in the 1970s after the oil shocks of that period, but the fears then raised, while valid enough in principle, were not realized because the world was still a long way from the midpoint of the depletion of its oil which more or less corresponds with peak production. Twenty years later, we are now close to midpoint: so, this time the shock will be for real.

It is an immensely important subject which deserves to be more widely understood. That is why I have written this book. I have tried to make it non-technical and readable. I don't want to sound didactic in expressing the views I have but rather to invite debate and discussion. I am therefore especially grateful to those who have contributed the interviews which are contained in the book. They express a different slant on the same subjects based on different experiences. There is an autobiographic streak in the book. Its purpose is not to indulge in an ego trip but rather to explain the experiences that led to the conclusions. I desire to be objective in the analysis and to state the premises.

Absolutely nothing is straightforward as you will soon discover if you read on. For example, I have only recently grasped the role that double-taxation treaties had on the price of oil. There are probably many more strange relationships that I have not spotted.

In tables, graphs and appendices, I have provided my current assessment of what the world's endowment of *Conventional* oil is. The numbers are quoted as computed which gives the impression of greater than justified accuracy. There is one thing that can surely be said about these estimates: they are wrong. The issue is how wrong? In a sense, they challenge those with counter views to come out and state them with details and evidence. Silence amounts to tacit acceptance. There may be surprises in individual countries, but it would be a mistake to imagine that all surprises will be positive. The numbers tell the story and deserve to be considered closely. Another related and very important issue is to understand the difference between *Conventional* and *Non-conventional* oil:

I will leave you to read on to do that.

The World is not running out of oil. Or rather, not for a long time. What it is running out of is cheap oil, and soon. It will still be cheap to produce, but it will be expensive to buy because it will be increasingly scarce and controlled by a few Middle East countries. This introduces an enormous political factor to the issue. I wonder if the US government would fire missiles at Iraq if it properly understood the resource constraints.

Some may think that the end of cheap oil-based energy, which has fuelled our consumer age, is a Doomsday message. The *Coming Oil Crisis* will be just that because the transition will not be easy, but I sometimes think that the World needs a change in direction in any case. From the ashes of the oil crisis may arise a better and more sustainable planet. It must at least become more sustainable as Mankind lives out his allotted life-span in the fossil record. Whether or not it is better depends on how well we manage the transition. We don't have long to prepare.

I hope that you will find what follows useful in planning your future. More than that, I hope that it will encourage you to plan: the crisis is imminent. Those who anticipate can do well from the economic and political discontinuity; those who react can survive; but those who continue to live in the past will suffer. I hope, too, that governments will somehow get the message and not leave everything to the open market that is ill-equipped to deal with the depletion of a finite resource on which we have all come to depend. Even if they cannot bring themselves to prepare, perhaps this will at least help them know what hits them when the crisis strikes, as surely it must.

Milhac, France.
8th February, 1997

ACKNOWLEDGMENTS

This book is the outcome of a cooperative effort to which many people have contributed. Some were already acknowledged in *The Coming Oil Crisis*, but many new contributions have led to this work.

I would like particularly to acknowledge the following:

Jean Laherrère has been a great help throughout, providing insights, many graphs, encouragement and friendship. He has been patient with someone almost devoid of mathematical skills.

Tom Jamison, Robert Harris, Ken Chew, Jean-Michel Frautschi and Bogdan Popescu of Petroconsultants in Geneva stimulated and encouraged the preparation of *The Coming Oil Crisis*, being generous in releasing their precious data and insights, as well as contributing to its publication. It has been a pleasure to cooperate with them.

Walter Ziegler, Alain Perrodon, Jean Laherrère, Richard Hardman, Leik Woie, Buzz Ivanhoe, Clive Needham, Ron Swenson, and Richard Duncan contributed greatly with their interviews.

Roger Bentley of Reading University has taken an active interest, now over many years, and provided many useful comments.

It is a pleasure to record the contributions of the many experts from the ASPO organisation, especially Klaus Illum. Peter Gerling and Professors Kjell Aleklett, Ugo Bardi and Rui Rosa.

Arne Raabe and Rory O'Byrne have contributed with the distribution of the ASPO Newsletter.

Sarah Astor and Jim Meyer at ODAC have contributed in raising awareness of the subject

Anders Sivertsson from Uppsala University did sterling work building a new database.

Chris Skrebowski, Editor of the Petroleum Review, has been a great source of inspiration, providing much invaluable information and insight.

Jack Zagar and Jeremy Gilbert have helped in many ways, taking on speaking engagements in North America.

Julian Darley has raised the flag in Canada, while Ali Samsam Bakhtiari has brought his wisdom from Tehran. My old friend Ray Leonard reported from Moscow.

The wide interest from the media has been a stimulation. David Strahan made two impressive television programmes for the BBC, and Amund

Prestegaard is at work on another. It has been a pleasure to respond to the gentle voice of Tamal Mukherjee of Radio Singapore, who calls from time to time.

I also appreciate the many e-mail correspondents from all over the world who maintain a lively debate, bringing insights and information. Walter Youngquist deserves special mention for his frequent letters. Many have become old friends without ever having been met in person.

Bill Hughes and his team at Multi-Science deserve every credit for their interest and support in this third publishing venture.

But it is with great sadness that we have to record the passing of Buzz Ivanhoe and Bill Pace who contributed interviews to the earlier book.

Bobbins put up with a lot of doomsday chat, making an invaluable contribution with editing and proof-reading. Oddny was forgiving when we "snakkered om olje".

HARRY WASSALL
FOUNDER OF PETROCONSULTANTS

THIS BOOK IS concerned with how much oil there remains to produce, and its depletion pattern. No one would be able to answer these questions but for the work of Harry Wassall who died on November 25th 1995. I met him not long before he died, and discussed the issue of oil depletion with him. It was a subject of great concern to him, and it is therefore a pleasure to dedicate this work to him.

He was an American citizen born in 1921, the son of a New York commodities trader. He graduated in geology at the University of Tulsa in 1942, and joined the Navy to serve out the war in the Pacific on board a Landing Ship Transport.

In 1946, he joined Shell as a field geologist before moving to Gulf Oil, accepting a foreign assignment to Cuba, where he worked as a field geologist from 1950 to 1956. There, he married the beautiful Gladys, a well-born Cuban lady. He enjoyed the life of the island, and when Gulf wanted to transfer him, he decided start his own consultancy, Harry Wassall & Associates. In addition to normal consultancy work, he started the Cuban Scout Letter, a newsletter covering industry developments in the country. He was by nature an entrepreneur and enthusiast, and soon added to his endeavours a drilling as well as a chemical company.

The Scouting Newsletter had expanded through much of Latin America, and a nucleus staff had been recruited by the time Fidel Castro's revolution brought enterprise in Cuba to a standstill in 1959. Harry resolved to extend the newsletter's coverage to the Eastern Hemisphere, and opened a small office in Geneva to build the business. The Wassalls themselves moved to Madrid in 1962, where he developed various consultancies, not all related to oil.

He then dedicated himself to building up the Geneva office, which grew from strength to strength. It was time for a new name: Petroconsultants. The scope of operations expanded rapidly, with a whole range of products beyond the newsletter, which was now global is scope.

I well remember, when working on international new ventures for Amoco, that the first port of call in evaluating a new proposal was always the Petroconsultants material to find out what wells had been drilled; to look at the concession maps, and find out the background.

Fig. 0-1. Harry Wassall

Petroconsultants established for itself an enviable reputation, performing an essential service for the industry, with which it maintained first class connections. Harry saw the growing potential of the computer, and his company was one of the first to computerize its database. It secured contracts to manage the major companies' data on a confidential basis.

Over the years, the Company expanded its network of contacts around the world, including the former Soviet Union

This was a great achievement that owes almost everything to its founder, Harry Wassall. He was an entrepreneur in the best sense of the word; a man of courage, insight and imagination. But above all, he is remembered as a man of great kindness, humour and intense loyalty to his friends and staff.

As the world reaches the peak of its oil and gas production, knowledge of the resource base becomes an increasingly essential issue for planning the future, with colossal political and economic implications.

It is a pleasure therefore to dedicate this book to a man who contributed so much to building the foundation for all realistic estimates of future oil and gas supply.

On Harry Wassall's death, Petroconsultants was acquired by IHS Energy of Denver, Colorando, and fell under new management, leading to many changes in its staff and manner of doing business.

For
Bobbins
Jack and Julia
Anne and Patch
Simon and Oddny
Emma, Clara and Oscar

Chapter 1

INITIATION

THIS BOOK IS ABOUT the World's endowment of oil, asking how much has been found to-date and how much remains to be found and produced in the future. It is a very important subject, considering that cheap oil-based energy has been the lifeblood of the economy over the best part of the last Century. Its influence has been enormous: it has driven the way people live; what they expect; and how they order their lives. Yet, oil has to be found before it can be produced, meaning that production has to mirror discovery after a time lag. The fact that discovery has been in relentless decline for forty years delivers a chilling message about the future production profile.

It is an intensely political subject, and it is a romantic story: "black gold" has turned many rags to riches, for individuals and for nations. It has spawned jealousies and arrogance, and has been a key factor in several wars: the Gulf War and Iraq Invasion being the most recent. The great American economic miracle, with its mindless consumerism, owes much to its endowment of oil: the country would have been much less of a world power without it. Yet, American production peaked in 1971,[1] and is now in terminal decline without hope of reprieve.

Yes, it peaked in 1971. No one particularly noticed at the time because there was plenty to import: the filling stations operated as before. But it was a critical and fundamental inflexion: in other words, a discontinuity. In fact, the 1971 peak in production reflected an earlier peak in discovery in the 1930s that also went un-noticed. We can now look back and see the importance of these events: they mark the beginning of the end. Oil will not last forever: you can't grow new crops of it. It was formed but rarely in the long geological history of the Earth, and only in a few places under most exceptional conditions. In fact, the great bulk of it came from just two brief epochs of extreme global warming, about 90 and 150 million years ago. It is decidedly a finite resource, being consumed at a rate that has been increasing at about two percent a year. In 2004, the World consumed almost 30 billion barrels, but found much less than half that amount.[2] A barrel holds 42 US gallons (or 35 British Imperial gallons), which is several tankfuls for an average car. We use a lot of oil, which is vital for transport and agriculture, which means food.

The 1971 peak was a local discontinuity in the United States, the World's most mature oil country, but it did not much affect that wealthy nation, which was able to import increasing amounts of oil from other less depleted countries. Imports have been rising ever since: such that a country that once supplied the World now imports more than sixty percent of its needs on a rising trend. But, what about the other less depleted countries? Can they go on supplying the World forever? Many of them have also passed the same inflection to falling production as the United States experienced so long ago. That is one of the questions that I will endeavour to answer in this book. The short answer is no, but it is not as simple as that. Another related question is how much will it cost. Will it be subject to profiteering? At the time of writing, prices have recently soared above $50, nudging historical peaks.

You would think that it was a fairly straightforward issue to resolve; surely economists have access to data banks that hold the answers; surely the great international oil companies know what the situation is; surely governments are now planning what to do – unfortunately: if they do, they are not telling us. The subject is clouded in mystery. As you begin to dig into it, you find a maze of conflicting information and disinformation, as well as imprecise and confusing definitions. There are few hard and fast facts. Judgment and experience are therefore called for to choose a path through the minefields. That is an apt analogy, for there are colossal vested interests with motives to distort and confuse: oil is money, and money is power. There is a great deal at stake. Even I, an old man in an Irish village, have received the attentions of the U.S. Intelligence Service.

But, judgment itself is part analytical and part intuitive, and in the latter regard is subject to bias. Many people's intuitive knowledge of oil is based on the weekly trip to the filling station. It has been there for as long as they can remember; and their intuitive judgment tells them that it is likely to continue to be there for the foreseeable future. My intuitive knowledge of oil, by contrast, has been based on looking for the stuff. In the course of my life, I have evaluated hundreds of prospects around the World, and my intuitive judgment, built on that experience, tells me that only very few prospects fully meet the geological criteria to succeed. I stress *fully* because a prospect depends on many factors. One weak link in the chain means an expensive dry hole. These factors are themselves not easily determined, as they depend on geological circumstances far beneath the Earth's surface, whose precise nature can be interpreted only with difficulty. The unravelling of geology relies on observing what can be observed, and then using geological principles to make a logical extrapolation from the known to the unknown. I will follow this soundly based scientific principle in this study.

It makes sense therefore to evaluate the subject through the eyes of an explorer, and recount how his knowledge and appreciation of the situation evolved. Perhaps one should say rather how this explorer came to grasp the nettle. So, I will adopt an autobiographical framework around which to weave the story. Along the way, I will interview others who can contribute their knowledge; and I will digress frequently into political and other matters that lie outside my own professional expertise. I will even risk exploring Conspiracy Theory. I want to try to make it a readable sort of detective story. Explorers are taught to look for clues.

Before proceeding, I should draw attention to the Appendices at the back of the book which provide tables of data, as well as graphs illustrating production and discovery for the World, the Regions and principal countries. This information may help resolve any specific queries that arise in the course of reading the text.

TRINIDAD

The *Regent Springbok* tanker tied up at the loading jetty at Pointe-à-Pierre, Trinidad on 9th April, 1958. It had arrived in ballast from England to load petroleum products from Texaco's giant refinery with which to supply the company's marketing chain in Europe. In addition to its normal crew, it carried a supernumerary crew of about twelve passengers, who were Texaco staff coming out as new recruits, or returning from the home leave to which they were entitled after three years' service. I was one of the new recruits coming out to his first job, fresh from Oxford University, where I had just completed my D.Phil. thesis in geology.

Texaco was a major American oil company that had recently bought the British-owned, Trinidad Leaseholds Ltd., which continued to be British-staffed. Trinidad itself had, not long before, become independent after many years of British colonial rule, and seemed to be a happy multi-racial society of Negroes,[3] Indians (originally brought in as indentured labour to work the sugar estates after the abolition of slavery), Chinese, Lebanese and others of French, Portuguese, Spanish, British and mixed extraction. There were splendid old white Trinidadian families with names like *de Verteuil, Rostow, Fernandez*, who made up a colourful and dynamic element, although vaguely conscious that they were an anachronism whose days were numbered. The average age was fourteen, promising to make it a very crowded island.

To be more accurate, it was not quite my first job, as I had spent a year in Borneo as a member of an Oxford University Expedition, mapping the remote Usun Apau Plateau in the interior of Sarawak,[4] later taking a brief assignment with the Colonial Geological Survey to write up the results. Rocks I knew, but precious little about oil.

Once the personnel formalities were over, I was taken to the Geological

Office, a long, low, green building with mosquito screens in place of windows. It stood beneath a spreading *samaan* tree on a slight elevation above the refinery with its storage tanks. I was introduced to Ken Barr, the Chief Geologist, a silver-haired and gentle-voiced man, who welcomed me to his department.

Before long, he said that he wanted me to meet Dr Kugler, the legendary father of Trinidad oil, who had had a key role in developing the island's oil industry since his arrival in 1913. We set off down the corridor, and soon

I began to hear an animated conversation in German: it was Dr Kugler, discussing some aspect of Trinidad geology with Karl Rohr, a fellow Swiss field geologist. I was ushered into the great man's presence. He was slight of build, and had a sort of rolling gait as he got up to greet me. His silver hair was swept back above a chiselled face. He was dressed in khakis and a white shirt as was normal in the tropics in those days. He turned his one good eye on me, and began to quiz me about my experiences in Borneo, immediately revealing a great breadth of knowledge, not only about its geology, but its people and natural history. It did not take long to realize that here was a remarkable man. He was to

Fig. 1-1 Dr K.W. Barr, Chief Geologist of Texaco Trinidad Inc.

have a lasting influence on my career, and indeed life.

So, began my career in the oil business. I spent only two years under his tutelage, but it was enough to instil in me a scientific interest in petroleum geology. At Oxford, I had had sybaritic tendencies, and, after the Borneo experience, had not at all wanted to return to the arduous life of a field geologist in the tropics. But under Hans Kugler's influence, I found my feet, becoming dedicated to my new profession.

My first assignment was a stratigraphic study of the Cretaceous rocks of the Central Range, as exposed in the Biche River, where I established a base in a disused police station. The next assignment was a mapping project in the Rock River area of the Trinity Hills. It was thick, tropical rain forest, virtually devoid of natural outcrops. So, the work involved controlling teams of Indians, drilling augur

Fig. 1-2. Dr Hans Kugler, father of Trinidad geology, who greatly influenced my career

holes from which samples of bedrock were recovered. I would sit in the forest under a huge beach umbrella, draped in mosquito netting, while I described the samples in my notebook; labelling them; numbering them; and plotting the location of the augur hole on a map. Typical entries in the brown field notebook read something like:

> *Cb238. Clay, greenish-grey, micaceous, silty, slightly calcareous, with carbonaceous fragments.*
> *Cb239. Clay, light grey, micaceous, with silty calcareous streaks.*

The work continued for several months, and the samples were studied in the Geological Laboratory under John Saunders,[5] a micro-palaeontologist. His job was to extract micro-fossils, called *Foraminifera*, which allowed him to date the sample by reference to a palaeontological zonation. Trinidad's geology is extremely complex and difficult, being made up of highly folded and faulted sequences of rather monotonous clays. It was only by means of their fossil content that the different formations could be identified. Mapping the strata so recognized then made it possible to unravel the structure, and thereby identify possible reservoirs and traps for oil.

Once the determinations were made, I set about the interpretation, and

Fig. 1-3. A mobile land rig for repairing wells

Fig. 1-4. My good friend, John Saunders, a micropalaeontologist with whom I worked closely

little by little, found how the pieces of the jigsaw fell into place. The juxtaposition of two normally separated form-ations, identified by their micro-fossils, could be traced from one augur line to the next, pointing to the existence a thrust fault,[6] which in turn could be correlated with a dip discontinuity[7] in a well to the north. In the well, the thrust was overlain by some water-bearing reservoir sands, which were missing at the surface. It meant that they had died out somewhere between the well and the surface where I had been mapping. I wondered if perhaps oil had migrated up-dip to accumulate in the truncation. But, I had also identified a cross-fault that offset the thrust. Exactly where the truncation was located was hard to predict, but my best guess was that it lay just outside Texaco's lease in an adjoining one operated by BP. In other circumstances, such information would have been a closely-guarded secret to be commercially exploited, but in Trinidad in those days, the companies maintained a friendly cooperation. When Dr Kugler had studied my maps, he sent me down to Palo Seco to show them to Sam Wilson, the Chief Geologist of BP's Trinidad operation. In due course, he tested the idea with a well, and found oil. My first step into the oil industry had been a success, but finding oil for another company did not exactly earn medals from the management.

Trinidad days passed happily, not only during working hours but in the friendly society of an expatriate oil camp, which somewhat resembled an expensive country club, save for the sulphurous smell from the refinery. Weekends were often spent water-skiing, or on the beaches of the beautiful Northern Range beyond Port-of-Spain. Rum-punch parties and Carib beer were there in abundance for the benefit of a young bachelor. At one such party, he met the beautiful Bobbins Ludford, a secretary who had stepped ashore from one of the subsequent tankers that brought staff from London. We were soon to be married.

Then, one day in September 1959, Dr Kugler called me into his office to ask if I would like to be transferred to Texaco's operation in Colombia. He had been talking to the Manager for Western Hemisphere Operations in New York, who was impressed by Trinidad's scientific approach and wanted to apply it throughout the Company. Kugler had also greatly impressed the Chairman of the Board, Augustus C. Long,[8] when on a visit to Trinidad. So, I was entrusted with the missionary work to bring palaeontology to the Texans running the Colombian operation. Thus, opened one of the most colourful and interesting chapters of my life.

But before we leave Trinidad, let us look back from today's vantage point, and see what its oil resource situation was, and has become. At the time, none of us had such a perspective, being preoccupied with detailed day-to-day projects and studies. There was no idea in the back of our minds that the resources were finite, or had a particular distribution. These subjects, forming the theme of the book, will be developed in later pages. Trinidad had had a long oil history. Columbus had caulked his ships with tar from the famous Pitch Lake, a natural seepage, and oil operations had already commenced during the 19th Century. Forest Reserve, the largest field, was found in 1914, followed by numerous other small onshore fields, most discovered before the Second World War. Efforts then turned offshore, first to be rewarded by the discovery of the Soldado Field in 1954 in the Gulf of Paria, a landlocked embayment that separates Trinidad from Venezuela, before attention turned to the Atlantic and Caribbean shelves. Trinidad

Fig. 1-5. Dr Hans Kugler, as I remember him best: observing nature in the rain forests

had been in fact the gateway to Venezuela which belongs to the same geological province. Overall, discovery peaked in 1968, ten years before production did so in 1978, but in detail the country has had four exploration phases: an early onshore effort; a second offshore in the shallow Gulf of Paria; a third in the Atlantic; and finally the fourth in the Caribbean to the North. I will later evaluate the significance of these patterns, but meanwhile will introduce the subject with two graphs, Figures 1-6 and 1-7. Most of Trinidad's onshore oil had been found before 1930. The term *Ultimate* means the total amount that will have been produced when production ends. A *wildcat* is an exploration borehole

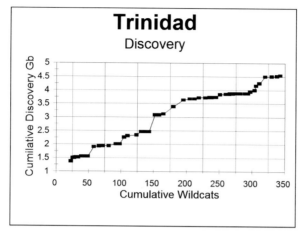

Fig. 1-6. Discovery pattern of Trinidad: the larger fields in each exploration phase are normally found first.

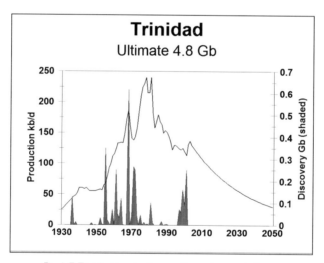

Fig. 1-7. Trinidad production profile and annual discovery.

on a new untested prospect that, if successful, results in a new field. The plot of cumulative discovery against *wildcats* tends to be hyperbolic,[9] or a series of hyperbolas, each related to a particular phase of discovery as a new geological province was opened up. Projecting the curve gives an idea of what the *Ultimate* will be, and how many boreholes it will take to get there. The concept of an *Ultimate* recovery is an important one, because production will one day end. The curves of discovery are flattening everywhere (see Appendix III).

Fig.1-8.An offshore platform in the Gulf of Paria, which gave Trinidad the second of its four lives

COLOMBIA

I left Trinidad at 7 am on October 1st 1959 aboard a DC-3 of Lineas Aereas Venezuelanas, and flying at an altitude of no more than a few thousand feet traversed the spectacular Coast Range of the Andes before landing at Maiquetia, the airport of Caracas at the foot of the mountains. From there, a modern highway, full of fast and seemingly huge American cars, led to the Venezuelan capital, Caracas. It was an exciting place, very different from sleepy Trinidad with its British colonial atmosphere. After a stopover, I continued on a Constellation airliner to Maracaibo before heading into the interior of South America, catching occasional glimpses of vast stretches of forest cut by meandering rivers. At last, we dropped through the cloud-base to land in a broad upland valley, in which lies Bogotá, the capital of Colombia, at an altitude of 8300 feet. It was a cool grey day, very different from the tropics I had left behind. An ancient Chrysler taxi, with several hundred thousand miles behind it, brought me to the Tequendama Hotel. The driver was wearing a poncho, known in Colombia as a *ruana*, and a black trilby hat. Entering the hotel was like stepping back into Europe, and I was delighted.

Now, began a very new and exciting experience: to live in this Andean city, where hung a pervasive smell of eucalyptus smoke; and to work on the magnificent geology of Colombia, cut, as it is, by three ranges of the Andes. The centre of Bogotá dated from the 16th Century, when it had been the Vice-Royalty of the Spanish Empire, known as Gran Colombia, covering what is now Venezuela, much of Central America and Ecuador.

The inner suburbs were built in red brick, mainly before the Second World War, and were somewhat reminiscent of the North Oxford I remembered. Farther north, developed the modern tree-lined suburb of El Chico, where the wealthy lived in large Spanish-style houses, surrounded by high walls and wrought-iron fences. Ancient buses of several rival lines raced each other along the roads, passing the occasional trails of donkeys, carrying fire-wood, which were led by a *ruana*-clad Indian. Above the town, rose an impressive range of barren hills, capped by the Virgin of Monsarrat.

Fig. 1-9. Bogotá, a charming Andean town in the days before drugs

Next day, I made my way to Texaco's office on the 14th Floor of the Edificio Seguros Bolivar, one of the few skyscrapers in the commercial part of the town, not far from the hotel. There, I met the Chief Geologist, Ken Bishop, a tall laconic American, who reminded me rather of one of those figures you see in American war films, stooping with an intent expression as he beckons his troops to follow him into action. The Exploration Department consisted of about twenty young Americans, who had been mobilized as what was termed a Task Force in one of the Company's periodic new thrusts to explore Colombia. It had been established in the country for many years, operating the Velasquez Field in the Magdalena Valley, but had not found much else. I was assigned to a small team evaluating the Sabana de Bogotá area, a high intra-montane depression within the Andes, which was characterised by some huge outcropping anticlinal features. They formed impressive oil prospects, provided they contained satisfactory reservoirs and source-rocks, the details of which were not then known. Pat Maher[10] from Minnesota, Ray Robbins and a quiet spoken Texan, Carl Henderson, were my new colleagues on the project. They probably did not know quite what to make of this *Limey* who had arrived in their midst, but they were open and friendly.

I started work reading reports to become familiar with the geology. I also started to understand the Company's style of operation. It was a far cry from the gentle academic hand of Ken Barr in Trinidad, where we had all worked together as dedicated geologists and friends on a more or less equal standing, despite our different ages, status and experience. Now, I witnessed something of the hierarchical atmosphere of American business: there were bosses and there were employees, for whom the pay-check was the prime motivation. It was not such a bad system, as you knew where you stood, which is more than was always the case with a European company, as I was later to find out with the Belgians. I did not mind the pay-check either: I had joined the then legendary dollar payroll, and was earning what seemed an astronomical amount, far above my Trinidad salary. Much as I enjoyed colourful Colombia, I often thought nostalgically of Trinidad and particularly of Bobbins Ludford to whom I soon sent a telex

"Letter asking for your hand in mail stop please answer soonest stop love Colin".

She did, and we were married by Adrian Clarence Buxton, the British Consul, on December 19th, 1959. In later years, when I was in management, I often found myself in the role of little more than a telex operator, feeding the bureaucracy of a large company. I sometimes reflected how appropriate it was that I should have also proposed by telex.

Before that happy day however, the Company had what was called a *Wise Box*, at which every project had to be presented with much theatre to two executives from New York, Dashy Bode and Bill Saville.[11] The former appeared to have extraordinarily primitive notions of geology, and thought that mapping was simply a matter of measuring the dips of the formations and constructing geometric cross-sections. He seemed oblivious of the subtleties of stratigraphy: how formations thicken and thin; and change their compositions, reflecting variations in the environments in which the rocks were laid down. He failed to recognize the importance of palaeontology by which to date the formations, and thus correlate them from one area to another. This, I thought, was the moment to don Kugler's missionary hat, and propose that the Company should now apply palaeontology to improve its lamentable lack of stratigraphic knowledge.

Fig. 1-10. Bobbins collecting fossils in Colombia

The message was not at all well received: in the hierarchy, it was not thought fitting for a junior geologist to challenge an executive, least of all a *Goddamn Limey*. The message of course carried an implied criticism, which is never appreciated: a subtler approach would have been more successful. It was a lesson I never managed to learn.

Probably as a consequence, I was soon shipped down to do field work in the Magdalena Valley to map an area of Miocene non-marine strata, conveniently devoid of fossils. My base camp was a bullet-pocked, estate house, near the Carare River, which was being used by a seismic crew. In a later chapter, I will explain this exploration technique that involves letting off explosive charges and recording the echoes reflected back from subsurface formations. The bullet holes were from the attentions of bandits who infested the area, and from whom a detachment of Colombian troops was supposed to protect us. Each morning, a small helicopter would carry me out to a clearing in the forest on one of the seismic lines, where I would be joined by Hernando Patiño, the surveyor, and my field party. It comprised about twelve tough Colombians, led by their Capitaz, Abel Alberacín, who camped out in the forest. The days were spent laboriously following overgrown streams, clogged with secondary growth, in the hope of finding an outcrop with a dip to be measured: that is to say, to measure the degree to which the strata are inclined. The men had to hack a path through the undergrowth with machetes. Mosquitoes abounded leading me to sew a flap on the back of my hat to protect my neck from bites, which made me look something like a French Legionnaire. One day, the supervisor from the Velasquez Field came out to visit me: he was a splendid Texan by the name of Wayne Marrs, whose opening words as he stepped from the helicopter were:

"Stud, where did you get that bonnet?"

The project was a futile exercise, since the structures of interest lay beneath a thrust-plane, such that the dips at the surface had barely any relevance anyway. Later in the office, I discovered that the Company already had a comprehensive report on the area, acquired from Shell years before, when it was still under virgin forest and outcrops were more plentiful. But scientific study was not the style of the Company at the time: reports were neither written nor read under the Task Force mentality.

It took several months to complete this mapping. I then moved south to the Rio Guaguaqui to map the thrust-belt separating the Magdalena Valley from the Eastern Andes. It promised to be a more interesting project, and soon I was collecting ammonites from Cretaceous shales. This time, we were on our own without helicopter support, having to rely on dugout canoes to transport us up the rivers. My notebook is full of sketches of tropical flowers and birds. I felt myself, however humbly, to be in the best tradition of the classical explorers of South America, like Humboldt or Darwin himself. But after several weeks of work, I woke one day with a fever and yellow eyes. When resting in camp brought no improvement, I sent some of the men to make their way on foot to the Velasquez Field, some thirty miles away, to get help. At last, a helicopter arrived and flew me to the oilfield from whence the Company DC-3 brought me back to Bogotá. I had contracted jaundice and amoebic dysentery, the normal hazards of tropical field work.

The cure was rest at home and Vitamin B injections. It was a welcome break, and it was good to see Bobbins again after my long absence in the jungle.

When, a few weeks later, I had recovered sufficiently to return to the office, I found that Hans Tanner, a Swiss geologist, had returned from vacation to supervise the evaluation of the Sabana de Bogotá and Llanos Foothills, to which I was reassigned. Hinting that he shared my view of the

Fig. 1-11. Field work in the Magdalena Valley

Fig. 1-12. Field camp in the Magdalena Valley

Company's scientific weakness, he instructed me to commence a thorough stratigraphic investigation of the eastern flank of the Cordillera Oriental. We needed to know what reservoirs to expect below the surface structures that had been mapped, and how deep they would lie. That could only be worked out from a stratigraphic evaluation.

Now, began what was to become the most rewarding chapter of my geological career. The field party was reassembled, and we set forth to map the road section from Bogotá to Villavicencio, a town in the foothills of the Andes at the edge of the Llanos plains that extend far into the heart of the continent. To begin with, we drove out each day from Bogotá, but when the distances became too great, rented places to stay in colourful villages along the way. I can remember waking to the sound of mules trotting along the cobbled streets: it was a scene from medieval Spain.

Hugo Cely, a Colombian trainee geologist, helped Hernando run the planetable and alidade survey, and the men with stadia rods were dispatched to mark the outcrops, which were abundant. We started in the Tertiary lake-bed deposits around Bogotá, before coming to the impressive scarp of the Upper Cretaceous, Guadalupe Sandstone, now one of the producing reservoirs in the Cusiana Field – of which more later. The sequence was only mildly deformed, making it possible to map accurately every formation. Furthermore, it was richly fossiliferous. For example, the oyster, *Exogyra squamata*, proved a useful marker for the Cenomanian stage, which could be traced across country. Careful search was rewarded by a rich fauna of ammonites. As we progressed, we came to another massive sandstone scarp of Albian age, which I saw might be a deeper reservoir for oil, although it appeared to be too indurated to have adequate porosity and permeability, at least within the mountains. At the weekends, I returned to Bogotá to organize and photograph the fossil collections, which were identified by Hans Bürgl, an Austrian palaeontologist at the

University, with whom I became a close friend. Eventually, the stratigraphy was worked out in detail with the identification of all the classic European stages: *Maastrichtian, Santonian, Campanian, Turonian, Cenomanian, Albian, Aptian, Barremian, Hauterivian, Valanginian, Berriasian and Tithonian* – the names are etched in my memory.[12]

This is a good moment to introduce in a few words the geological column and explain the meaning of terms like Cretaceous and Jurassic.

Fig. 1-13. Hugo Cely surveying outcrops with the planetable and alidade

Fig. 1-14. Saddling up the mules

The rocks laid down during the last 600 million years of the Earth's history are ascribed to four systems: Palaeozoic (230-600 Ma); Mesozoic (65-230 Ma.); Cenozoic (62-2 Ma.) and Quaternary for the last two million years. Rocks older than 600 million years are attributed to the Precambrian or Proterozoic (Figure 1-15). These systems are further sub-divided, each stage being recognized by its fossil content. The ammonites, for example, died out at the end of the Cretaceous Period, 65 million years ago. The absolute ages in millions of years of most of the intervals have now been determined by studying the decay of radioactive elements. Once the first traverse had been completed, I turned to the next road northwards that led through a more remote section of the Andes to Gachalá, famous for its emerald mines. Comparing the two sections, and using the fossils for correlation, began to show the stratigraphic trends: how particular formations were thinning or thickening, and subtly changing in composition. This knowledge would prove essential to predicting conditions in the prospective areas.

From Gachalá, we hired mules and followed an ancient *camino real*, the trails followed by the runners of the Inca Empire, to cross the Farallones de Medina, a massif rising to over 4000 m. I will never forget emerging into the Llanos foothills and riding muleback into the small and remote town of Medina. It was a scene from a Wild West film. As we rode into town, we

Fig. 1-15. A Cretaceous ammonite

Millions Years	ERA	SYSTEM	Characteristic Life
		QUATERNARY	Homo sapiens
2			
		Neogene	
23	CENOZOIC		
		Paleogene	
65			Mammals
		Cretaceous	
141			Dinosaurs
	MESOZOIC	Jurassic	
195			Ammonites
		Triassic	
230			
		Permian	
290			
		Carboniferous	Coal-forming forests
345			
		Devonian	
395	PALAEOZOIC		Fishes
		Silurian	
425			
		Ordovician	Trilobites
500			
		Cambrian	
600			Molluscs
	PRECAMBRIAN		
4600		Formation of the Earth	

Fig. 1-16. Geological column

could sense the hostility, and almost feel the rifles being aimed at us from the low adobe houses around the square, to which horses were tied up. A gang of rough characters riding into town usually meant only one thing. It was a tense moment before we established our credentials as *petroleros*, not *bandoleros*.

I had become a dedicated geologist, fascinated by Colombia's geology in all its majesty, and I loved this colourful country with its still evident Spanish old-world culture. Although I was so often away in the field, Bobbins and I enjoyed our new married life at our small apartment in El Chico. We were wealthy, at least by our standards, and had many new friends amongst the expatriate community[13]. Bogotá was full of splendid European restaurants from the Balalaika, where a barrel-chested White Russian sang; to La Zambra, where its owner, Juanillo, gave renderings of *La Flor de las Clavelles*, before cutting off the tie of one of his guests to add to a collection hanging from the ceiling. Then, there was the *Bella Suiza*, a typical Swiss inn at Usaquén, where we would often go for a Sunday lunch of *wienerschnitzel mit roesti*, washed down by locally brewed German beer.

Julia, our daughter was born on January 22nd, 1961. It was a splendid life. Professionally, I felt that my missionary work for palaeontology had been vindicated, and I was disappointed to find myself so out of tune with the company I worked for, who did not seem to appreciate, nor give credit for, what was being achieved in scientific terms.

Our tour of duty now came to an end, and we flew home to England for a three-month leave. As the days passed, I increasingly dreaded returning to the baleful environment of the office, which I mistakenly attributed to it being an American company. I decided to approach BP to see if they had an opening, thinking that I would fit better into a European company, such as I had known in Trinidad. To my great delight, Norman Falcon, the legendary Chief Geologist, who, with J.V.Harrison, had made the initial surveys of Iran, offered me a job to return to Colombia, where BP was then starting up an operation under a joint venture with Sinclair Oil & Gas. It was too good to be true.

With hindsight, I can see that my reaction to Texaco was an immature, emotional one, most of all to attribute what I perceived to be its weaknesses to its American base. As I was later to learn, all companies can fall into bad hands in the ups

Fig. 1-17. I seem to have had a penetrating gaze

and downs of their existences. Everyone in the corporate jungle faces his own tensions and pressures. They had more to worry about than some young geologist in Colombia with a bee in his bonnet about fossils.

A SECOND BITE OF THE CHERRY

Fossils were not a priority in the new company either, where force of circumstances pushed me into a much more commercial role. As a newcomer, BP had very little acreage in Colombia, and was therefore interested in securing access to existing concessions. Companies were often willing to assign a 50% interest in a concession in return for the drilling of a test well. Many of the World's oilfields have been found by other than the original owners of the concessions in which they lie. The concessional system in Colombia at the time awarded initial rights as *Applications*, which could later be converted to full *Concession* status, carrying drilling obligations. These rights changed hands frequently as companies acquired and released acreage subject to shifting ideas and strategy, often brought about by changes in the tax regime. My job became that of an evaluator of such offers and opportunities. Most I rejected, but occasionally would come across an interesting possibility. Indeed, had BP accepted all the recommendations, it would soon have become the largest company in the country. It had already found the Provincia Field, drilling on Esso acreage, to which could have been added the Rio Zulia and Orito fields. They in turn would have led to other promising finds along the mountain front, extending across the border into Ecuador.

I would interview the company concerned for information, armed with tracing paper on which I would try to extract the maximum of information from their maps by vaguely surreptitious means, and then rush back to the office to write up an appraisal, fearful of being forestalled by another company. It sharpened my wits and made me more of an oilman than a geologist. It was a very valuable experience. It also gave me a comprehensive knowledge of Colombian geology, distilled from the work of all of these companies, many of whom had been working in the country for years. In those days, before the advent of fax and internet, reports had to be written and mailed to head office. It was good experience, because the process of writing the report greatly facilitated the logical analysis.

It was a very pleasant office staffed by an agreeable group seconded from BP and Sinclair.[14]

We were to remain in Colombia for another four years until the spring of 1966. Despite the reminiscences, this is not an autobiography but an account of the evolving understandings of a young exploration geologist in relation to World oil resources, the theme of the book. There is no need therefore to recount all the adventures and incidents that made up this happy chapter of our lives, which included the birth of our son, Simon, on

Fig. 1-18. Freddie Martin visits me in the Cauca Valley

October 29th, 1962. It is worth mentioning, however, that my final year was spent compiling a regional report on the geology of Colombia, which anticipated in fact aspects of the then evolving geological revolution of Plate Tectonics. It recognized the presence of several major transcurrent faults, caused by lateral movements of segments of the Earth's crust, and proposed a structural evolution in terms of a westerly movement of the Continent, overriding earlier deposits at its leading edge. The report correctly identified the Llanos Basin as the most promising province in the country, a conclusion that was not to be vindicated until 1983 when the Occidental Oil Company drilled the Cañon Limon giant discovery.[15] BP did not act upon this recommendation in 1965 when the entire basin was there for the taking, but belatedly succeeded in buying into a concession in 1988 that yielded the giant Cusiana Field, not far from the town of Medina into which I had ridden on a Texaco mule twenty-five years before. It tends to confirm that there are fewer technical surprises in exploration than is commonly imagined.

In addition to assessing the merits of the different basins that make up Colombia, the report also assembled information on the size of fields and made some prognosis as to future discovery. It demonstrated an awakening understanding of resource constraints: the limitations of many of Colombia's basins were already evident. In this regard, it was a useful step in the growing appreciation of this critical subject. Yet, although I could see the local limitations, the World still seemed a very large place, and I had at the time no feeling of any global constraint.

In the same way as we took a preliminary look at Trinidad's production profile, let us also look at that of Colombia in Figures 1-19 and 1-20.[16] It too had had an early start when General Virgilio Barco let the early concessions, and when the Tropical Oil Company developed the giant La Cira-Infantas Field in the Middle Magdalena valley in the 1920s. The remote Llanos Basin, with its crop of giant fields, was not developed until late because of its very remote location on the wrong side of the Andes for the purposes of export. Like Trinidad, Colombia has now passed its peak and production is falling steeply.[17]

These days in Trinidad and Colombia characterized an epoch in exploration that has virtually ended. It was onshore, and involved studying the rocks at outcrop: whether in tropical rain forests, the oil lands of the United States or the deserts of the Middle East. The geologists who did this

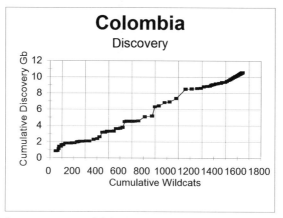

Fig. 1-19. Discovery pattern of Colombia: the late stage giant fields in the Llanos were not unanticipated

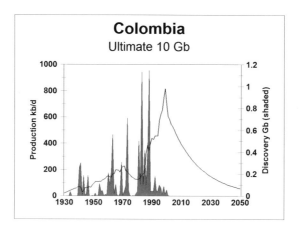

Fig. 1-20. Production profile of Colombia, which has now passed peak

work had a pioneering role, often working far from base and without supervision. It is not surprising that they were individualists. It was a slower pace of working, involving the gradual accumulation of

Fig. 1-21. Field work in Colombia

Fig. 1-22. On horseback in the Western Andes

information with a hand lens and notebook. We worked outdoors and in a natural environment, and probably gained an intuitive knowledge of the rocks amongst which we spent our days. As I read professional papers to-day, I am often struck by how artificial they seem, however elegant the intellectual hypotheses expounded. The pioneering explorers also faced many practical daily decisions of where to camp, how many mules they needed, and how to get the best from their labour force. No courses were offered in these subjects in those days: it was more or less a matter of common sense and a growing confidence built on experience. Today, companies are obsessed by safety policy to the extent of having inspectors making sure that no one uses the stairs in the office without a hand on the rail: it seems a far cry from working in remote bandit country in Colombia without means of communication.

Exploration has moved on from the outcrop to rely on seismic surveys and borehole data, displayed on the computer screen, and manipulated by enormous computing power. Oil geologists now rarely see rocks, and spend their lives glued to their screens in corporate offices under supervisory hierarchies. It must be much less fun, but yet, even now, prospects are still to be found only in the heads of explorers.[18] The days of poking holes into large structures are over, so that ever more ingenuity will be required to identify the smaller and more subtle traps, which are all that remain. The need for stratigraphic insights that characterised my days in Trinidad and Colombia has become ever more important. Old Kugler was right.

Forgive me, if I sound a bit nostalgic for the "good old days", but perhaps it is useful to explain the foundations upon which this study is based.

Until now, a geologist's mission has been to develop Nature's resources for the benefit of Man. He has been only too successful in doing so, and as this book will try to explain, must now adopt a new mission: to explain just how finite these resources are, in the hope that Man will come to understand that he must make better use of them. Geology, unlike engineering, is a descriptive science that imbues a sense of detachment and a long sense of history. We cannot change history, nor the resource limitations of the rocks formed millions of years ago. It can be said that oil is no more than a liquid rock, whose occurrence is controlled by Nature. However skilled the engineer is at taking the oil out of the reservoir, he is no good at all at putting it into the reservoir. The economist and manager can do nothing without this basic ingredient of economic life: oil has had a crucial role over the past century. When geologists tell us that we are about to face a major change as the resources of available oil dwindle, we had better pay careful attention.

ENERGY

Why was I, and others like me, employed to look for oil? The answer is because it was a potent source of energy, which is what makes the World go round, whether we speak of industry, transport, trade or agriculture, which means food. It also played a critical role in the creation of the financial capital that resulted, as well as the economic theories and practices to manage investment. The term *energy* comes from the Greek *en* meaning *in,* and *ergon* meaning *work*. It had already been recognized by Aristotle, 300 years before Christ. But, it was not until the development of thermodynamic theory in the 19th Century that the physical principles were understood. The coming of the steam engine and the first railways in England in 1825 brought revolutionary developments in the use of energy, which had a great effect on the way people lived. They moved from the countryside to the towns, often to work in depressing circumstances, and received little of the benefit of the new energy, which effectively introduced capitalism as we know it. Since then, the use of energy in all its forms has expanded greatly. I remember visiting my aunt's farm in Ireland just after the Second World War. They had fifteen farm labourers, horses to pull the carts and plough the fields, and a pony trap in which to do the weekly shopping. A paraffin-driven generator provided electricity for lighting. Now, the farm is connected to the mains electricity, having tractors and sophisticated harvesters but hardly any labour. It grows Christmas trees for the German market instead of wheat.

The new sources of energy, which made this transformation possible, involved colossal national investments. It is estimated that more than ten trillion dollars of investment was spent to furnish the World's energy needs over the last three decades.[19] The provision of such energy itself consumes energy on a vast scale, whether we speak of oil production, coal mining or nuclear power, not to mention the electricity generating station itself. It may take up to five years for the plant to produce more energy than its construction consumed. A modern coal-fired power station, yielding one gigawatt of electricity, consumes 10 000 tons of coal a day, but also gives 600 tons of ash and 200 tons of sulphur dioxide, as well as much carbon dioxide released to the atmosphere. We are running out of space for the waste.

Furthermore, all materials and foods require substantial amounts of energy both to produce them and deliver them to market in usable form. The production and use of energy also carries adverse environmental effects in the form of acid rain, greenhouse gases, nuclear waste, not to mention the blighting of the landscape in many ways. It may also be affecting the climate, leading to a new phase of global warming, when the ice caps melt and the sea-level rises to submerge many populous lowlands. The usage of energy varies greatly from country to country, depending on the degree of so-called progress. Pre-industrial countries, such as India, use less than 0.5 tons of petroleum equivalent per capita per year, whereas the inhabitants of the most profligate user, the United States, consume twenty times more. They use twice as much energy as do the Europeans but their per capita income is only fifty percent higher. It seems inconceivable that the growth of usage in the extravagant countries can continue, nor that the less industrialized countries can catch up, even assuming that that was desirable.

Source	%	Output	Use
Renewable		Electricity generation	Commercial, industrial, household heat, light and power
Biomass	n.a.		
Wind	—		
Wave	—	Direct fuel	Transport
Solar	—		Agriculture and fisheries
Hydro	3		
Non-renewable			
Geothermal	—		
Nuclear	7		
Gas	23	Waste heat and wasteful usage (about 70%)	
Conventional Oil	40		
Non-conventional Oil	—		
Coal	27		

Fig. 1-23. Energy input and output

Some theory

For the physicist, energy is a constant in a closed system. It is not consumed but simply transformed from one form to another: heat to movement or movement to heat. A calorie is the amount of energy required to raise one gram of water 1°C. It is equivalent to 4.184 *joules*, which is another measure of energy, defined under the S.I. system as the kinetic energy of one watt per second. Power is measured in terms of the energy produced or consumed over time in *watts*, being one *joule per second*. There is also the distinction between kinetic energy, which relates to objects in motion, and potential energy, such as the power of a compressed spring.

In fact, heat and mechanical energy are related phenomenon under the principles of modern physics. The laws of thermodynamics govern the efficiency of energy transformation into useful work and heat. A wind-driven generator captures some seventy percent of the energy input, whereas one fired by heat captures only about forty percent. What is not used is released as wasted, and even damaging, heat to the environment.

Food

Food provides both nutrition and energy. The average person requires 2500 kcal per day, or three kWh, to keep him going: that is to say, the equivalent of running a small domestic electric heater for three hours. The continuous power consumed is about one-sixth of a horse-power.

The most primitive tribes of hunters and gatherers expend about one unit of energy to recover sixteen units from the food obtained, but they need plenty of forest to wander through. Modern agriculture reduces the relationship to about one-to-four, through the use of mechanization and fertilizer, which consume energy. It is even down to one-to-two in some very intensive production.

Most food today is processed, packaged, transported, stored, refrigerated, cooked and advertised, which may consume as much as seven times the energy content of the food itself. Oil provides much of the energy needed. Even where I lived in the depths of rural France, there was an almost continuous distant throb of tractors at work, while here in the village of Ballydehob in the west of Ireland, huge trucks and tractors often block the small roads.

Fertilizers and pesticides are in ever greater use. Some are made from hydrocarbon gas, with more energy being consumed in their manufacture, transport and application.

Fisheries are very heavy oil users, but again there is scope for improved efficiency, with drift netting using much less than trawling.

Transport

People are on the move all the time, and goods are shipped all over the globe in ever increasing quantities. Transport uses about twenty percent of the energy consumed in developed countries. The efficiency ranges widely: a bus uses about one-third as much as a car per passenger moved. Furthermore, there are gas-guzzlers and efficient light-weight small cars.

Air travel has proliferated greatly: jet trails are visible somewhere in the sky almost everywhere. The oceans are traversed by tankers, container ships and cruise-liners, all powered by oil. The military too burn it up. The US military alone consumes as much as 5.6 billion gallons a year[20] to no useful end.

Transport depends heavily on oil, by far the most convenient fuel. About sixty percent of the oil produced goes in transport. The huge waste of energy incurred by stalled traffic in over-crowded streets is appalling. Many towns are now close to gridlock, which itself tells us that this particular profligacy has reached a limit, notwithstanding the pending decline in the essential fuel that drove it.

Sources

The energy required for food, heat, mechanical power, and producing materials comes directly or indirectly from the sun and natural nuclear activity. Much comes from fossil fuels comprising coal, peat, oil, and natural gas, all of which are present on the planet in finite quantities, having been formed in the geological past. They are obviously subject to exhaustion, as this book will consider in respect of oil and natural gas.

In addition, there are renewable sources of energy, including biomass, hydro-electric, wind, wave, tidal and solar sources,[21] which have produced little so far, but must become much more important in the future. The scale of the World's dependence on fossil fuels is alarming and dangerous. Oil makes up forty percent of traded fuel. Its production is now virtually at peak and set to decline irreversibly: yet, few have grasped the implications. Governments remain in denial, and tend to resent being told because it exposed them to the risk of criticism for failing to act on the information available to them.

There is a need for some form of accounting to properly understand the energy content of things and activities, so as to better see how vulnerable we may be, not only to shortage as such, but to increased price. In due course, and in the not too distant future, it will be necessary to allocate or ration energy supplies, not necessarily by price alone. Already Australia set an example by questioning the wisdom of a new road to the Sydney Airport for the Olympic Games.[22] It would not be difficult to contrive effective fiscal measures aimed at energy efficiency. For example, it is worth knowing that in energy terms importing 1 kg of meat is equivalent to 1.3 kg of gasoline. Recycling metals uses much less energy than does primary production: iron, one-third; copper, one-tenth; and aluminium, one-sixteenth. The scrap dealer is a more valuable member of society than is often appreciated.

Electricity is a particularly useful form of energy, but it is not at all an efficient usage. About one-quarter of the energy used in an industrial country is dedicated to producing electricity, but the output is a fraction of that because of the losses in generation and transmission. The terrible waste of pointless street lighting is very evident. Flying over Europe or America by night shows glittering lights from horizon to horizon: all for no real purpose and burning up a non-renewable resource. Better building design could reduce the needs for heating and air conditioning greatly.

Energy is a critical factor in the World's future, and oil is the most vulnerable to early exhaustion. It is time to think about the consequences.

Energy Concentration

In considering energy, it is important to realize that what really matters is energy concentration. All of it, apart from nuclear energy, comes ultimately from the Sun. Looking at a mediaeval walnut mill in France, I wondered how the thing could work on the little trickle of a stream, which was all that flowed through the valley. The answer was that the stream filled a reservoir providing a sufficient head of water to drive the mill for a day or two before having to be refilled. It was a form of energy concentration. Look at a log burning on a fire. The tree, from which it came, had been concentrating sunlight for twenty years while it grew. We are able to concentrate that into heat for an hour or two while the log burns. Fossil fuels are even more concentrated forms of sunlight that fell upon the Earth millions of years ago.

In early years, Man used human and animal muscle power, itself derived from food grown in sunlight. Imagine for example the work of the shipwrights building the *Santa Maria,* in which Columbus sailed to America. They had to cut down the trees by hand, transport them by horse, and saw them into planks by hand: it took a long time. Now, the same work could be achieved rapidly by the use of power saws and drills, using highly concentrated energy.

As we progressively use up the concentrated energy of the past, provided by fossil fuels, things will have to slow down and become more sustainable. This represents a daunting change of direction, which will likely be accompanied by great tension.

NOTES

1. The USA peaked in 1971 with 9.65 million barrels of oil a day, according to Oil & Gas Journal data for oil only. The inclusion of Condensate and/or Natural Gas Liquid gives a slightly different date.
2. The exact amount is uncertain but some reports suggest that no more than about 7 billion barrels of Regular Conventional oil were found.
3. The term Negro has for some strange reason acquired a pejorative overtone in some circles: here it is used with nothing but great respect. To me it sounds much better than the anglicized Black.
4. Campbell C.J., 1956, Geology of the Usun Apau Area, East Sarawak; Brit. Borneo Geol Surv. Ann Rept
5. I can't list all my other associates, but should nevertheless record my indebtedness to them all for their happy comraderie and cooperation in those early days: they included George Higgins (dec'd), my supervisor; Max Cater, Vernon Hunter, Jimmy Frost, and Howard Bennett, not to forget Polly, the Secretary, the team of draughtsmen, and Watty the fossil collector.
6. See Appendix 1 for a short glossary of geological and oilfield terms.
7. Dip means the inclination of the strata.
8. See Texaco (Chapter 3).
9. Those of you who are mathematical will have no difficulty in knowing what hyperbolas and parabolas are. Simply stated, they are plots in which the degree of curvature gradually diminishes. Several examples are illustrated herein.
10. He went on to form the Lone Wolf Oil Company in Denver and become a millionaire, before the oil glut of 1985 wreaked havoc with his endeavour.
11. I don't know exactly what their roles were. Bill Saville had been a field geologist in Colombia and was an impressive character: a tall, gaunt frame with a moustache, who could have played the part of a sheriff in a Western movie. He had greater sympathy for what I proposed, but in the corporate politics was probably unable to act. I, of course, had no insight into corporate politics at the time. Another figure in the office was Don McGirk, a pleasant, distinguished-looking man in a tweed suit, who was the Exploration Manager, but seemingly not part of the Bishop-Bode axis.
12. I later published a paper with Hans Bürgl on this key section, which is one of the best exposed in South America (see Campbell, 1965).
13. Our particular friends included the Barry's whose paths we crossed many times later; the Ross's also from Texaco Trinidad, and the Tanner's.
14. Moose Bernard, a larger than life American, was in charge, while Freddie Martin (dec'd) was Exploration Manager: a quintessential upper class Englishman with a nervous cough and a love of night life. Others were Bill George, the Deputy Manager, who later helped found Nordic-American; Richard Hardman (later a director of Amerada and President of the Geological Society); and David Walker (later to become Chairman of the British National Oil Company). Jim Spence (dec'd) and John Harrison were also there when I first joined. The draughtsmen, López, Obando and Prieto, and the secretary Magda Luz Gutiérrez, were especially close, helping me to produce my reports, which in the days before photocopying and word processing were reproduced on ozalid skins.
15. The report identified the prime tract in the Llanos Basin with the following words "We would therefore rate the block of acreage bounded by the Venezuelan border, the Tame and La Heliera wildcats and the Rio Ariporo as of prime interest." This is precisely where Cañon Limon was subsequently found.
16. Colombia has had a rather exceptional discovery pattern with three hyperbolic trends: an early one around the La Cira-Infantas discovery in the Magdalena Valley; a second in the 1960s with Orito, which opened up the southern Sub-Andean Basin extending into Ecuador; and a third with the Llanos discoveries, highlighted by the Canon Limon and later the Cusiana Fields.
17. As, in fact, predicted in *The Coming Oil Crisis*.
18. Wallace Pratt, an American expert, put it well in 1952 when he said "Oil is found in the minds of men".
19. Knoepfel H., 1986, gives a first class discussion of energy issues.
20. Hardy D.R. et al., 2003 *DOD Future Energy Resources* ; Dept. of Defense, Washington
21. One could include geothermal energy, which is renewable where it comes from volcanic rocks, but exhaustible where coming from a buried aquifer.
22. This is an interesting case. Mr Brian Fleay has written a book "The decline of the age of oil", based on

an identical reasoning to that developed here. It captured public imagination and led to strong opposition to the construction of a road called The Eastern Distributor, which had been planned as infrastructure for the Olympic Games in 2000. It may be one of the first examples of where concern about future oil supply has led to a major public outcry (see Macleay, 1996).

Chapter 2

OIL: CHEMISTRY, SOURCE AND TRAP

IF YOU DON'T KNOW much about where oil comes from, don't be ashamed. Until quite recently even professional explorers had only a rather hazy idea about the exact origin of oil. In the last Chapter, I described my experiences as a young geologist looking for oil in Trinidad and Colombia. I knew of the classic view that most of the oil in northern South America came from the *La Luna Formation*, a sequence of petroliferous Cretaceous shales and muddy limestones, but I had difficulty in understanding how this deep-seated source could charge the overlying Eocene reservoirs in the Middle Magdalena Valley without filling intervening sandstones. To overcome this difficulty, I proposed that the oil was generated in the shallow sequences themselves, despite their non-marine origins. It seemed a reasonable explanation at the time but it was no more than a hypothesis, which in fact proved to be wrong. Now, geochemistry can specifically identify not only source but the date of generation. It has revolutionized the search for oil, permitting a much improved assessment of what the World's oil endowment actually is. It is of cardinal importance, and I will try to explain it in this chapter. It may be rather heavy going for the generalist, but it is important to understand at least something about the physics and chemistry of oil and gas at the outset.

Before coming to explain the origin of oil as now accepted by virtually all explorers, I should dispose of an alternative theory that was been advanced by the late Thomas Gold,[1] an academic, who proposed that oil and gas come from the deep interior of the Earth. If this were so, petroleum would be present in almost infinite quantities. This theory apparently endorsed by Soviet scientists, and has been again advanced by J.F. Kenney[2] and others. The flat-earth economists,[3] who reject the notion of finite limits beyond market forces, are prompted by their vested interest to promote the theory. When the Swedish government bowed to environmental pressures and suspended the production of nuclear energy in the wake of Chernobyl, it was persuaded at huge cost to test Gold's theory. Two wells were drilled to about 7000 m at Siljan, the site of a meteor crater in an area of granitic rocks, lacking[4] conventional prospects. Some indications of oil and asphalt were known in the vicinity, but are thought to be due to the local distillation of oil from Ordovician black shales, due to the heat of the meteor impact.[5] It is claimed that the wells encountered some indications of oil, but a degree of mystery surrounds the critical samples. If valid, which is far from certain, probably such indications either came from lubricants used in drilling the well, or from the shales into which the granite body was intruded. Virtually no professional oilman takes the hypothesis of an inorganic origin of oil seriously. I will leave it at that beyond mentioning recent research[6] that has determined that virtually all known reservoirs have a temperature of between 60 and 120 °C. This is simple observation not relying on scientific theory. Drilling deeper does not therefore find more oil or gas.

THE CHEMISTRY OF PETROLEUM

Organic chemistry is so termed because it is concerned with the essential element of living organisms; namely carbon. This is not the place to go into the subject in any depth: suffice it to say, that oil and gas, together with other related substances such as asphalt or bitumen, are hydrocarbons made up of carbon and hydrogen in various molecular combinations. In fact, we are concerned with a family of such molecules which can dissolve intimately in each other, depending partly on the ambient temperature and pressure: they are, so to speak, born, transformed, degraded and disappear, having in this respect a sort of life cycle that mirrors their organic origins. The term *petroleum*, meaning rock oil, is strictly speaking a misnomer given the organic origin of oil, but is nevertheless a useful generic term for oil, gas and related substances.

We are concerned with three hydrocarbon families:

1. Saturated hydrocarbons
Saturated hydrocarbons are paraffins (or alkanes), which are quantitatively the most important, making up 50–60% of most oils. They form a linear molecular chain with the general formula C_nH_{2n+2}, and occur in three states:

$$gas \quad - \quad with\ C_1\ to\ C_4$$

$$liquid \quad - \quad with\ C_5\ to\ C_{15,}$$

$$solid \quad - \quad with\ above\ C_{15}$$

So-called n-paraffins, with odd numbers of carbon atoms, are synthesized in living organisms. Such molecules are found in oil, being true biological markers inherited from the living organisms from which they were derived. C_{15}, C_{17} and C_{19} characterize microscopic organisms, including algae, whereas molecules of above C_{21} typify plants. These chemical links give the game away, showing that oils come primarily from algae. Another group of molecules, the iso-alkanes with a branched structure, include pristane (C_{19}) and phytane (C_{70}), and is derived from chlorophyll in living organic material. They also demonstrate the link.

2. Unsaturated hydrocarbons
Unsaturated hydrocarbons comprise the aromatics which have a ringed molecular structure. They are so named because of their pleasant smell, although the naphtheno-aromatic sub-family is commonly associated with sulphur compounds, giving them the exceedingly unpleasant smell of bad eggs. Benzene (C_6H_6) is the simplest aromatic hydrocarbon.

3. Resins and asphaltenes
Resins and asphaltenes are complex compounds with high molecular weights, rich in nitrogen, oxygen, sulphur, nickel and vanadium. They are mainly the products of the chemical alteration of ordinary oils.

It is of course not as simple as this. For example, the carbon itself occurs in two isotopes ^{12}C and ^{13}C, the proportions of which can be used to help identify whether the oils were deposited in marine or non-marine conditions.

THE PHYSICAL PROPERTIES OF PETROLEUM

As I have explained, petroleum can occur as a solid, liquid or gas, partly depending on the ambient temperature and pressure, and each phase may contain dissolved elements of the others. Oil, thus, commonly contains gas; and gas contains liquids.

Natural Gas is divided into two types: *dry gas*, consisting mainly of methane, and *wet gas*, which contains liquids, such as propane and butane, and is normally found in close association with an oil accumulation. The isotopic composition of methane generally reflects the degree to which it has been subjected to rising temperature and pressure on burial. Some natural gas deposits contain hydrogen sulphide (sour gas), nitrogen and carbon dioxide, depending on depositional conditions and the effects of alteration. Small amounts of helium and argon are also sometimes present.

Methane is relatively soluble in water. Natural gas is highly compressible, such that its volume may expand by a factor of 200 to 300 on being brought to the surface, which incidentally means that it can provide a valuable drive mechanism to expel associated oil from a reservoir.

Hydrates are special deposits of methane in an ice-like solid condition that are found in Arctic and deep oceanic environments. Some hopes have been entertained for exploiting such deposits, but they are unlikely to be fulfilled as the methane, being held as granules and laminae in a solid matrix, has no opportunity to migrate and accumulate in commercial quantities. The alleged large size of the deposits was based on a particular seismic reflector, which was later found to be an artefact, caused by temperature boundaries in the overlying seas.

Oil is a liquid hydrocarbon but generally contains varying amounts of gaseous and solid hydrocarbons in solution. These phases may separate naturally, and can be extracted by processing. Oils come with many different characteristics. Density, mainly reflecting the chemical composition, is one important property, which in the Anglo-Saxon world is traditionally measured under a scale set by the American Petroleum Institute, with most oils being in the range 15° to 45°API (0.9–0.7 S.G.). The heavier oils are rich in resins, asphaltenes and sulphur, whereas the lighter oils tend to contain dissolved gas. Heavy oils are generally dark brown or green in colour whereas the light oils may be almost as clear as a refined product. Another important property is viscosity, the inverse of fluidity, which generally increases with density, and decreases with the dissolved gas content and at higher temperature. It is measured in centipoise,[7] ranging from 1 cP to more than 10 000 cP. This property is related to pour point, which is linked to the paraffin content. High viscosity crudes, especially waxy ones of non-marine origin, become pasty and solid below about 10°C. For example, the heat generated by a high through-put is needed to prevent the oil in the Alaskan pipeline from solidifying.

A third important property is solubility: namely the ability of the several fractions to mutually dissolve in each other. In particular, large amounts of gas can be dissolved in oil. It is measured as a gas-oil ratio (GOR) which may be as high as 6000 cubic feet per barrel (1000 m^3/t). The ratio varies inversely with density and rising pressure. Where conditions approach the bubble point, the gas separates to form a gas cap above the oil accumulation. The dissolved gas increases the volume of the liquid, and a so-called *Formation Volume Factor* has to be applied to convert volumes of oil in the reservoir to those at the surface where the gas comes out of solution.

High Density Oils may be classified into three groups : *Bitumen*, defined by viscosity, *Extra-Heavy Oil* with a density greater than 10°API, and *Heavy Oil*. There is unfortunately no standard industry definition for the upper threshold of *Heavy Oil*, adding to the confusion.[8] Canada and Venezuela have high cutoffs at respectively 25 and 22°API, but this study takes 17.5°API as a pragmatic lower limit of normal production. The sulphur content may be as high as ten percent.[9] Heavy oils have various origins but most commonly are normal oils, from which the light fractions have been removed by water-leaching, oxidation or microbial degradation. Huge

Fig. 2-1. The Pitch Lake of Trinidad, a natural seepage that has long been a source of asphalt

deposits of heavy oil and bitumen occur in Eastern Venezuela, Western Canada and Siberia, forming important resources for the future.

Solids comprise methane hydrates, already described, and a whole family of complex bitumens and asphalts,[10] such for example as found in the famous Pitch Lake of Trinidad.

That, in a nutshell, describes the physical and chemical properties of petroleum: it is a slippery substance in more senses than one as we shall see as this account unfolds. Its diversity in a way reflects the diversity of the life from which it was derived and the conditions to which it was exposed thereafter.

Having briefly covered its properties, I will now turn to consider its formation in Nature. It is another very complex subject, but one that has to be understood, at least in general terms, to gain an appreciation of the reasons why the World's endowment is so finite.

THE FORMATION OF OIL AND GAS

As those on diets soon come to learn, living organisms are largely made up of proteins, carbohydrates and lipids (fats). Lipids are abundant in algae, especially the *Botrycoccus* family, and in diatoms, and are also present in plants, being found in pollen, cutin, chlorophyll and caretonoids. The sunlit upper waters of seas and lakes supported abundant life, especially micro-organisms, including these algae and bacteria, which naturally flourished at times of global warming. The surrounding lands were dominated by plant life. The organisms and plants had their life-cycles ending in death. The remains of those that lived in lakes and seas sank directly to the bed of the sea or lake to form the basic ingredient of oil, while on land, leaves and plant remains were washed by rivers into lakes or seas and form a source of gas.

In the summer of 1988, the Italian tourist industry faced a serious setback. The beaches along the Adriatic Coast became covered with evil-smelling slimy masses; and offshore, the fishermen reported that their nets were being clogged with the same stuff. Unusual weather conditions had led to a prolific flowering of algae, which absorbed so much oxygen from the sea as to poison not only the algae itself but marine life generally. Apart from what was washed onto the beaches to dismay the tourists, the organic debris eventually sank to the sea bed.

I am myself familiar with other examples of places where marine life proliferates excessively. Fishing in the Humboldt Current off Ecuador was such an experience. The current runs northwards along the coast of South America, and is affected by offshore winds that drive the surface waters westwards to be replaced by up-welling deep cold water along the coast. This deep water is heavily mineralized, providing important nutrients to support the proliferation of micro-organisms on which the whole food chain is built. The grey sea with mist banks, caused by the cool water, was full of game fish, while the fins of numerous sharks could be seen circling our boat. The sky was as full of sea birds, living off vast shoals of anchovies. The sea and sky merged into one huge organic soup. The sediments being deposited beneath this current have been investigated,[11] and do indeed form potential source-rocks for oil, although not in the life-span of *Homo Sapiens*. Note in passing that this same life stock was responsible for the great deposits of *guano* in Peru and Chile, on which Europe depended for agricultural fertilizer prior to the advent of artificial fertilizer made from hydrocarbons and by electrolytic methods. Likewise, the Norwegian fjords, in which we used to sail, sometimes turned milky white due to the proliferation of algae during the long hours of daylight in the northern summer. There were even reports of so-called "red tides" when algal flowerings took so much oxygen from the waters that they turned a reddish hue.

Looking into the geological past, it is clear that there were only periodic explosions of such life in places where conditions were exceptionally conducive, as in the present day examples quoted. The critical factors were sea temperature due to latitude, global climate and local circumstances, as well as the supply of nutrients. In practice, abundant hydrocarbons were generated only in tropical latitudes: remembering that some areas that were previously in the tropics subsequently moved to higher latitudes under Plate Tectonic displacements. The configuration of the Tectonic Plates affected the distribution of ocean currents, having a major consequential influence on climate. Most such movements were northerly which explains the paucity of oil in the Southern Hemisphere.[12] These brief and rare events are responsible for the oil trends as we know them, and we do now know almost all of them. It transpires that the bulk of the World's oil comes from just two short epochs of extreme global warming, 90 and 150 million years ago. Clearly, the abundance of the organic material was the first essential ingredient, but it was not in itself sufficient, for the organic debris had to be both preserved and concentrated before it could become a commercial source of oil or gas. The total organic content of a source-rock may exceed ten percent in ideal conditions. For these and other reasons to be explained later, only a very small fraction of the Earth's original endowment of biomass has been available to form oil and gas.

The fleshy parts of dead fish settle quickly to the sea bed, but the microscopic plankton, which forms the main source of oil, settle at a rate of no more than about 100 m per week. Only about two percent reaches the sea-bed of shallow seas or lakes, and perhaps one tenth of that survives in the deep oceans, because it is oxidized before it gets to the bottom. Furthermore, much of what does reach the seabed is destroyed by bottom-living organisms. It is only in the depths of stagnant marine troughs and deep lakes, where the oxygen content is low, that large amounts of the organic material can be preserved. It is also necessary for the material to be concentrated, as occurs in areas receiving little other sediment. Present day analogues are the Black, Baltic and Caspian Seas, the Gulf of California and Lake Maracaibo.

The World's most prolific oil province, at least so far as *Regular Conventional Oil* is concerned, is the Middle East.[13] The organic material responsible for it was formed in warm Jurassic seas and accumulated in stagnant sink holes and lagoons within a broad carbonate platform that received only limited amounts of sediment washed in by rivers draining the surrounding low-lying lands. Another prolific province is the North Sea, where organic material accumulated in stagnant rifts towards the end of the Jurassic Period, 150 million years ago, when the North Atlantic was beginning to open. A third example is provided by the rift lakes that developed along the west coast of Africa and Brasil, as the South Atlantic opened during the early Cretaceous.

The point is that these conditions were met only very rarely, both in time and place, which explains why prolific oil and gas deposits are restricted to only a few well defined trends. To confirm this, you have only to look at an oil map of the World, which shows clusters of oilfields separated by huge barren tracts that lacked the essential conditions.

The rate of sedimentation plays an important part in the process. If sedimentation is too rapid, the organic material becomes disseminated, whereas if it is too slow, its preservation is impaired. Clay minerals are also involved, helping to fix the organic material.

So far, I have spoken mainly of oil, which is associated with restricted marine or lake environments characterised by an abundance of planktonic and algal organic debris. By contrast, gas is associated with more brackish environments, as found in deltas, where ligneous and humic material from plant-life predominates. In some areas, the two environments overlap, giving both oil and gas source-rocks.

WALTER ZIEGLER: International Explorer

Q Walter: you come from a family of well known geologists. Can you tell us a little of this background?

A *My father was a doctor in Wintertur in Switzerland. He was a keen naturalist, and instilled in all four of his children an early scientific interest, and an appreciation of nature, particularly botany and geology. He also had a philosophical streak, and taught us to first observe, and then to try and explain, and comprehend. He wanted logic not magic. It influenced our choice of career: but all the same, it was strange that there should be three geologists in one generation.*

As children, we often went to our grandfather's summer house in the Jura, a beautiful range of beech clad mountains and green valleys. The shiny yellow-white limestone cliffs formed spectacular rock arches in the gaps where the rivers broke through. We later understood that they were structures termed anticlines, and still later that oil is trapped in anticlines. Ancient castles and ruins stood on these cliffs guarding the passages from one valley to another in this spectacular country. We couldn't fail to notice the alternating strata with the hard limestones standing out from the soft shales. We began to have an intuitive feel for geology, which was further encouraged when we found fossils: ammonites (Ammon's Horn), belemnites (Devil's fingers) and other shells, which our mother often pointed out.

A family friend, Freddy Senn, was an oil geologist, who impressed us all with his stories of travel in distant lands, including Burma and South America.

Fig. 2-2. Walter Ziegler, an international oil explorer who took part in the North Sea discovery

When he died unexpectedly on his way to Morocco, his widow gave us a book from his library "Geologie der Schweiz". Its pages were full of cross-sections, maps and panoramas which allowed us to identify and understand the geology of the country we knew.

Naturally, we also went to the Alps, and broadened our knowledge with minerals, igneous and metamorphic rocks, as well as the sight of the great overthrust nappes, exposed on the mountain side. My brother, Peter, won a school prize with his collection of minerals. I had a teacher called Dr Peter Walter with the prominent "beak" of the true scientist, who had a great interest in geology. During my last summer vacation, I hiked with him from the Helvetic Alps through the Aar and Gotthard Massifs to the Engadine, traversing a great section of Alpine geology. The results of my exams were less than impressive, but my mind was far away in the Alps. I remembered particularly a chance encounter at an altitude of 3402m on the Rheinwald Horn, where we met Wolfgang Leupold, a professor of geology at Zurich University, who gave us an impromptu lecture on the geology of the range we could see below us, stretching from the Po Valley in the West to the Tauern Mountains of Austria in the East. I was stimulated to learn more and understand how this great mountain chain came into being.

In the autumn of 1947, I started a general science course at the Swiss Federal Institute of Technology (ETH), but I did not enjoy it and was glad to be called up for Military Service. I was a driver and took the opportunity to learn as much as I could of the geology of the areas where I was stationed, especially the Tessin. I had a normal army experience for two or three years, alternating with university semesters, and ended up as an Artillery Lieutenant. It was the beginning of the Cold War, and we understood that we had to be ready to defend our country lest it suffered the same fate as Czechoslovakia where in the spring of 1948 a democratic government was overthrown by Russian-backed communists.

Gradually my interests at University concentrated on geology, and I was fortunate to fall under the influence of Professor Rudolf Staub, a great Alpine geologist, who instilled enthusiasm for the subject and taught us how to think, observe and interpret, and step beyond that to speculate about the unknown. Although I had inscribed in a course of Engineering Geology, I came to realise that working on foundations and tunnels would be rather restrictive, and resolved to try to become a petroleum geologist and see the world. To this end, I spent 2½ years on a higher degree, studying an area of flysch in the Grisson under Professor Staub. I enjoyed the good comraderie of the post-graduates under Capo Staub, as we affectionately called him. We learnt a lot from each other and at times had too much to drink, which didn't harm us either. I finished at the ETH in February 1955 at the great age of 27. I felt I needed to escape from the somewhat closed society of Swiss academia and the highly regulated social structure of the country. I yearned for freedom and adventure, not creeping socialism.

Q When you graduated, your brother followed the traditional path to Shell, but you went to Canada to join Esso, now ExxonMobil, the world's largest oil company. Can you say something about these early days and how your career evolved?

A *My first job was as a photogeologist with the Institute Francais du Petrole, where I was engaged in mapping the Sahara under Daniel Trumpy, a famous retired Shell geologist. Never have I been so cold as that winter in Paris in 1955/56. Gas, which had to be piped from the Saar coalfields, was in short supply, and the pressure was barely sufficient for heating or cooking for days on end. France was then facing its retreat from Empire, having suffered the defeat of Dien Bien Phu in Viet Nam. There were frequent strikes, high inflation and a general malaise. Army trucks provided transport, but despite all the hardships, Paris still held a great attraction for me.*

I was offered a job by Shell as a geophysicist, but that did not appeal to me, and instead I followed up a suggestion from friends in Canada, who told me that Imperial Oil, an Esso affiliate, was looking for geologists. I successfully applied.

I quit my job, and in due course set off on a great adventure: by train to Liverpool to board the Saxonia to Montreal and on to Calgary. A few days later, I was sent on my first mission to join a field party of three geologists, a cook, helicopter pilot and mechanic, working on the frontal ranges of the Rockies by the Athabasca River. The helicopter was a magic carpet that swept us up the mountain side to the outcrops we wanted to study, although it was often a hair raising experience when buffeted by mountain winds. However before long the carpet broke down, and we had to hire a string of pack animals from a backs-woodman called "Old John the Swede" who slept under his stinking saddle blanket. It was an exciting life of riding, climbing, measuring geological sections and living under canvas, and we saw lots of wildlife: moose, wapiti, Rocky Mountain sheep, goats, mules, deer, coyote, wolves and partridges. When the field season ended with freezing rain and snow flurries, it was time to head back to base.

I then joined the research group in Calgary where amongst other things, they were studying how to measure the chemical alteration of hydrocarbon source-rocks. Diane Loranger had noticed the progressive loss of ornamentation on fossil ostracods with increasing depth of burial, and Frank Staplin had observed how fossil pollen became darker in colour. These lines of research were to provide critical means of determining where and when the source-rocks would give up their oil: it was a glimpse of a new understanding.

I continued to spend the summers on field studies full of colourful and interesting experiences, and I also widened my experience of basin studies, wellsite work and becoming generally more of an oilman. Then in the early

1960s came bad times. The industry was suffering one of its periodic downturns, and the geological department was reduced from 40 to 20. The Company had decided that it was not interested in gas, and preferred to import cheap Venezuelan oil rather than explore the foothills of the Rockies. It was to be a recurring theme.

But then a new play opened up with the discovery of Swan Hills and Judy Creek, where the objectives were hard-to-find reefs. We were back in business with all the excitements of a boom. I was promoted to Area Geologist in Edmonton, and now had for the first time a more commercial role, making deals and competing in lease sales. It was great fun.

Q Clearly, you integrated into the Canadian environment, but still you were a European at heart with a wider perspective. Esso evidently recognized this, when they invited you to transfer to their international operations. What was your reaction?

A On January 2nd 1964 I was told that I was to be transferred to Esso's Geneva office to study Triassic reefs in Austria. It was totally unexpected, and I could not believe my good luck. I had greatly enjoyed my time in Canada, but the limitations of the Alberta Basin were already evident. The new frontiers of the Arctic and the offshore were only just awakening, and the thrust of exploration was turning overseas, especially to Libya. I welcomed the chance to broaden my horizons.

Q What were the highlights of this new assignment?

A The first was a cultural shock: I had almost forgotten how to speak French. But then another stroke of good luck was a reorganization that transformed what had been a research office into a European Frontier Exploration office. My new mission was to integrate a preliminary North Sea study into a regional context. I made numerous trips to Germany to collect information. Next came another transfer: this time to Spain where a new office to evaluate Africa had been established under Dave Kingston, one of Esso's so-called "Rover-boy" geologists with whom I hit it off particularly well. I started collecting information with which to prepare the basic basin maps, and also made a brief trip to Morocco to investigate an exploration idea.

I was then recalled to Calgary for what turned out to be a splendid and stimulating period of work in a very congenial environment. I was able to catch up on the latest progress in geochemistry and the early steps towards seismic stratigraphy on which great strides were being made. But despite the enthusiasm, I think at the back of our minds we were beginning to see clearly the limits of exploration. There were fewer new ideas to be tried, and often we would find that many of the promising concepts that did develop in our minds turned out on further investigation to have been already tested.

Before long I was back overseas to be based in Spain on African studies. I began travelling in earnest with numerous trips all over Africa: Mozambique, South Africa, Tanzania, Kenya, Cameroons, Gabon, Rio Muni, and Fernando Po. I could see at first hand the full misery of post-colonial Africa: camps of starving refugee children, violence, stone-walling petty bureaucrats, and the general degradation of a once fine people under ill-absorbed Western influences. Technically, I soon realized that much of Africa was effectively non-prospective apart from a few key areas along the Equatorial Atlantic Coast, where most of the possibilities were already controlled by other companies.

In March 1969, began my final assignment to Walton-on-Thames in England to act as Chief Geologist in a newly established East Atlantic Study Group. It was a bad experience: we drilled some forty dry holes along the coast of NW Africa at a cost of more than $100 million. At the same time, the Company found itself in the grip of restructuring as various affiliates were consolidated and absorbed into what eventually became Exxon, which at the time we unkindly dubbed as the "sign of the double cross". The ranks of middle management swelled with ill-experienced and sometimes abrasive individuals from the domestic organization in the United States. They had been successful in the Gulf Coast and now wanted to conquer the world, which they thought of as a great extension of Texas, marred only by grasping foreign governments. The Viet Nam War was in progress.

Africa was assessed "to death" under the new administration, but the record shows that between 1972 and 1983 the Company had failed to find a single barrel of oil (save for Chad on a Conoco farm-in, which only now starts production) despite a search costing millions of dollars. Africa lacked the pervasive source-rocks of the Gulf Coast and so a "Gulf Coast" approach to exploration did not work. The Company had the expertise to understand but those in charge had grown up in a different environment and did not apply it rigorously. The thing rolled on under its own momentum at huge cost.

Later, the Company did learn from this mistake, and instituted a programme of training to very high technical standards. Exxon courses became renowned for their excellence in the industry, but in many respects it was a case of closing the stable door after the horse had bolted. There was much less left to find.

Finally in the summer of 1983, the Europe-Africa regional office was packed on to a 747 and shipped to Houston where it reopened in diminished form.

Q The North Sea was opening up then, and you published one of the first syntheses of that area. Did Esso play a fundamental role there?

A *In 1970, after the N.W.Africa fiasco, I joined a North Sea Study Group with the task of putting together a regional synthesis. We were in partnership with Shell as operator in the United Kingdom. The southern North Sea gas province had already been developed as an extension of the giant find at Groningen in Holland, and in the early 1970s some people thought that there was little further potential. However, when we were preparing for the UK Fourth Round of Licensing, Pat Kelly, one of our geophysicists, drew our attention to a regional seismic line that a contractor had shot east of the Shetlands. It showed "tilted fault blocks" below the Cretaceous, which reminded him of the El Morgan Field in the Gulf of Suez. Shell/Esso tested the idea with Well 211/29-1 in May 1971, and held the positive results in the greatest secrecy as a basis for bidding on surrounding blocks, many of which turned out to hold major oilfields. We did not have a close daily contact with Shell but followed its every action with close attention, contributing constructively as the ideas evolved.*

During these days, I compiled the regional picture within a plate-tectonic framework, making analogies with other comparable settings around the world. I authored one of the first published accounts of North Sea geology.

When the oil crisis of 1974 hit, I was hustled off to evaluate coal potential around the World, the Company having decided to move into that business. In the spring of 1975, still another new study group was formed to evaluate the Circum-Atlantic in the hope that the eastern US shelf might provide much needed new oil supplies. I was called in to advise, but on the basis of my evaluation was neither enthusiastic nor popular for explaining yet again that no amount of ingenuity can compensate for the lack of active source-rocks. Our team's advice was ignored, and several expensive wells were drilled with predictable results. By now, we really understood the essential factors needed to make an oil prospect, and we had all the technology to provide the raw data. The sad fact was that there was much less left to find. The US East Coast was as poorly endowed as its counterpart in N.W.Africa, which we had already explored to our cost.

After other assignments in Libya and Africa, we faced in 1983 the final dismantling of the once mighty global Esso Exploration organization. The world evaluation had effectively been completed. It showed that there was nothing left to do beyond a mopping up operation and a single office was enough to handle the ever smaller projects that remained.

I was offered a senior position in this new organisation but preferred to stay in England, partly for family reasons. A chance meeting led to an offer to join Petrofina, a Belgian oil company, that was then seeking to expand. I accepted, and after a period in the U.K. office, moved to Brussels where I ended my working career. It was a new company with a very different character. We tried our best looking for by-passed ventures, but the world was the same and I could not find what was not there to be found.

Q So you had a very varied career in many parts of the world, and you were able to observe the unfolding situation from a very privileged position in the world's largest oil company, having access to a colossal database and call on the most advanced technical expertise. How did you assess the situation in resource terms?

A *I regretfully came to the conclusion that it was a war of lost causes. There were thrilling moments as during the pioneering days in Canada and later in the North Sea when we opened up a new province, but much of it was a catalogue of frustration. We knew what we were doing, and, with the best will in the world and the best expertise available, we simply could not make a purse out of a sow's ear. There are only about thirty provinces world wide as prolific as the North Sea, that is to say having an oil endowment of more than ten billion barrels, an amount itself enough only to supply the world for six months. The key to the North Sea was the prolific Upper Jurassic source-rocks. Every successful basin relies on a well defined source and an appropriate thermal history.*

Q Time and again you have stressed the importance of source. What do we know about the world distribution of source-rocks.

A *Each successful basin has its source system, by which I mean not only the rock itself, but the timing of generation and migration. It is a thermokinetic process whereby the rocks give up their oil and gas on critical exposure to heat. In addition, the preservation of the oil and gas formed is vitally important: all oilfields leak over time. So, we had to search for structures that had been charged in the relatively recent geological past.*

Q So how would you sum up the conclusions of a career spanning fifty years of exploration in many parts of the world?

A *Above all I would say that it was great fun. I saw a lot, learned a lot, lived in many different countries, and met a great number of varied and interesting people from many backgrounds. It was my good fortune to live during an epoch when petroleum geology became a science, subject to rigorous scientific discipline. It is well capable of answering questions about the availability of hydrocarbon resources on which the modern economy depends. Our studies have confirmed beyond any doubt that the globe has a decidedly finite potential for oil exploration. The implications are colossal. The World must finally come terms with the fact that changes in the way it lives are imminent. It has no option but to adjust to resource limitations. "No more candy for the kids!" The game is nearly over.*

Q One final question: can I ask you to use your worldwide experience to show is where the key provinces are: not a present day map but one showing the continents in the positions they then occupied.

A Yes indeed, oil is concentrated into a few provinces, for well understood reasons, and we can plot where they are. I don't think that there are many, if any, new ones to find considering how extensively the world has now been explored, and considering that we now know so much more about the factors responsible.

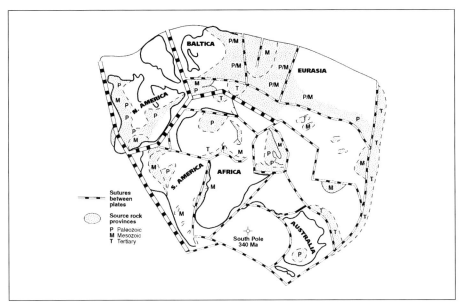

Fig. 2-3. The main oil and gas provinces of the world in their original setting along the sutures marking the break-up that led to the present continents

THE CONVERSION OF ORGANIC MATTER INTO OIL AND GAS

In a sense, the organic material had a life after death as it is gradually converted by chemical reactions into oil and gas. Once it had landed on the sea- or lake-bed, it was buried beneath layers of generally fine-grained sediment, washed in by rivers from the surrounding lands. Physical and chemical reactions commenced almost immediately, being primarily driven by bacteria and other micro-organisms that can continue to attack the sediments until they are buried to depths of several hundred metres. The main effect was to create a strongly reducing environment. The bacteria split the complex molecules, releasing carbon dioxide and methane, leaving behind an insoluble residue known as *kerogen*. It occurs in three important types: *sapropel* yielding oil; *vitrinite* giving gas; and *inertinite* that provides neither. It is also classified into three categories, known as Types I, II and III, according to the path of chemical change.

Types I and II with a high hydrogen/carbon ratio are oil prone, whereas Type III, with a low ratio, is gas prone. It is all rather complicated.

The temperature rose with burial below the sea-bed, and the resulting chemical activation tended to break down the complex molecules into more simple structures, commonly releasing methane in the process. Three stages of alteration are recognized: *diagenesis* at shallow depth; *catagenesis* at moderate depth, normally in the range of 2000 to 4500 m, when the bulk of the oil was generated; and *metagenesis*, at greater depth where the oil was cracked[14] to gas. Coaly material, called *vitrinite*, changes its reflectivity on alteration, which means it can be used as a sort of thermometer to indicate the degree to which the rocks have been "cooked" in terms of both temperature and the time to which they were exposed to it. *Vitrinite Reflectance* values thus allow us to track the course of alteration for both hydrocarbons and coals, as shown in Figure 2-4.

Stage	Vitrinite Reflectance	Hydrocarbons	Coal
Diagenesis	<0.5	Biogenic gas Early gas	Peat Lignite Sub-bituminous
Catagenesis	0.5-1.0 1.0-2.0	Oil Gas-Condensate	Bituminous Coals
Metagenesis	2.0-4.0	Methane	Anthracite
Metamorphism	>4.0		

Fig. 2-4. The alteration of organic material into hydrocarbons and coal

As temperature rose over the critical range, more and more *kerogen* was converted to oil, but as the process proceeded there was less and less left to convert. The term *oil window* is used to describe the depth-range of generation, which generally does not extend over more than about 1000 m. The peak itself is even more important, as it takes place over a relatively short time span. It has a dynamism of itself, which I will consider more in relation to the migration of oil. It is critically important to determine the structural conditions that were present at the time of peak generation because they will primarily determine where the oil moved to and which traps were filled.

HOW OIL MIGRATES AND ACCUMULATES IN OIL FIELDS

I have explained how organic matter was converted to *kerogen*, which in

OIL & GAS GENERATION

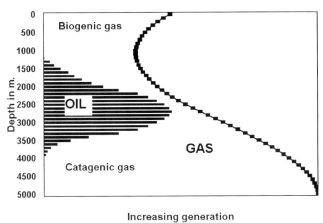

Fig. 2-5. The oil window: the depth at which oil is generated

turn yielded oil and gas on being buried to a critical depth, normally between 2500 and 4000 m. It was preserved solely in certain environments, in which, for various reasons, only fine-grained sediments were deposited. Such rocks, with their high organic content, generally consist of dark-coloured clays and muddy limestones, often having a petroliferous odour. They are called *source-rocks.*

The expulsion of oil from the source-rock, under the circumstances described, is termed *primary migration,* whereas its subsequent movement into a reservoir is called *secondary migration.* Chemical reactions at the critical temperatures broke down the large *kerogen* molecules to form the smaller molecules of oil and gas. This resulted in a moderate expansion of volume because the density of oil is 0.7–0.9 g/cm^3 compared with 1.15–1.35 g/cm^3 for *kerogen.* The differential in the case of gas was even greater.[15] This expansion led to an increase in pressure, which was a critical factor in oil migration that needs to be explained.

At a given depth in a sedimentary basin, the weight of the overlying rocks is partly carried by the stress transmitted directly through the network of grains making up the rocks, known as *vertical stress,* and partly by the fluid pressure in the pore spaces. The generation of oil and gas, which entered the pore space, therefore increased the fluid pressure both absolutely and relatively to the *vertical stress.* The *kerogen* in the source-rock was commonly concentrated into a sequence of individually thin layers, which became fluid on being converted to oil. The pressure in such fluid intercalations rose until it exceeded that of the stress imposed by the overburden weight. Since the lateral rock stress is less than the vertical stress, the excess pore pressure led to the formation of vertical fractures, along which the oil bled off to relieve the pressure, thereby closing the fluid layer until the grains again come in contact to carry the overburden. If you sit on an air-cushion not strong enough to support your weight, you will experience something of the same sort of phenomenon.

Once the oil and gas were forced out of the source-rock, the main driving force for further movement was buoyancy, because the density of oil and gas is less than the water that naturally filled the pore space. The strength of the force is determined both by the height of the oil column in the rocks and the counter capillary pressures either between the oil and rock grains directly, or between the oil and a thin film of water that coats the grains, as is usually the case. The capillary resistance to flow is related to the throat-size of the passages between the individual pores. The throat-size in an un-cemented sandstone is large enough to allow the oil through easily, in which case the sandstone is said to have adequate *permeability*. Rising damp in an old building demonstrates how seemingly solid stone has both porosity and permeability, allowing water to gradually flow though it. So it is in underground reservoirs, save that the pressures are much higher. The pore throat-size in shale is low, typically having a permeability of about a *nano-darcy* (10^{-9} D) - permeability being measured in units termed *darcys* – and, in such cases, the movement of oil is prevented. Micro-fractures, however, provide a pathway by which the oil can cut across such impermeable layers, many of which have normally to be breached in the course of its migration.[16]

Knowledge of the process of rock-fracturing is routinely gained from drilling operations. Ideally, the drilling mud is weighted up to a level sufficient to exert enough pressure to preserve the borehole: if the mud weight at depth is too high, it will force its way into the formation, fracturing it in the process, whereas if it is too low, the formation may encroach on the borehole to grip the drill bit. The rocks at depth become quite plastic under the stresses to which they are exposed. One of the main challenges to efficient drilling is to properly match the mud-weight to the changing subsurface formation pressures: failure to do so can result in a dangerous blow-out or "stuck drill-pipe".

In many basins undergoing active subsidence, the critical fracture pressures are commonly found in and around the oilfields, suggesting that oil migration is in progress at the present time. In fact, in some instances, oil, or more commonly gas, from depth can to some extent continuously replenish the reservoirs. In other cases, where the original basins have been subjected to subsequent uplift, the formation pressures are much lower, approximating to the hydrostatic head of the water in the formation. In indurated sequences, geological faults can form migration paths for oil, but in cases where the sediments were still un-compacted at the time of faulting, the faults are smeared with clay, and form seals rather than conduits for oil.

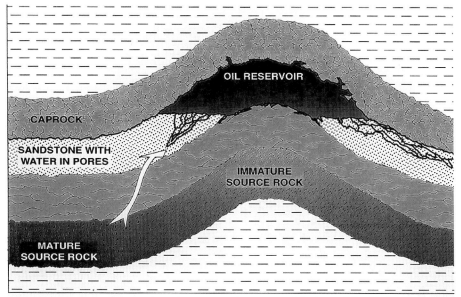

Fig. 2-6. Oil migrates from the source-rock into an anticlinal trap (after Bjørlikke)

In cases where there were no intervening reservoirs and traps, oil and gas migrated vertically along the fracture systems created in the process of migration to dissipate or to escape at the surface, giving rise to seepages which are found in many petroliferous basins. In other cases, the migrating hydrocarbons encountered a conduit, such as a porous sandstone, and then flowed through it laterally and upwards under the influence of buoyancy.

Again, if this conduit led directly to the surface, the oil and gas escaped. The huge degraded deposits of western Canada are an example. If however the rocks in the basin had been folded and faulted, the hydrocarbons accumulated at the highest point of such traps. There is an infinite variety of circumstances: sometimes the trap lies directly above the source, as for example in the case of the Ekofisk Fields of Norway. In other cases, it lies far away, perhaps at the margins of a basin as in Eastern Venezuela, where oil is found at shallow depth far from where it was generated. The catchment area for traps is colourfully referred to as the "kitchen", namely where the source-rocks are cooked to give up their oil and gas. The traps themselves are subject to various pressure regimes as they become charged with oil and gas.

Figure 2-8 is a cartoon of the different types of traps that can hold oil in a sedimentary basin bordering a continent. In the real world, it is unusual to have so many different types of trap in close proximity.

The degree to which the traps could hold their charge depends on the sealing quality of the overlying rocks, and the pressure differential. Salt

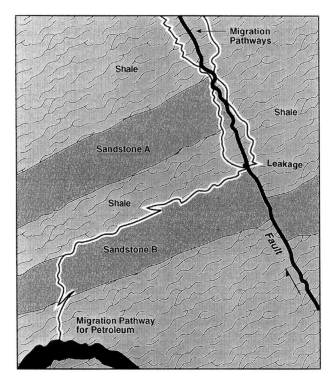

Fig. 2-7. Oil may also be trapped by faults, some of which leak (after Bjørlikke)

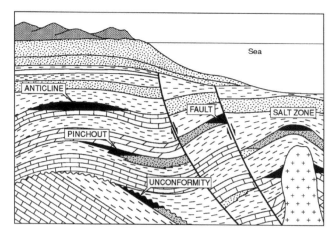

Fig. 2-8. A stylized illustration of diverse oil traps

and anhydrite are particularly efficient seals, and their widespread occurrence in the Middle East is one of the factors responsible for its rich endowment. Shales and clays also served as seals that could hold the hydrocarbon charge until a critical pressure was breached. When that was passed, the seals fractured, as described above, and the hydrocarbons vented from the weakest point in the trap, either to collect in a shallower trap, if one was present, or to escape at the surface. Since gas tended to segregate from the oil and collect at the top of the trap, the gas escaped from an imperfect seal more easily than did oil. No seal has complete integrity, and some degree of seepage is inevitable over time. It explains why the occurrence of oilfields becomes progressively rarer in older rocks. It is still another factor that limits the resource. The process of oilfield destruction has been a very important one that is not always appreciated.[17] At shallow depth, the petroleum is prone to leakage, biodegradation and weathering. At great depth, it is cracked to gas. However, oil that has leaked from one accumulation may in some circumstances collect at shallower depth. For example, the oil in the West of Shetlands discoveries, off the United Kingdom, evidently migrated early into a larger structure termed a "holding tank", from which it later leaked and re-migrated into other structures that had formed in the meantime.[18] Much was naturally lost in course of each re-migration, but it does explain some otherwise anomalous discoveries. It is estimated that the median age of the World's oilfields is 35 Ma (Oligocene), which is quite recent in geological terms. Older fields are preserved only where they lie in quiet tectonic settings and where they are endowed with very effective seals, normally salt.

In summary then, oil and gas are hydrocarbons derived from organic material: oil coming primarily from algae that lived in seas and lakes; and gas from plants whose remains were washed into them. Only a small fraction of this biomass was preserved from oxidation and destruction, and it accumulated in only a few places where conditions were right. Such organic material as was preserved was buried below sediments that were subsequently washed into the seas and lakes, and it was heated on burial by the Earth's heat flow. Chemical reactions converted it to a material known as *kerogen*, which in turn yielded oil and gas on further burial. Peak generation was reached at a certain depth of burial and lasted for a comparatively short period of geological time. The oil from this short-lived charge filled such structures as were in communication with it. The organic material and any hydrocarbons formed were progressively destroyed when depressed beneath the *oil window*. Only a very small fraction of the oil generated, perhaps about one-millionth, actually found its way into exploitable oilfields. It has been calculated that the World's entire available supply of *Regular Conventional Oil* would do no more than fill Lake Geneva in Switzerland.

This may all seem very complicated for the non-specialist, and indeed there are many more factors than covered in this summary. But it is important to have at least an inkling of what is involved to understand why it is such a finite natural resource.

NOTES

1. See Gold T., 1988.
2. Kenney J.F., 1996, describes a Russian theory that oil comes from primordial material deep in the Earth, suggesting that the international industry is mistaken in its understanding and should now look in crystalline rocks and beneath volcanoes. Most experienced explorers would dismiss the idea. In any event, we should not rely on such an untested hypothesis for future supplies.
3. Professors P. R.Odell and M.A.Adelman, together with the latter's former student M.C.Lynch, are the high priests of this persuasion.
4. Krayushkin V.A. *et al.*,1994, provides the original paper behind Kenney's paper.
5. Vlierboom F.W. *et al.* 1986 gives an explanation of the Siljan drilling in Sweden.
6. See, Nadeau P., Bjorkum and O.Walderhaug, 2005. Evidently, this temperature range is controlled by several phenomena including pore pressure, fracture pressure, sealing conditions and diagenesis, making it the optimal zone for oil accumulation. It may be thick or thin, deep or shallow, depending on local circumstances, but it has great significance in defining the finite nature of the resource.
7. In the S.I. system, it is measured in mPa.s or milliPascal- second.
8. The official (UNITAR) classification terms oil with a gravity below 10o API Extra Heavy Oil.
9. The oil from the Tengiz Field bordering the Caspian contains as much as 16% sulphur.
10. Bitumen (tar) is a term applied to oils with a viscosity above 10 Pa.s (10 000 Cp), and asphalt is a term for bitumen with a viscosity above 20 Pa.s.
11. Demaison G.J. and G.T. Moore, 1988, have investigated the sediments being deposited by the Humboldt current. Demaison emanated from the French school of geochemistry but worked for Chevron in California.
12. Klemme H.D. & G.F. Ulmishek, 1991, provide a very valuable insight into the world's source rock realms, which formed the basis for the early assessments by the US Geological Survey.
13. In fact, the most prolific provinces, if oil that has been subsequently degraded is included, are in northern South America; Western Canada and Siberia, where huge tar sand and heavy oil deposits occur. A very effective seal is as important to the Middle East as is a source.
14. The term "cracked" means that the bonds holding the atoms together in oil molecules are broken down at high temperatures to yield simpler gas molecules.
15. Bjørlykke K., 1995, explains the complex process of oil migration in a very readable paper.
16. Heum O.R., 1996, provides further information on modern concepts of oil migration.
17. Macgregor D.S., 1996, has published an excellent discussion of this subject including a listing of giant fields.

Chapter 3

THE PIONEERING EPOCH

T HE LAST CHAPTER was rather heavy going technical stuff, but important reading all the same if one is to gain an appreciation of the physical constraints to oil production. There are not great underground caverns full of oil waiting to be tapped. It accumulates in the minute pore space of rocks under complex geological circumstances, which only rarely come together in the right combination. One weak link in the geological chain converts a promising prospect into an expensive dry hole. I explained how flush oil generation occurred when the critical temperature was reached in the source-rocks, and how it progressively consumed the amount of organic raw material available to convert. The result was a classic bell-shaped curve of generation, starting at zero, rising to a peak and then falling back to zero. This is a pattern which will recur in this story in different contexts: it is well to get used to it, for it is immensely important.

The World has now virtually reached peak production facing the onset of terminal decline, as I will discuss later. Indeed, the calculations suggest a peak by 2006. As many as 54 out of the 65 principal producing countries have already passed peak and are into decline. Some have been in decline for many years, while others are too close to it for them to have fully appreciated their position. We can learn much from the history of discovery and production prior to peak which has to be mirrored in inevitable decline.

While many of the details are fuzzy, no one can doubt the generality of the picture of the oil age in Figure 3-1.[1] It can be usefully divided into four epochs: *pioneering* to 1950; *growth* to 1970; *transition* during the early years of this Century and then *decline*. The transition is further sub-divided to include what we may call the *Crisis Years,* which promise to be just that. Identifying the precise date of the peak itself is not that important although it attracts much attention. What matters more is the recognition that the First Half of the Age of Oil, which was characterised by growing production, is giving way to the Second Half when production declines without reprieve, along with all that depends upon it.

In this chapter, I will describe the Pioneering Epoch when most of what we know about oil was worked out; when most of the technology needed to extract it was developed; and when most of the major oil-bearing

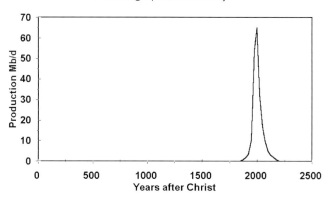

THE AGE OF OIL
A fleeting epoch in history

Fig. 3-1. The age of oil: a fleeting epoch in history

provinces were found. There have been important subsequent advances, but the back was broken during the *Pioneering Epoch*.

Oil and gas from surface seepages have been known since long before the time of Christ. Moses' basket of reeds was caulked with tar; Nebuchadnezzar's fiery furnace and the Biblical burning bush were, it may be assumed, located on gas seepages; and the eternal flames that were worshipped in antiquity were flammable oil-impregnated clays near Baku on the shores of the Caspian.[2] I remember seeing Papuan natives in the highlands of New Guinea, when I was there in the 1960s, daub their bodies with oil collected from seepages when they gathered for their ritual *sing-sing* dances, a custom that had being going on for time immemorial.

It did not take much to move from skimming oil off the pools into which the natural seepages ran, to start digging shallow pits to collect it more easily. The early Burmese were the most advanced in this regard, using bamboo pipes to case shallow wells and transport the oil. The Romans and Chinese were also exploiting oil before the time of Christ, mainly for medical purposes. Torches of burning tar were sometimes used in naval warfare. There has also been a very long history of digging wells for water and salt brine; salt having been a very valuable commodity for preserving food in the Middle Ages and before. The early Chinese are credited with the development of a form of rig for drilling wells, consisting of a heavy stone on the end of a rope that was repeatedly raised and dropped, slowly punching a hole into the earth. It was the precursor of the *Cable-Tool* with which the early oil wells were drilled.

A trade in oil had been established both in Europe and the United States long before wells were drilled specifically for it:[3] there was much demand for domestic illuminants derived from lard, whale-oil, camphene, as well as oil from seepages and coal workings. The distillation of cannel coals in Scotland in the 19th Century was particularly noteworthy as it opened the process of oil refining. The growing demand for this lamp-oil, coupled with the decline in whale oil due to over-whaling,[4] were the stimulants for the search for crude oil. The technique for drilling wells was already well established in North America in the production of brine, from which to extract salt needed for preserving meat in the days before refrigeration. There were even cases where such wells encountered gas, which was used to fuel the salt works.

The kerosene lamp was a great revolution in the way people lived, adding a useable evening to the working day, especially in rural areas. A second and greater revolution came on July 3rd 1882, when Karl Benz in Germany powered the first automobile, or horseless carriage, with the internal combustion engine, which had been invented by Nicholas Otto a few years before. It itself was an evolution of the steam engine that introduced the fuel directly into the cylinder, which was more efficient. At first, it used carburetted benzene distilled from coal but soon turned to gasoline refined from crude oil. The automobile developed an unquenchable thirst for oil.

Fig. 3-2. In 1882, Carl Benz powered the first automobile with the internal combustion engine
(Courtesy of Mercedes Benz)

THE BIRTH OF THE OIL INDUSTRY

The birth of the oil industry is generally attributed to the famous well drilled for oil in 1859 by the self-styled Colonel Edwin. L. Drake at

Titusville, Pennsylvania, although it is now claimed that F.N.Semyenov actually got there first with a well on the Apsheron Peninsular, near Baku on the Caspian, eleven years before.[5] The first *production* statistics in Romania go back to 1854. It does not really matter who wears the crown, for in any event the oil industry grew rapidly in the succeeding years in both Pennsylvania and on the shores of the Caspian.[6] Much has been written on the early history of the oil industry, and it does indeed make colourful reading. The oil pioneer, Beeby-Thompson,[7] arriving at Baku in 1898, for example, wistfully notes that:

> *"The Caucasian women, about whose beauty I had heard so much, were disappointing, for all the most lovely creatures had long ago been sold to Turkish harems, leaving the least attractive for reproduction".*

Here, I will be content to summarise only the highlights of this exciting epoch, concentrating mainly on the discovery of oil; and the companies and people that were responsible.

Drake's well, which encountered oil at a depth of 67 feet[8] in an Upper Devonian sandstone, led to the first great oil exploration boom as the shallow reservoirs were tapped by an army of pioneers and speculators descending on the oil lands of the Appalachian Basin of the United States. Stills were erected nearby to make kerosene, which within the remarkably short span of two years was already being exported to Europe. Fortunes were made, and many were lost. Prices fluctuated wildly: from its outset, the industry has been plagued by "boom or bust". The reason is the special depletion pattern of oil, which

Fig. 3-3. In 1859, Drake's well in Pennsylvania launched the American oil industry (from Pennsylvania Historical & Museum commission Drake Well Museum Collection, Titusville Pennsylvania)

Fig. 3-4. Early oilfields at Baku, another important source of oil.

flows rapidly under its own pressure from the wellbore, as soon as it is tapped, in a manner very different from the laborious process of mining coal with a pick and shovel.

This is perhaps a good moment to introduce the notion of *Resources* and *Reserves*, a cause of much misunderstanding. A coal deposit covers a wide area so that the *Resource* is huge, but it is mined only where the seams are thick and accessible, with the quantities in reach of the mine shafts being described as *Reserves*. It is a matter of concentration, such that if prices rise or costs fall, more becomes viable and converted to *Reserve* status. This has given rise to the notion of the resource pyramid of near infinite size. It is beloved of economists who dream of more difficult and costly resources becoming available as the market dictates. However, oil is different because it is concentrated by Nature into a few places having the right conditions. It is either there in profitable abundance, or not there at all. The oil-water contact in the reservoir is abrupt, so it is not a matter of concentration. There is a certain polarity about oil that distinguishes it from other minerals. That said, extracting the heavy oil and tar deposits is indeed virtually a mining process and obeys the same rules.

The early explorers drilled by guesswork, although they soon began to develop an empirical understanding of geology, identifying the trends where drilling succeeded. The State of Pennsylvania did appoint a geologist to investigate; and his report of 1865 observed that oil tended to accumulate in anticlines, where the strata are folded into an arch. It went on to state that

> *"Petroleum is a bituminous liquid resulting from the decomposition of marine and land plants".*[9]

In other words, the essential geological constraints had already been understood in America within six years of the first discoveries. It did not take them long to figure out the basics.[10]

Looked at with to-day's eyes, it is rather surprising that these ancient rocks in Pennsylvania, which had been deposited over 350 million years ago, should still be oil-bearing so close to the surface. No seals are perfect, and this was a long period of time over which the oil could have seeped away. Probably, the explanation is that, although the source of the oil is ancient, generation was not achieved until much later after long exposure to low temperatures.

The Appalachian Basin is an interesting example of depletion. The oil boom was already over by 1900 by which time some 183 oilfields had been found, yielding an ultimate recovery of 1.33 billion barrels.[11] It is in fact quite a small province, notwithstanding its early importance, as its total contribution would be enough to supply the World's present demand for little over two weeks. It is noteworthy that it still has reported reserves of some 20 million barrels, showing how production declines exponentially, such that old fields continue to produce a few barrels a day for a very long time during the tail-end of their depletion, and how ever smaller fields continue to be found even in mature areas. There is an old well in a museum that continues to produce a few gallons a day. This emphasises why it is much more important to address the issue of the peak between the growth and decline of production than to worry about running out altogether. Today, we hear a great deal about technological progress, with much play made of the importance of horizontal wells and multi-lateral wells, in which several branches are drilled from a single initial borehole. It is sanguine therefore to learn that both techniques were already being used in Pennsylvania before the end of the 19th Century,[12] albeit with by no means today's technical sophistication.

STANDARD OIL

In the same year as Drake drilled his well, a man by the name of John.D. Rockefeller went into partnership with a newly arrived British immigrant, Maurice Clark, to establish a trading company in Cleveland, Ohio. It was an opportune moment with the Civil War of 1861–65 creating a demand for goods of all sorts, and it was an opportune place connected by two railways and the Great Lakes navigation system. The new firm soon turned to trade in kerosene that before long led it into the refinery business. In this way, started the great Standard Oil Company which grew to become the World's largest corporation, being run on ruthless business lines. It was the precursor of the modern company with its bureaucracy, driven solely by the motive of return on investment. It was a marketing company: a market it sought to control by fair means and foul. It succeeded in doing so by placing a stranglehold on oil transport both by securing the pipelines and negotiating special rebates from the railways. Ownership of the tank-cars also gave it a particular advantage. The wild fluctuations in oil price, which arose as new discoveries flooded the market, were anathema to its orderly plans. Even in these early years of the industry, a need for regulation had already arisen: Standard Oil was in fact exercising a function no different from that subsequently applied by the Achnacarry Agreement,[13] the Texas Railroad Commission[14] or OPEC.

Standard was reluctant to enter the hurly-burly of exploration, although it eventually did so in the 1880s when another new oil province was found in Indiana. Its motive was to protect its existing market from competition. By then, the Pennsylvania fields were beginning to decline, the first example of oil depletion. By 1885, the State Geologist of Pennsylvania had already stated:

Fig. 3-5. J.D. Rockefeller, ruthless founder of Standard Oil, and philanthropist
(Courtesy of Exxon Corporation)

Fig. 3-6. Henry Flagler, Rockefeller's associate who had a key role in the creation of Standard Oil
(Courtesy of Exxon Corporation)

"the amazing exhibition of oil is only a temporary and vanishing phenomenon - one which young men will come to see come to its natural end"

He was both right and wrong in his prognosis: he was right about the area he knew but wrong insofar as he did not know how much oil would be found in new areas. Now, a century and a half later, we have a much better idea of that issue, and can confidently repeat his words on a global scale.

Standard Oil's ruthless capitalism was much reviled by the independent oil producers who regarded it as a creeping octopus that would eventually ensnare and devour them. It was particularly disliked in Texas and the southern

States, still smarting from the Yankee victory in the Civil War. Pressure against it grew, until the Government in 1911 was forced to break it up under anti-trust legislation, a democratic response to an overweening monopoly. After the break-up, some of Standard's daughters, Esso, Chevron, Mobil, Amoco, Conoco, Sohio and Arco, to name the largest of the thirty-seven, grew to become some of the World's most important oil companies in their own right. Esso became Exxon and has now merged with its sister Mobil, almost rediscovering the essence of their progenitor, Standard Oil.

Although they did practice successful exploration throughout the World, in some cases pioneering new projects in the best traditions of the explorer, it can be said that they owed most of their growth to their great financial strength, inherited from Rockefeller's empire, which allowed them to swallow their weaker competitors. I think that the major companies have always been traders at heart, with making money being their sole objective: nothing wrong with that of course – save when one comes to consider the depletion of a finite resource that should perhaps be governed by more sophisticated criteria to better respect the value of this irreplaceable material to Mankind.

Standard Oil's control was centred on the eastern United States, but new oil lands were being explored in the south. California came in first with several important discoveries before the end of the 19thCentury. Union Oil (Unocal) was the dominant producer, which managed to preserve its independence until recently, when it was acquired by Chevron-Texaco for $16 billion, paying almost $10 a barrel for its reserves. The complex geology of California where the oil occurs in strongly folded and faulted Tertiary rocks, partly affected by the famous San Andreas Fault, prompted a greater attention to scientific geology here than elsewhere, and the companies in California took on increasing numbers of professional geologists to guide their efforts. Later, Standard Oil did move in, and its affiliate, Standard of California, now Chevron, has had an exceptional reputation in exploration, in due course bringing in Saudi Arabia.

Oil was found in Texas when a water-well at Corsicana unexpectedly encountered oil in 1893. It was followed in 1901 by a spectacular blow-out at Spindletop, near Beaumont, in which seventy-five thousand barrels a day gushed high into the sky. It was a mighty roar that heralded another oil boom, opening up a new province. It transformed the economic life of the United States, and indirectly contributed to its world political power – now in decline as depletion bites. Discovery followed discovery as the new oil belts were drilled up, but eventually it peaked in the early 1930s. Today, only a few million barrels are found each year, and they in very small fields. The total endowment of Texas, again resulting mainly from the early discoveries, is about 60 billion barrels, equalling the North Sea.

Standard Companies in 1911	External merger & liquidation	Today
SOC (California) SOC (Kentucky)		CHEVRON
Swan & Finch	liquidated 1965	
SOC (New York) Vacuum Oil Co		MOBIL
Waters Pierce	dissolved 1940	
Borne, Scrymser Co		Borne Chemical
Galena Signal Oil Co Cumberland Pipe Line Co Southern Pipe Line Co	Ashland 1917	Ashland Oil Co
Chesebrough Mfg Co		Chesebrough-Ponds
The Crescent Pipe Line Co	liquidated 1925	
Prairie Oil & Gas Co Atlantic Refining Co	Sinclair 1917 Richfield 1955	ARCO
SOC (Kansas) SOC (Indiana) SOC (Nebraska)	Pan American 1930	AMOCO
Washington Oil Co	liquidated 1976	
SW Pennsylvania Pipe Lines National Transit Co South Penn Oil Co Eureka Pipe Line Co		Pennzoil
Union Tank Car Co		Trans Union
Colonial Oil Co	liquidated	
Indiana Pipe Line Co Buckeye Pipe Line Co Northern Pipe Line Co New York Transit Co	Pennsylvania Co '52	Penn Central
Anglo-American Corp SOC (New Jersey)		EXXON
Continental Oil Co.	Maryland 1917 Du Pont	Du Pont (Conoco)
SOC (Ohio) Solar Refining	BP 1938	BP
The Ohio Oil Co	US Steel	US Steel (Marathon)

Fig 3-7. Standard's daughters

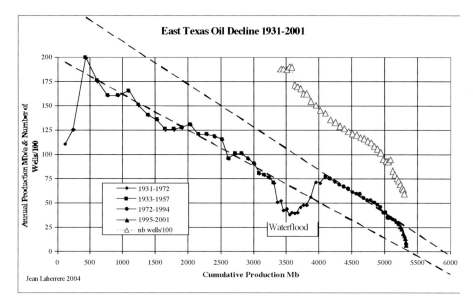

Fig. 3-8. Depletion of Texas

The discovery at Spindletop spawned two more major companies, Texaco and Gulf Oil, both to be later taken over by Chevron. The acquisition of Gulf followed an unsuccessful assault by a clear thinking corporate raider, Boone Pickens.[15] He correctly realized that its past was worth much more to its shareholders than its future. It is a reality that most other major oil companies have faced, with varying degrees of frankness, explaining the wave of mergers that have occurred over recent years. Exxon merged with Mobil, Chevron with Gulf and Texaco; BP with Amoco and Arco; Total with Elf and Fina. Only Shell stood aloof, apart from the minor acquisition of Enterprise, itself run by ex-Shell men, which explains why in 2004 to was forced to downgrade its reserves, causing a financial furore which cost the Chairman his job. Looking for oil in the ground was evidently no longer delivering adequate results despite the Company's technical excellence.

TEXACO – A HARD-NOSED COMPANY

The Spindletop discovery well was drilled by a one-armed, self-educated mechanic, named Patillo Higgins, who was backed by a Captain A.F. Lucas, an immigrant from Yugoslavia. They eventually sold out to what became Gulf Oil. Another man on the scene was Joe Cullinan, known as "Buckskin Joe" for his abrasive manner. He set up the Texas Fuel Company in Beaumont to trade in oil and oilfield equipment. He was backed by New

York and Chicago investors, managed by a retiring German immigrant, by the name of Arnold Schlaet,[16] who exerted a considerable influence on the Company. In 1906, the name Texaco with its logo of a green "T" overlying a red Texas Star was registered. Buckskin Joe knew his job, running the business with an autocratic style that did not endear himself to his investors who eventually ran him off. Some say that the Company retained a characteristically autocratic manner over its remaining life, which perhaps owed something to its founder.

Fig. 3-9. Joe Cullanin, "Buckskin Joe", (1860–1937) founder of Texaco

As the business grew, Texaco moved its headquarters to New York, and expanded its upstream operations throughout the United States, being more successful than many of its competitors. Because much of the U.S. market was controlled by the Standard group of companies, Texaco paid especial attention to overseas markets. In 1936, its global marketing strategy prompted it to join forces with Chevron in the Eastern

Fig. 3-10. Spindletop on fire: the field that opened the southern United States

Fig. 3-11 Captain Rieber (b.1882 in Norway), the Chairman with Nazi friends

Hemisphere, creating the Caltex Group. With hindsight, this can be seen as one of those rare transcendental corporate decisions with incredible and unforeseen consequences transforming a company. As a result, it gained access to Chevron's rights in super-rich Saudi Arabia, becoming a founder member of Aramco. In 1942, Caltex also found the Minas Field in Sumatra, the largest in Indonesia, which could not, however, be developed until after the Second World War. The oil supply from Caltex was to meet Texaco's needs for a long time to come, which probably explains its own rather indifferent performance as an international explorer. It had no need to find more, save for strategic reasons to reduce its dependence on Saudi Arabia.

The mate on the first tanker to leave Spindletop with Texaco oil was Torkild Rieber, a tough Norwegian seaman, born in Voss near Bergen in 1882. This dynamic man went on to build up the Company's tanker fleet, and by 1935 had risen to be Chairman of the Board. He was a ruthless businessman, intent on building a market in Europe, where he admired the efficiency of the growing Fascist movements, which he supported. He was by no means alone in such a view: more European businessmen than later cared to admit it shared this respect for power and decisiveness. In the Spanish Civil War, he supplied General Franco's forces in contravention of the U.S. Neutrality Law, and did so on credit, being eventually rewarded with a dominant share of the Spanish market for much of Franco's life. This connection brought him into contact with leading German Nazis with whom he struck deals to swap tankers built in Germany for supply of oil, which continued to be made available after the Second World War broke out. Some of the tanks that pushed the British Army into the sea at Dunkerque in 1940 may have been fuelled by Texaco. His close ties with the Germans were exploited by the latter's Intelligence Service which established a presence in Texaco's New York office. This group sent intelligence about convoy sailings to Britain, disguised as corporate patent numbers, through the Company's cable office to Germany, and sought also to influence American politicians to the German cause. The plot was unmasked by British counter-intelligence, and the Captain was eventually forced from office after America came into the war.[17] Texaco's black and red livery and its red star did seem to have a slightly Germanic touch to it.

Texaco, like most other American companies, was operating in the cut-throat environment of the United States. Oil production was rising, and competition for market was fierce. It occupied most of the Company's attention. Growth was achieved primarily by buying up competitors, with the major companies holding portfolios of partly owned affiliates. The deals were often complex and contentious, and some of the directors of the swallowed companies remained on the Boards of the affiliates representing minority interests. Litigation was the order of the day. Texaco's half interest in Caltex was the jewel in its crown. Robert E. King, based in New York, was its well respected Exploration Manager. The corporate history[18] makes no mention of Texaco's own Exploration

Fig. 3-12 Augustus C. Long, (b.1904) a post-war Chairman, with an autocratic style

Manager: perhaps there was no such title. In the post-war years, Texaco in its own right achieved one of its few international exploration successes when it found the Orito Field in Colombia in 1963, which led to a string of finds in neighbouring Ecuador, in a joint venture with Gulf. It also had a modest success in the UK North Sea. Otherwise, it seems to have followed a policy of obtaining its reserves by acquiring other companies, including Seabord in Venezuela, Trinidad Leaseholds, and Getty: the latter landing it in an expensive law suit with Pennzoil. Later, tax regulations led to the consolidation of affiliate interests, and Caltex was absorbed by its parents, who were later to merge as already mentioned.

An interesting development in the marketing area was the Company's decision in 1985 to sell a one-half interest in its refining and marketing organization in thirty-three U.S. States to a Saudi company, after Aramco was expropriated.[19] It was a noteworthy link between a producing government and a downstream organization that made sense for both parties. It is surprising that it did not set a precedent for other major companies. Perhaps only the hard-nosed Texaco was willing to let a cuckoo into the nest.[20] It would not be surprising if there had been a strong Saudi shareholding in Texaco.

In an article[21] published in the 1990s, the Company explained a new strategy of concentrating on a few key fields, returning in fact to its US homeland where it aimed to buy out its partners and conduct highly efficient operations with a slimmed-down staff. This policy was a clear hint of the decline being forced upon it by falling discovery, which eventually led to its demise through merger with Chevron.

The climactic moments in Texaco's life were its spectacular birth at Spindletop, opening up Texas; its decision in 1935 to join forces with Chevron in the Eastern Hemisphere that delivered Saudi Arabia; followed in turn by the loss through sequestration of that country's production in 1979. It seems never to have quite had its heart in international exploration, although from time to time has made rather transitory forays. The most successful of these was opening up the Sub-Andean basins of southern Colombia and Ecuador, which involved some stirring exploits in the Amazon jungles. It was the weakest of the former sisters, and eventually failed, as some analysts[22] had foreseen.

OTHER EARLY OIL PATCHES

The early development of the oil industry in the United States had a lasting world influence: the large American companies expanded overseas taking their business and technical culture with them. In technical terms, the industry's American roots has left its legacy, including for example its units of measurement: the traditional well-casing sizes of 9⅝″ and 13⅜″ are still in almost universal use as are such colourful drilling terms as *roughneck*, *rat-hole*, *kelly bushing*, not to mention a piece of equipment, delightfully known as a *Donkey's Dick*. But the United States was not by any means the only pioneering oil country. During the 19th Century, there were already developments in many places including: Baku in Russia; Borneo; Burma; Sumatra; Romania; Poland; Trinidad; Peru; and Mexico. Of them, the most important was Baku on the shores of the Caspian, a backward and poorly administered territory on the southern fringe of the Russian Empire. Oil extracted from hand dug pits had been a State monopoly, but in the early 1870s, the area was opened up to private capital. One of the first entrepreneurs to arrive on the scene was Ludwig Nobel, a member of an inventive Swedish family that made a fortune out of producing dynamite. It is now remembered by the Nobel Prize, which it endowed.

The World's first tanker, the *Zoroaster,* was designed and operated by the Nobels to transport oil on the Caspian, the development being prompted by the absence of sufficient oak to make traditional barrels. The Rothschild Bank later came in to finance a railway to Batumi on the Black Sea, opening an export route. They in turn were followed by Shell, which started exporting Baku kerosene in tankers to Europe and the East. That company later merged with Royal Dutch, which had been pioneering oil production in the Dutch East Indies (now Indonesia), to become the giant Royal Dutch/Shell, or simply Shell as it is known today.

The Baku oilfields lie on the ancient delta of the Volga River in front of the Caucasus. The geology is characterised by complex folds and faults, with multiple reservoirs.

Seepages of both oil and gas were abundant. A peculiarity of the geology was the presence of curious features, known as mud-volcanoes, which are also found in Romania, Colombia and Trinidad. They are mounds of mud, as much as several hundred feet in height that form over

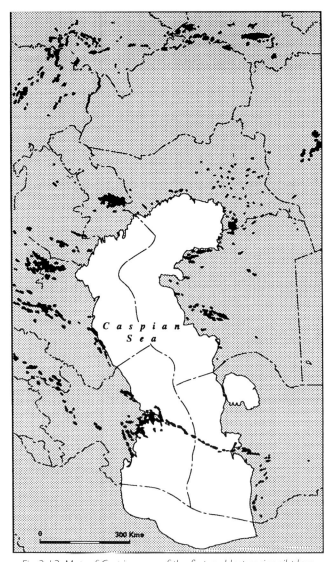

Fig. 3-13. Map of Caspian: one of the first and last major oil plays

active gas seepages. They sometimes explode and catch fire.

Beeby-Thompson[23] describes how he commenced geological investigations of the area, but no doubt most of the fields were found by hit-or-miss. It was evidently easier to hit than to miss in this prolific area that in 1900 produced as much as 75 million barrels from 1700 wells less than 1000 feet deep. It was a violent place of banditry and strife, with appalling operating conditions. No less a figure than Josef Stalin was a workers' leader, masterminding strikes and disturbances in Baku in the early years of the last Century. Such pressures spread and culminated in

Fig. 3-14. A sailing tanker: the bark "Brilliant" that carried a cargo of 30,000 barrels of oil (Courtesy of Exxon Corporation)

the Bolshevik rising of 1917, which, to put it mildly, transformed the World. It was not to be the last occasion on which oil shaped human destiny: the most important being about to strike as depletion bites in earnest during the years ahead. The Bolshevik Revolution effectively brought the Caspian oil boom to a close. Hopes of a second boom came after the fall of the Soviets by opening the offshore, but the results have not lived up to expectation.

The greatest oil province of all, the Middle East, was also attracting attention as the 19th Century drew to a close, but to do business there was difficult. Most of the area was controlled by the Ottoman Empire with its decadent and corrupt Sultan spending much of his time in a harem surrounded by eunuchs. The rest was in the hands of the Shah of Persia, whose authority barely extended beyond his own capital.

The Germans became interested in building a railway from Berlin to Baghdad as part of a foreign policy initiative aimed to[24] catch up with the colonial expansion of France and Britain in other parts of the World. As a land power, it recognized the military mobility afforded by railways, which it conceived would be more effective than the slower British sea-power. It secured to this end a concession in Anatolia and Mesopotamia, which included mineral rights for twenty kilometres on either side of the track, presumably as a source of building stone. Its engineers soon reported the numerous oil seepages which they encountered in the Mosul area of what is now Iraq.

At about the same time, the head of Persian Customs, General Antoine Kitabgi, hearing of the growing oil interest in the vicinity, resolved to see

Fig. 3-15. Marcus Samuel, a London merchant, who built Shell Oil
(Shell International Photographic Services, London)

Fig. 3-16. Henri Deterding, dynamic head of Royal Dutch, which merged with Shell in 1907
(Shell International Photographic Services, London)

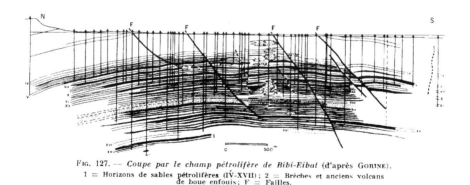

FIG. 127. — *Coupe par le champ pétrolifère de Bibi-Eibat* (d'après GORINE).
1 = Horizons de sables pétrolifères (IV-XVII) ; 2 = Brèches et anciens volcans de boue enfouis ; F = Failles.

Fig. 3-17. The Bibi-Eibat Field near Baku, characterized by multiple thin reservoirs

if he could let an oil concession in his country. After one or two false starts, he managed in 1900 to bring it to the attention of William Knox D'Arcy, an entrepreneur who had just returned to London from Australia where he had made a fortune in gold-mining.

He saw the possibilities, and eventually secured the rights: a £20,000 signature bonus to the impoverished Shah being the main inducement.

Fig. 3-18. William Knox D'Arcy (1849-1917) whose initiative in Iran founded what is now BP.

Drilling commenced in 1902 under appalling conditions, with summer temperatures of over 43°C. in the shade, and success was slow in coming. D'Arcy was becoming overextended, although encouraged when the British Government started to take an interest in his project. Britain had always sought to deter Russian expansion into the Middle East to protect its trade route to India and the East, and oil became a new factor. D'Arcy's immediate problems were, however, partly solved when the old established Burmah Oil Company, based in Glasgow, agreed to take a share in his company.

In January 1908, after six long years of travail and disappointment, the third well was spudded at Masjid-i-Sulaiman (The Mosque of Solomon) in the Zagros foothills: it was pretty much the last throw of the dice. By May, the well was down to 1000 feet without results, and a cable to suspend operations was received from London. The local manager, G.B.Reynolds, however, decided to continue until he received written confirmation. His initiative was rewarded at 4.30 a.m. on the morning of May 26th, when the well blew out throwing a jet of oil fifty feet into the air.[25]

In World resource terms, it was a climactic event. As Dr Ziegler explains (see Chapter 2), the World contains no more than about thirty significant petroleum systems, that is to say provinces with the unique set of geological circumstances to yield prolific oil. This discovery in 1908 found the largest.[26] In terms of such petroleum systems, as opposed to

Fig. 3-19. The well that opened the Middle East in 1908 and changed the World

individual fields, almost forty percent of the World's total oil supply was found by this well, although that would not of course become known until much later. It was undoubtedly a turning point in history.

Another turning point of a different sort was about to unfold in 1914 with the outbreak of the First World War. Britain's last major naval engagement had been the Battle of Trafalgar in 1805, a critical action in the Napoleonic wars. At the height of Empire, the Navy was the corner-stone of Britain's power and prestige, but by the turn of the Century it had become more of a symbol, with polished brass, holystoned decks and smartly dressed crews, than an efficient fighting machine. It was just this pageantry that so impressed the Kaiser, himself an honorary admiral, when he came to take part in his favourite sport of yacht racing at Cowes.[27] "Why", he asked himself, "does Germany not have such impressive battleships, dressed over all in flags and illuminated by night, to stand guardians for forthcoming yacht races at Kiel?" From this harmless vanity began the Anglo-German arms race, as each new German warship had to be matched by a British one to maintain the balance of power. Gradually the emphasis changed from the pomp and splendour of the Marine band on the quarter deck to actually making the thing a lethal weapon, able to out-speed and out-gun its competitor. The maverick Admiral Fisher was dedicated to this transformation. He realized that his ships would have to convert to oil fuel to obtain the required performance, but his proposals met with resistance. For the first time, but certainly not the last, as this book will explain, the issue was *security of oil supply*. Britain had no oil of its own, and was reluctant to rely on American oil, or even Shell oil with its Dutch connection. What it needed was its very own supply that it could control. Winston Churchill, then the First Lord of the Admiralty, concluded that Persian oil was the answer for two reasons: first, to supply the Navy; and second, to strengthen a British presence in the Middle East to deter the threat of German or Russian expansion in that area. The Government took up a fifty-one percent interest in the Anglo-Persian Oil Company, later to become British Petroleum, or BP, royal assent being granted six days before the First World War broke out.[28]

The war opened with cavalry charges, as plumed Uhlan lancers galloped into action, but ended with tanks driven by internal combustion engines running on fuel refined from crude oil.[29] Oil became the great new driving force of the World, changing the meaning of the term, horse-power. But in terms of oil resources, perhaps the most significant feature of the First World War was that Turkey found itself on the losing side. Had it been an ally, or remained neutral, events would have turned out very differently.[30] As it was, Britain had a motive to encourage Arab nationalism,[31] which effectively resulted in the creation of the state of Saudi Arabia out of a tribal desert. It was to have far reaching economic and political consequences that have still to be played out, as this book will discuss in later chapters. The defeat of the Turks led to the break up of the Ottoman Empire into new administrations, of which Iraq, Kuwait and Saudi Arabia are the most important. Not far below the surface, was the division of the region's oil rights to the three victorious allies, Britain, France and the United States. Recent events in Iraq echo history.

The first solution for the division of the oil rights was to share them. This was achieved by the formation during the 1920s of the Iraq Petroleum Company, owned by Shell, BP, Compagnie Francais des Pétroles (CFP, now Total), Mobil and Esso, not to forget the legendary Calouste Gulbenkian (Mr 5%) who put the deal together (see below). This group had what is called an Area of Mutual Interest (AMI) agreement that prohibited independent activities by the partners in the area of the former Ottoman Empire. It became known as the Red-Line Agreement covering all the productive Middle East territories outside Iran and Kuwait,[32] and was the cause of bitter conflict for a long time to come.

Chevron, which was not restricted by the Red-Line Agreement, took up

Fig. 3-20. An anticline in Iran, an obvious oil prospect
(Photograph courtesy of British Petroleum)

rights in Bahrain in 1929, to be later joined by Texaco, and struck oil there two years later. This find, coming from Tertiary sandstones at fairly shallow depth, was itself comparatively modest, but it was nevertheless of immense importance, for Bahrain lies only a few miles off the coast of Saudi Arabia. Up to that point, interest in oil had been concentrated on the huge folded structures of the Zagros Foothills in Iran and Iraq that were obvious surface features visible for miles around.

Many geologists, seeking analogues for this familiar type of prospect, were at first sceptical of the platform province to the west of the Persian Gulf, where the strata were largely obscured below sand dunes and, where seen, were flat-lying or, at most, shallow dipping. At first sight, it seemed to lack adequate structure to provide large traps for oil. So, the discovery of oil on Bahrain on the edge of this new province carried immense implications, which were at once recognized. Chevron began negotiating for rights in Saudi Arabia, partly through an eccentric and disaffected Englishman, by the name of Harry St. John Philby. He was trading in Jidda, where he had the Ford agency, and was, remarkably enough, no less than the father of the infamous British Cold War double-agent, Kim Philby. King Ibn Saud, himself a British protégé from the War, was desperately short of money, and Chevron clinched the deal in 1933 with delivery of thirty-five thousand gold sovereigns that were shipped to Arabia in seven boxes aboard a P&O liner. It was a substantial and risky investment at the

USA	18 Mb	64%
USSR	3.4	12%
Mexico	1.9	7%
Venezuela	1.2	4%
Dutch East Indies	0.7	2%
Rumania	0.7	2%
Iran	0.6	2%

Fig. 3-21. Oil production in 1935

time, for no one could have imagined that Saudi Arabia would become the World's most prodigious oil province, with an ultimate endowment of about 275 billion barrels, fifteen percent of the World's total.[33] Chevron, when it later found that it lacked the resources to develop the area single handed, brought in Texaco, as already described, followed in 1947 by Mobil and Esso, the latter two in flagrant disregard for the famous Red-Line Agreement arising from their commitments to the Iraq Petroleum Company. These four companies formed the Arabian-American Oil Company (Aramco), the emphasis being on the second word. In pre-war days, Ibn Saud, absolute ruler of a feudal and primitive country that was little more than his private estate, effectively became an American satrap. The further evolution of this remarkable and extraordinary situation has yet to unfold with or without the House of Saud.[34] I will return to the issue in later chapters.

While rights to Saudi Arabia were being negotiated, BP and Gulf turned attention to Kuwait, which lay at the head of the Persian Gulf, outside the Red-Line Agreement area. They eventually decided to join forces rather than compete for the territory, and signed a lease for it in 1933. It completed the primary carve-up of the Middle East.

Although by far the most important, the Middle East was by no means the only oil territory being explored and developed during the first half of the last Century. Most progress was in the Western Hemisphere, especially in the United States itself, which was already becoming a fairly mature province, but also in Venezuela and Mexico, where impressive finds were made. Shell, which had rather missed out on the carve-up of the Middle East, took up a strong position in the Western Hemisphere, competing successfully with the major

Fig. 3-22. King Ibn Saud of Saudi Arabia, who sired a thousand princes

American companies. Generally smaller scale operations were also taking place elsewhere. By 1935, twenty-five countries were in production, of which the seven largest are shown in Figure 3-21. The United States was producing sixty-four percent of the World's needs. The Middle East was barely represented: Iran, the largest producer, was in seventh place with only two percent. The other countries were, in decreasing order: India (incl. Pakistan); Poland; Peru; Colombia; Argentina; Trinidad; Japan; Sarawak; Brunei; Iraq; Canada; Germany; Egypt; Sakhalin; Ecuador; France; Italy; Czechoslovakia and Bolivia.[35] \What a different world it was!

Seven major companies, comprising Shell, BP, Esso, Mobil, Chevron, Texaco and Gulf, later dubbed the "Seven Sisters" by Enrico Mattei, the Italian oilman, had already brought World supply under their control. With most oil coming from the United States, security of the supply was not a serious issue, although not out of sight. The Soviet Union was closed to foreign companies, and those with rights from the pre-war days in Baku were formally expropriated in 1928. Mexico also ousted the foreign companies ten years later as a nationalist movement, somewhat akin to the Fascist influences of contemporary Europe, gained political ascendancy, believing perhaps with some justification that foreign influences were becoming excessive. Its oil industry was placed in the hands of a State enterprise PEMEX, the first example of a State oil company that was later to be copied widely. These expropriations were harbingers of what was to come.

CALOUSTE GULBENKIAN: OIL MAN PAR EXCELLENCE

"If you can't bite the hand that feeds you, kiss it"

This immortal dictum, with which he is credited, sums up one of the most remarkable oil men of all time and also underlines a common feature of international oil deals: the joint venture between several companies, each having an undivided interest in a concession. The term *Undivided* means that the owner had a share of the total,

Fig. 3-23. Calouste Sakis Gulbenkian, oilman (Gulbenkian Foundation)

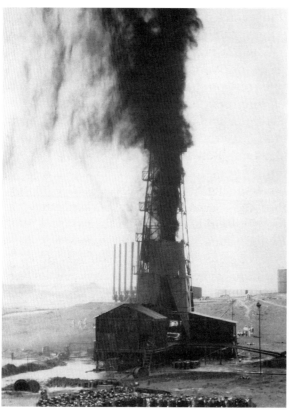

Fig. 3-24. The discovery well near Kirkuk, October 1927
(Photograph courtesy of British Petrol)

such that the concession was not physically divided. The manage-ment of undivided interests calls for compromise and nego-tiating skill, as families who jointly inherit property know to their cost, and Gulbenkian needed both to hold onto his 5% in his brainchild and life's work, the Iraq Petro-leum Company. Calouste Gulben-kian[36] was the son of a prominent Armenian business-man in Istan-bul, who had made a fortune as an importer of Russian kerosene. He was educated in France and England, where in 1887, at the age of nineteen, he secured a first-class degree in engineering at London University. Although he had some aspirations to further studies, evidently having serious intellectual leanings, his father sent him off to Baku to learn about oil, the foundation of the family fortune. On returning, he wrote several well-received articles on his experiences in *Revue des Deux Mondes*, a respected French journal, which established him as something of an oil expert in the Turkish capital. When the Government became aware that oil seepages had been encountered in Mesopotamia (now Iraq) by a German group, backed by the Deutsche Bank, which was surveying a proposed railway, it asked him to investigate. He produced a comprehensive report, which fired his imagination to somehow secure rights to this oil himself. It was a finally successful quest that was to occupy him for many years.

But in 1896, the Sultan embarked on a ferocious campaign of what would now be called ethnic cleansing, in which large numbers of Armenians were massacred. The Gulbenkian family was forced to flee to Egypt. A chance meeting on the boat with a fellow refugee, Alexander

Mantashov, who was an influential figure in Russian oil, introduced him to the right circles in Cairo. They included Sir Evelyn Baring of the then famous, but later discredited, British family bank. This contact proved useful when he moved to London in the following year to represent Russian oil interests. He soon became a close associate of Henri Deterding, the head of Shell.

In 1910, the Sultan was overthrown by the Young Turks who were bent on modernizing the country. They invited a group of London financiers and Gulbenkian to establish a national bank. Gulbenkian then persuaded the bank to form an alliance with the Deutsche Bank and Shell to establish the Turkish Petroleum Company in London, Gulbenkian retaining 15% for himself. Later, British Government pressure brought about the entry of BP, reducing his stake to 5%. A concession for the Mosul oil lands was granted on June 28th 1914, a few weeks before the First World War broke out, whereupon the Deutsche Bank's interest was taken over by the British Custodian of Enemy Property.

With the advance of Allied forces, the future of the Ottoman Empire was settled by France and Britain under the so-called Sykes-Picot Agreement of 1916. It provided, amongst other clauses, that the Mosul area with its oil potential should fall within the French sphere of influence, but the British eventually recovered control provided that France could participate in the oil development. A further complexity was introduced when Woodrow Wilson, the American President, proclaimed at the peace treaty the worthy principle of self-determination, which was not what France and Britain exactly had in mind for the Middle East. Britain in particular reneged on the promises it had made for wartime expediency to create an Arab nation. This American pressure eventually led to the somewhat artificial creation of the State of Iraq. However altruistic the proposal of self-determination was as a general principle, it was not long before American oil companies with the support of their government were pressing for a share of Mesopotamian oil. Difficult negotiations continued over many years, involving issues as fundamental to Gulbenkian as the validity of the Turkish Petroleum Company's rights. He followed every step with immense care, preferring the written word of the telegram to the spoken one at a meeting, and somehow contrived to hold on to his five percent. In 1927, drilling commenced six miles north-west of Kirkuk, and at 3.00 a.m. on October 25th, the well blew out at a depth of only 1500 feet, flowing at some 95 000 barrels a day. Even so, more than a year was to elapse before Gulbenkian was ready to sign the final agreement for what became the Iraq Petroleum Company, to be owned by Shell, BP, and CFP of France, with 23.75% each; Mobil and Esso with 11.875% each,[37] and Gulbenkian with 5%. The numbers themselves speak of who did, and who did not, give ground. It had taken him thirty-five years to get there, but he made it by dint of attention to detail and remarkable tenacity.

By now, even without Iraq oil, he was a fabulously wealthy man living variously in the Ritz Hotels of London and Paris, attended by a string of very young mistresses, whom he apparently regarded as a medical necessity. Apart from business, his interest was the impressive art collection that he had put together.

Although his Iraq venture was his most spectacular success, he was also instrumental in getting Shell into Mexico and Venezuela, which were not inconsiderable feats in their own right. He worked closely with Henri Deterding until their relations soured – over their shared admiration of Lydia Pavlova, the wife of a Russian general, who eventually became the second Mrs Deterding. This was one of his few negotiations that failed.

When the Second World War came, he moved from Paris to Portugal, where he died in 1955 at the age of 85, leaving a vast fortune, a remarkable art collection and a charitable organisation, the Gulbenkian Foundation. He regarded himself as an architect – not of buildings, but of immaculate and elegant business constructions and negotiation.

In the same way as we may speculate about the different course of history had Turkey not joined the losing side in the War, we may be sure that the world of oil would have been very different had not Calouste Gulbenkian dedicated his life to it.

PIONEERING EXPLORATION

Exploration and production during the Pioneering Epoch was onshore, where in fact most of the World's oil lies. It began with primitive technology and understanding, but progress was remarkably rapid. The rotary rig was introduced to replace the cable tool; well logging was

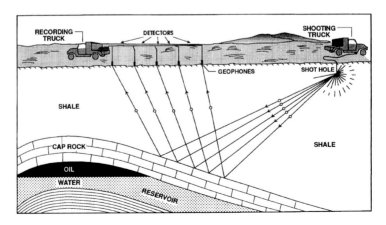

Fig. 3-25. Seismic surveying

developed to identify the formations and their oil content; and seismic surveys were brought into wide use to investigate geological structure at depth. All were established well before the end of the epoch.

It was the great age of field geology. Pioneering geologists on foot and muleback scoured the World for prospects, working under often arduous conditions far from base.[38] They had to be men of initiative, as well as rugged individualists. They searched, first, for seepages, and then for promising structures in the vicinity. But most of all they looked at the rocks to see if they had the essential characteristics and structural style to make an oil province. They developed a nose for what worked and

Fig. 3-26. J.V. Harrison, who mapped Iran with Norman Falcon, has the rugged countenance of the pioneer explorer

what did not, and they mainly got it right. Kingston[39] describes how Esso had a small group of such explorers that roamed the World from 1940 to 1960, searching out every nook and cranny. He comments on the ending of this chapter:

> *The field studies and overseas travel came to an end, and the roving assignments ceased* [in response to the move offshore]. *During the 1970s and early 1980s, the price of oil rose significantly, and the oil industry went on an exploration binge. Basins that we had repeatedly rejected in the past (we called them "dogs") were leased and drilled, mostly with negative results ... Some of our reconnaissance field work led to acreage acquisition, and to drilling, a few of which resulted in major oil discoveries in Europe, North Africa, and South America. Very often our regional studies resulted in negative recommendations – to stay out of basins because they did not look "oily" from a geological standpoint.*

It is a telling comment. It is not so difficult to determine if a new area holds promise or not, and the pioneering geologists were not often mistaken in their assessment.

In the 18th Century, there was a belief in the existence of a habitable "Great Southern Continent". But when James Cook[40] in 1769 sailed into the Southern Ocean and confirmed that it did not exist, the World realized that all the continents had been found. Later voyages of discovery found and mapped the smaller islands. Now, satellite imagery can identify the

Fig. 3-27. Geological field work: an almost lost exploration technique,
which found most of the World's oil
(Photograph courtesy of British Petroleum)

smallest rock. Oil exploration has followed a similar path: the "continents" and most of the larger "islands" were all found during the Pioneering Epoch.

The international fraternity was a small and closely knit one, despite the fact that individually they were working for competing oil companies.

I take 1950 to mark the end of the Pioneering Epoch, a convenient mid-century milestone, but it could as well be the Second World War. Oil was of course a critical factor to both sides. Germany was supplied by Romania and large quantities of synthetic oil distilled from coal (rising from 28% in 1939 to a peak of 62% in 1944), whereas the Allies were supplied largely from the United States.

In terms of the resources, about one-quarter of the World's ultimate endowment and many of the critical giant fields were discovered during the Pioneering Epoch. But, no more than 67 billion barrels had been produced by the end of the period, leaving plenty of reserves to be produced in the future. Production had been inexorably rising: to 3.8 million barrels a day by 1930 and 10.4 Mb/d by 1950.

The World was still perceived to be a large place, and few people doubted that the discovery of oil and gas would continue unabated. The Second World War had ended with a feeling of optimism for growing economic prosperity and a new international order for justice and respect. Oil, by now fully controlled by the major international companies, was

	Production kb/d	Cumulative Production	Reserves	Yet-to-Find
USA	5407	41.3	40.0	123.7
Venezuela	1498	5.5	38.2	38.2
FSU	758	6.7	23.2	210.1
Iran	664	2.4	31.8	85.9
Saudi Arabia	547	1.0	131.6	167.4
Kuwait	344	0.3	60.5	29.3
Mexico	198	2.5	3.9	45.6
Iraq	140	0.5	28.6	85.9
Indonesia	136	1.3	10.9	17.9
Colombia	93	0.5	1.3	8.2
WORLD	10477	67	394	1339

Fig. 3-28. The important producing countries in 1950. Note how much had been produced in the USA.

expected to fuel the wheels of industry. Things have not quite worked out that way.[41] The new Century has opened with an Anglo-American invasion of Iraq, which has cost the lives of over 100,000 innocent people and caused indescribable suffering. The invasion was made on the pretext that the country posed a threat to the region, despite indications to the contrary by UN Inspectors, whose findings were subsequently fully vindicated. Few now doubt that control of the region's oil supply was the underlying reason, with support for Israel, having extensive ties to the financial community, being perhaps an added inducement.

A SELECTION FROM THE ROLE OF HONOUR TO INTERNATIONAL PIONEERING GEOLOGISTS

E. Lehner (Trinidad)
H.G. Kugler (Trinidad)
D. Trumpy (Ecuador)
W.E. Humphrey (Mexico)
N.J. Sander (Saudi Arabia)
V. Oppenheim (Latin America)
J.V. Harrison (Iran)
N.L. Falcon (Iran)
H.D. Hedberg (Venezuela)
M. Steinekke (Saudi Arabia)
H.H. Renz (Venezuela)
O. Renz (Venezuela)
J.W. Bausch van

Bertsbergh(Venezuela)
L.G. Weeks (Australia)
H.R. Tainsh (Burma)
M.K. Hubbert (USA)
J. Ruthven Pike (worldwide)
A. Gannser (Venezuela)
E. Rod (Venezuela)
F.B. Notestein (Colombia)
T.A. Link (Brasil)
W.C. Hatfield (Colombia)
H. Loser (Peru)
J.W. Harrington (Latin America)
H.V. Dunnington (Iraq)

G.M.Lees (Iran)
De Boeckh H (Iran)
P.E. Kent (U.K.)
F.E. Wellings (Iraq)
J.D. Moody (worldwide)
F.C.P. Slinger (Iran)
A.N. Thomas (Iran, Canada, Libya)
R.A. Brankamp (Saudi Arabia)
S. Elder (Iran)
H.P. Schaub (Venezuela)
J. Weeda (Borneo)
E.H. Cunningham Craig (Trinidad)
R. Arnold (Venezuela)
G.A. Macready (Venezuela)
T.W. Barrington (Venezuela)
V. Benavides (Peru)
S. Papp (Hungary)
L. Eötvös (Hungary)
V. Zsigmondy (Hungary)
F. Bóhm (Hungary)
F. Pávai-Vajna (Hungary)
G. Csíky (Hungary)
H. Kirk. (Middle East)
G. Flores (Cuba, Sicily, Mozambique)

R.E. King (World)
W.E. Aitken (Colombia)
C.C. Wilson (Trinidad)
W. Saville (Colombia)
H.Widmer (Borneo)
H. Harrington (L.America)
A. Beeby-Thompson (worldwide)
W. Link (Brasil)
G. Noble (Iraq)
G.M. Mugoci (Romania)
I. Popescu-Voitesti (Romania)
G. Macovei (Romania)
G.Cobalcescu (Romania)
W.O. Leutenegger (worldwide)
Charlie Hares (USA)
Tom Wilson (Alaska)
L. Mrazac (Romania)
G.Grigoras (Poland)
Desheng Lee (China)
I. Lukasiewicz (Poland)
H.de Cizancourt (Poland)
W.D. Gill (Pakistan)
P. Bokov (Bulgaria)
V. Dank (Hungary)
T. Buday (Czechia)

and many unknown warriors

They found most of the world's oil basins with technology no more advanced than the hand lens and the hammer. The country of their most notable contributions is shown in parenthesis

NOTES

1. See also an interesting article by Haldorsen 1996.
2. Bilkadi Z., 1996, has published a beautifully illustrated book "Babylon to Baku" covering references to oil in antiquity: the earliest record being 3000 BC.
3. Fuller, J.G.C.,1993, gives a good account with historical insights.
4. Coleman J.L., 1995, The American whale oil industry: a look back to the future of the American petroleum industry; Non-renewable Resources 4/3 273-288.
5. Narimanov A. and A. Palaz, 1994, point out that wells had been drilled for oil in the Baku area before the famous well of Col. Drake in Pennsylvania.
6. In fact, Romania boasts of being the country with the oldest record of oil production: 275 tonnes in 1857. A foreign oil company had been registered there already in 1864. Production comes from

the foothills of the Carpathians, which were also investigated in early years in neighbouring countries including Poland (Popescu 1994, Jacob. 1938).

7. Beeby-Thompson A., 1961, gives a marvellous account of the early days of exploration and production around the world.

8. Yergin D., 1991, has written the seminal oil history which is essential reading; while Pees S.T., 1989, fills in the detailed memorabilia of the early Pennsylvanian fields.

9. Beers F.W. 1865 quoted in Pees S.T., 1989.

10. Possibly the only useful contribution of Kenney's paper on the inorganic origin of oil was the record of an observation by Academician Mikhailo V. Lomonosov in 1757, who stated "*Rock oil originates as tiny bodies of animals buried in the sediments which, under the influence of increased temperature and pressure acting during an unimaginably long period of time, transform into rock oil*". He got it right in 1757: who said that the Russians were not good explorers!

11. Energy Information Administration, 1990, provides useful data on the USA.

12. Pees, S.T., 1989.

13. An agreement between the companies to define their spheres of interest. It did not survive for long.

14. This organization, which controlled the movement of oil by rail, was used to prorate production to support prices.

15. Pickens, T. Boone, 1987, has written a candid and illuminating book about his assault upon a moribund Gulf Oil, with not always complimentary observations about the senior management of oil companies.

16. James, M., 1953, authored the official, and possibly expurgated, history of Texaco.

17. Farago, L, 1973, in a book on wartime espionage reveals the role of Texaco.

18. James, M., 1953.

19. Apparently, this situation arose from circumstances relating to the Pennzoil law suit. Texaco's suppliers began withholding credit, fearing for the Company's future, and it turned to Saudi Arabia for help. The latter provided the much needed credit, but the *quid pro quo* was access to Texaco's downstream.

20. US Shell also planned to join this venture with Saudi Arabia in a highly significant move.

21. Cazalot, C.P., 1996, in a rare interview explained Texaco's policy in the 1990s, which seems to be to concentrate on US opportunities.

22. As indeed anticipated in the First Edition of this work.

23. Beeby-Thompson, A. 1961.

24. British Petroleum, 1959, provides a delightful pictorial history of the company.

25. British Petroleum, 1959, *idem*.

26. In some regards, one could distinguish the Zagros foothills as a somewhat different petroleum system from the rest of the Persian Gulf and Arabian area, but that is probably splitting hairs at least in this context.

27. Massey R.K., 1991 writes about the events leading up to the First World War.

28. In a sense, BP became the first state-controlled oil company, although the State did not normally intervene as other than an investor. BP had difficulties entering Latin America because of this state-ownership which was not allowed under the constitution of several Latin American countries.

29. Clemanceau, the French premier in the war is quoted by Zischka as saying "*Une goutte de pétrole vaut une goutte de sang*"(a drop of oil is worth a drop of blood).

30. See Fromkin D.,1989, for an excellent account of the factors that led to Turkey's alliance with Germany, which were evidently much misunderstood by the British Government at the time. British policy had hitherto supported the maintenance of the Ottoman Empire to act as a buffer against Russian imperialism.

31. Lawrence T.E., a former archeologist, led an Arab rising against the Turks, promising recognition for an Arab state, a promise the Allies reneged upon in the Peace Treaty.

32. Kuwait had of course been part of the Ottoman Empire, although as a trading port it had strong external links, especially with Britain. It was excluded from the Red-Line Agreement by Gulbenkian on the famous occasion when he defined the area of interest with a red-pencil at a meeting. He probably did so to avoid difficulties with Britain.

33. In terms of Regular Conventional Oil, a term used here to clarify the meaning of the general term Conventional. It excludes oil from coal, shale, bitumen, Extra-Heavy and Heavy Oil, as well as

deepwater and polar oil, and liquids produced in gas-plants.

34. Aburish S. K., 1994, gives a revealing account of the corruption of a family that controls the world's supply of oil.
35. See Sell, 1938, for a listing of production data.
36. Yergin D., 1991; Tugendhat C & A. Hamilton, 1975, both give lengthy accounts of the life of Gulbenkian, which is also covered in an autobiography Pantaraxia. Other accounts are by his colourful son, Nubar, Portrait in Oil, and a biography by Ralph Hewins.
37. Other American companies were originally involved but dropped out.
38. Blakey, S., 1985, 1991 has published two well illustrated books on pioreering exploration.
39. Kingston, D. in Hatley, A.G., 1995, describes the life of a roving field geologist.
40. Boorstin, D.J., 1983 has written an excellent book of scientific discovery and exploration.
41. Kennedy, P., 1995, highlights the dangers of the global economy in a very perceptive work.

Chapter 4

GROWTH AND TRANSITION

GROWTH 1950–1970

IN 1950, WORLD OIL production stood at 10 million barrels a day (Mb/d), but within twenty years it had risen to 45 Mb/d, a staggering near five-fold increase. I call it the *Growth Epoch*. At first sight, you would be forgiven for thinking that this must have been a golden age for oil. But in fact there were darkening shadows, and the rapid growth was in a sense a reaction to a new uncertainty and insecurity. The companies had no intrinsic reason to flood the World with cheap oil, which, had they been assured of their future, would have been contrary to their long term interests. There are several reasons why they began to dig into capital, which I will explore in this chapter.

As was noted in the last chapter, the Soviet Union had formally expropriated the foreign owners of its oil in 1928, and Mexico did the same ten years later. They were ripples on the pond compared with the waves that enveloped the industry only one year into the *Growth Epoch*.

For several years prior to 1950, BP had been in dispute with Iran over the level of government take. The Iranians found it inequitable that they should receive royalties of £90 million while the Company registered a profit of £250 million, half of which went to the British Government, holding fifty-one percent of the shares. The well-worn cries of foreign exploitation rang in the Iranian Parliament and in the streets of Tehran. The most outspoken and impassioned voice was that of Mohammed Mossadegh, a frail-looking, seventy-year old, land-owning aristocrat, who harangued Parliament with theatrical tears and faints. He was often interviewed, stretched out in pyjamas on his iron bedstead, seemingly close to collapse, but was evidently far from that. The situation deteriorated when the Prime Minister, who proposed moderation, was assassinated, followed soon afterwards by the Minister of Education, who suffered the same fate. Events were spiralling out of control. In the face of these popular pressures, Parliament passed a law nationalizing BP's Iranian affiliate, and on April 28th 1951 elected Mossadegh as Prime Minister to implement the decision. He did. By September, a British warship, with the band playing "Colonel Bogey", evacuated the last British nationals[1] from

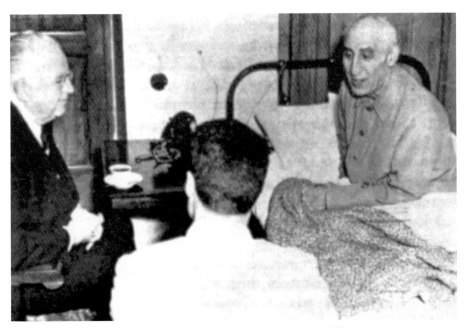

Fig. 4-1. Dr Mossadegh, the frail Iranian who ousted BP from Iran

Abadan, the Company's base since 1908.

It was a far cry from the gunboat diplomacy of an earlier epoch. A war-weakened Britain no longer had the stomach for Empire or the enforcement of contract. It was a retreat much welcomed by the American Government, under the cloak of anti-colonialist rhetoric, but with an eye for oil. A few years later, when the Iranian government was deposed in a coup organised by the CIA,[2] U.S. companies found themselves usurping, through their stake in the Consortium which had replaced BP, much of the position once held exclusively by that company.

At the time, this was seen as little more than an unfortunate chapter in a changing post-war world. After all, the socialist government in Britain was then bent on nationalising almost everything in sight. So, who should deny the Iranians the right to do the same? But in fact it had far-reaching implications, and led to shifts of attitude and policy, some of whose consequences have yet to be played out.

Let us reflect for a moment on BP's position before the Iranian nationalization. It had a concession that ran for sixty years from 1901. It relied on Iran for much of the oil it needed to supply its European and imperial markets that were gradually growing. It could easily balance supply and demand, and make long term plans, both for the duration of the concession and with the reasonable expectation that it could be extended. The Company knew what resources it had, and could plan an orderly exploitation so as to conserve them in a reasonable way. Such an

Fig. 4-2. Cheap imported oil replaced indigenous coal for electricity generation
Photo: Jørgen Schytee/Still Pictures

Fig. 4-3. Releasing a huge explosive charge for calibrating the refraction seismic survey that found Hassi Messaoud, Africa's largest field.
Photo: Jean Laherrère

attitude likewise influenced the American companies in their home country, where they owned the mineral rights outright. It had been an epoch of stability – even complacency one could argue – but at least it provided no motive for squandering resources.

All that changed after 1951. It was a searing experience for BP, which reacted by launching into a vigorous and remarkably successful exploration campaign around the World to find new supplies: an effort, culminating in Alaska and the North Sea. Unlike today, there were still major new provinces to open up. The fate of BP in Iran did not pass unobserved by the other major companies who relied heavily on Middle East oil. They realised too that they were becoming increasingly unwelcome tenants of unfriendly landlords: in ironic contrast with the socialist attitude to property in Europe where the tenant was favoured at the expense of the landlord. The companies concluded that they had only two priorities: to produce as fast as possible while they still owned the rights; and to find more sources of oil to lessen their dependence on a single supply.

To produce as fast as possible meant that they had to find new markets; and the main challenge of the epoch was to do just that, dumping cheap oil on the World. It soon created an energy dependent society, driving to

Fig. 4-4. Drilling in Venezuela
Photo: Shell International Photographic Services,
London

work from suburbia and buying consumer durables that could be transported around the World at minimal cost by cheap oil. Strawberries became available everywhere on every day of the year. Caesar salad was trucked from California to Toronto. The European market in particular was opened up, even to the extent of fuelling electricity generation by cheap oil imports at the expense of indigenous coal.

To find new sources of cheap oil was a greater challenge. It was obvious already by then that nowhere in the World compared with the abundant Middle East. Nevertheless, for strategic reasons as opposed to strictly economic ones, exploration was stepped up. Attention turned to Africa, a continent that had not been widely explored before. BP and Shell in joint ventures took up pioneering positions in East and West Africa, the latter soon yielding important discoveries in Nigeria. French companies turned to Algeria, where they were rewarded by the giant Hassi Massaoud discovery in 1956. A third prolific new basin in Libya was developed by both European and American companies, with major discoveries in 1958 and 1959.

Other useful finds were made in India by the State oil company, later culminating with the Bombay High field in 1974; as well as in Indonesia; Australia and Latin America. Meanwhile, behind the Iron Curtain, the systematic exploration of the Soviets was being rewarded by major discoveries of oil and gas.

In 1968, towards the end of the epoch, another frontier was crossed with the discovery of the giant Prudhoe Bay Field in Alaska, which was of enormous importance to the United States, giving it a second lease of life after discovery onshore in the Lower 48 States had peaked. The field was found by Arco which had leased the crest of the structure that contained the gas-cap, but BP's lesser bid yielded the flanks where most of the oil was located. Prudhoe Bay did much to compensate BP for the loss of Iran, and further cemented the Company's reputation as a first-class explorer. It was no mean feat for a newcomer from Europe to walk off with most of North America's largest oilfield.

These new areas were onshore, but attention also began to turn offshore. In fact, offshore extensions to fields had already been attacked by drilling deviated holes from the land, as in Trinidad and Peru. Some shallow-water prospects had been drilled from steel platforms erected on the sea-bed, as in Lake Maracaibo in Venezuela and the Gulf of Mexico.

What was needed was a mobile floating rig that could be used for truly exploration purposes where the cost of a fixed platform could not be justified. Engineering work in this direction had already been put in hand, and the first such floating rig, the *Breton Rig 20*, designed by John T. Hayward, was put into operation in the Gulf of Mexico in 1949. It consisted of no more than a barge with a drilling rig mounted upon it: the innovations being in the sea-bed wellheads and the connections with the barge above. This technology could be used in waters up to about 100 m in depth, but was very susceptible to wave conditions.

Further development led to the concept of the semi-submersible rig which was based on the ingenious idea of building the rig on two submerged pontoons that floated below the wave base, providing a stable platform irrespective of surface conditions.[3] The first such rig, *Blue Water No. 1*, came into operation in 1962. This new technology widened the scope of exploration to the continental shelves of the World within water depths of about 200 m. The fleet of such rigs grew rapidly so that during the 1960s semi-submersible drilling was undertaken in Borneo, Iran, Canada, West Africa, the United Kingdom, Norway and New Zealand. Designs were continuously improved such that, by the end of the epoch, wells were being routinely drilled several kilometres deep in the stormy waters of the North Sea.

Two alternative technologies deserve mention. One was the jack-up, in which the platform, having arrived on location, put down long retractable legs that rested on the sea-bed, and then jacked itself up above the waves. The Zapata Company of Houston dominated this market. It was managed by George Bush Sr, the former American President, who gained thereby a particular insight into the oil business

Fig. 4-5. The Ocean Traveller *rig drilled the first five wells in the stormy waters of the Norwegian North Sea*

PRODUCTION GROWTH
1950 & 1970

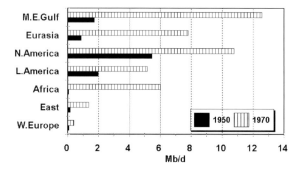

FIG. 4-6. 1950 and 1970 production by region

	Production Gb	Cumulative Gb/a	Reserves Production	Yet-to-Find
USA	3.5	96	45	64
FSU	2.6	32	106	102
Iran	1.4	14	79	27
Venezuela	1.4	26	34	22
Saudi Arabia	1.3	13	237	50
Kuwait	1.0	12	71	7
WORLD	16.8	193	949	608

Fig. 4-7. The most important producers in 1970

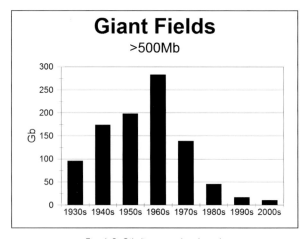

Fig. 4-8. Oil discovery by decade

that was later to influence American foreign policy. The second alternative was the drillship, being a conventional vessel with a rig mounted amidships, that was held in place either by anchors or thrusters. Global Marine of California pioneered this approach to meet the deeper water needs of that State.

Much play is made of technological progress, which some hold will provide a solution to the coming supply crisis. The technology of the semi-submersible rig did indeed make a major contribution by bringing the reserves of the World's continental shelves into reach for the first time, but I think it will prove to be the last such technological breakthrough having a significant global impact. Most subsequent technology has succeeded mainly in increasing production rate, which accelerates depletion, without adding much in terms of reserves. It is an issue to which I will return.

Five years after BP's loss in Iran came another nail in the sanctity of contract: the Suez Crisis. Anthony Eden, the new British Prime Minister, having previously resigned office in protest at the appeasement that preceded and perhaps made possible the Second World War, was resolved to react forcibly to Egypt's nationalization of the Suez Canal. An Anglo-French force was mobilized to defend the canal from a contrived Israeli attack. But America saw it as neo-colonialism and undermined the effort, such that the troops had to withdraw ignominiously a few weeks later. It was plain for all to see that countries could renege on contracts with impunity: likewise, the socialists of Europe had scant regard from the sanctity of private ownership. Whereas France and Britain had had long experience of overseas political administration and exerting their influence around the World, the United States, which was now in the driving seat, had only two policies: economic hegemony and curtailment of the perceived Communist threat, itself not without significance for its burgeoning arms industry. The fruits of this abnegation of the enforcement of international order were soon to be felt, especially by BP as the flag-bearer of an ex-colonial power. It was expelled from both Libya and Nigeria for political problems related to its mother country.

Production rose rapidly during the *Growth Epoch*, mainly because fields in the Middle East, especially those in Kuwait and Saudi Arabia which had already been found before and during the Second World War, were now brought into full production. Yet again, the classic depletion pattern manifested itself, as production in the giant fields rose rapidly to peak levels. It was the same problem of flush production that had so concerned Standard Oil during the early days.

Discovery too peaked although this was not evident at the time. In particular, the discovery of the critical giant fields peaked as shown in Figure 4-8.

The growth of oil production from new sources around the World held down the price of oil, and hence the revenues accruing to the principal producers. Their populations were growing, and their economies expanding. They needed more money. Since the main source of such money for most of them was oil revenue, they set about seeing how they could increase their cut. Such moves led to the creation of the Organization of Petroleum Exporting Countries (OPEC) in 1960. It was a move in which Venezuela secured the backing of the Gulf producers to try to exert more influence. I will return to the pricing of oil and the evolution of OPEC in a later chapter.

The period of growth drew to a close around 1970, and gave way to what can be termed the *Transition*.

TRANSITION 1970–2010

Many of the companies had been pumping as if there was no tomorrow: as it turned out, for many of them there was not to be one. Around 1970, radical changes began to be felt in many ways, leading to a new epoch, which I term the *Transition,* before *Decline* sets in during the early years of the 21st Century. The word *Transition* is perhaps too gentle for what will include years of abject *Crisis,* as foreseen in Chapter 8.

During the twenty years ending in 1970, production had been growing at almost ten percent a year. But, over next the three decades it averaged no more than two percent a year.

This fall in demand was brought about by a combination of factors. They included improvements in engine efficiency, as exemplified by the jet engine, but more than that, there was a sort of saturation in the industrial countries. You can only drive one car at a time, even if that is stationary in a traffic jam. The pattern differed from one region to another with the increase in Asia and Australasia being offset by an anomalous fall in the Former Soviet Union, due to its difficult economic circumstances after the collapse of Communism. On a per capita basis, consumption ranged from about 27 barrels per year in North America, to 13 in Western Europe and 3.6 in the rest of the World.

Gasoline has been the dominant product in the United States, and its consumption has been growing fast in both South East Asia and the developing countries. Fuel oil use, by contrast, has been in decline, especially in Western Europe.

As noted, the share of production coming from the five main Middle East producers had been rising rapidly during the preceding *Growth Epoch,* such that by 1973 they were supplying thirty-six percent of the World's needs. The World was dangerously vulnerable to this single source, as it was soon to find out.

This is not the place to review the Arab-Israeli conflict in more than a few words. In pre-war days, Britain had some sympathy for the idea for the

	Production Mb/a	Consumption Mb/a	Consumption b/capita	Trade Mb/a	Population M
N. America	1701	7794	25.6	−6093	305
East	1363	5793	2.8	−4430	2034
Europe	1982	5112	13.1	−3130	391
Eurasia	5222	3637	2.1	1585	1445
L. America	2800	2437	5.0	363	484
M.E. Gulf	7067	1384	11.3	5683	122
Africa	2715	930	1.2	1785	780
M.E. (Other)	1050	563	4.8	487	117
Non-M. East	17071	19859	3.5	−2788	5674
WORLD	24138	27653	4.6	2278	5979

Fig. 4-9. Oil consumption and production

Note: Production is Regular Oil only (balance coming from Non-Conventional & Refinery Gains).
Consumption and Population includes non-producing countries.

creation of a Jewish homeland in Palestine,[4] which was then a British Protectorate, provided that it did not adversely affect the indigenous population. It had not bargained for the large-scale immigration in the aftermath of the holocaust that brought the matter to a head. Jewish terrorist pressure finally led the British to leave in 1948. The State of Israel was proclaimed, soon to be the recipient of massive aid from the United States, which was being influenced by a strong Jewish lobby. The Palestinians and neighbouring Arab States were not exactly consulted, and it is not surprising that they objected to what was tantamount to an invasion and an exercise in what would now be called ethnic-cleansing. Whatever claims the Jewish people had to repossessing a homeland they lost 2000 years ago, it is obvious that the indigenous Arab community have become the victims of what transpired. A deep-seated conflict has developed ever since and shows little sign of abating.

Some years ago, I took a taxi in Reading, England, far from the Middle East, driven by a man, clad in loose-fitting white clothes, who described himself as a Pakistani Islamic Fundamentalist. He confided that, given the chance, he would be very willing to strap explosive to his waist and destroy a Synagogue. It was a strong reaction for a man living in gentle Berkshire, having had no direct connection with the conflict. He further confided that whereas the Israelis had modern American weapons, his group had time and people: an ominous message.

On October 6th 1973, a combined Egyptian-Syrian attack was launched on Israel in what became known as the Yom Kippur war. The idea of using oil supply as a weapon had been around for several years, and on October 17th, the Arab oil ministers, meeting in Kuwait, decided to use it. They resolved to reduce oil production at five percent a month, and to totally cut off supplies to the United States and the Netherlands[5]. By December, oil prices had risen to $17/b, up from about $2, around which level they had hovered for the past half century. It was what came to be known as the First Oil Shock. By definition, a shock is a short-lived affair, and the embargo was relaxed by the end of the year. Oil prices did not however fall back to their previous levels but fluctuated in the low 'teens.

Fig. 4-10. Sheik Yamani, oil minister of Saudi Arabia.
The World listened to his every word
Photo courtesy of Sheik Z. Yamani

A second oil shock followed six years later in 1979 when the Shah of Iran fell from power to be replaced by the Ayatollah Khomeini, who returned from exile in France to form a new government bent on strengthening Islamic values in the country. Panic buying drove oil prices up to almost $50/b, which is the equivalent of $85 in 2004 dollars.

These two shocks were warning signals, like the minor tremors that presage an earthquake. They prompted many people to become aware of oil resource constraints for the first time. Could the World continue to sustain economic growth indefinitely was a question that was being increasingly asked, as epitomised by the *Limits to Growth Study*.[6] Its general conclusions remain very valid but the immediate fears proved to be groundless because the new productive areas that had been found during the *Growth Epoch*, especially in Alaska and the North Sea, were now coming on stream. Flush production from the giant fields, found early in the exploration process, was beginning to flood the market. It was a repetition of the same old problem that had plagued the industry since its birth. It is important to remember that these new provinces had already been found in the 1960s, and were not a response to the oil shocks, even if higher prices did accelerate development. As a result of this new production, the share of

World supply from the five Middle East producers fell, and was down to about thirty percent by the time of the Second Oil Shock, which was accordingly a short-lived affair.

Meanwhile, there was another important development that transformed the industry during the early years of the *Transition*: in a word, sequestration. OPEC had substantially failed to regulate the business either by imposed posted price or later quota, and the producing governments now turned towards sequestration: Iraq in 1972; Kuwait in 1975; Venezuela one year later; and finally Saudi Arabia itself in 1979. Furthermore, State companies, often with privileged rights, arose in many countries under the socialist movements of the day, except of course in the capitalist

Fig. 4-11. The Ayatollah Khomeini, who returned Iran to Islam
Photo: Associated Press

environment of the United States. In Britain, rights previously freely negotiated were clawed back by what was cynically described as a voluntary transfer to give to a new state-owned British National Oil Company (BNOC); and in 1973, Norway established Statoil, as we shall see in Chapter 12.

Sequestration did not work as its proponents had hoped: they had forgotten one vital ingredient, namely tax which distorts most rational behaviour. Previously, the companies had been paying tax at 50% on the basis of "posted price", which was substantially above the price actually being realized in the market. In effect, the take by host governments was closer to 80%. However, by virtue of the double taxation treaties which were in force, the companies were able to claim the tax paid to the producers as a charge against the tax payable to the consumer governments. In effect, the consumer taxpayer was paying the tax to the producing governments. This happy hidden subsidy ended after sequestration, when the producing governments found themselves having to sell their oil on the cold open market. It really is remarkable to see how the companies actually made so much of their money by juggling their tax obligations.

With the expropriations, the major oil companies lost ownership of most of their prime sources of supply, such that by the end of the last Century their share of World reserves was no more than about seven percent. The Seven Sisters, then down to Six on the demise of Gulf Oil, no longer dominated the world of oil. The concept of property is a deep one, ranking close to life itself as the expression "life and property" confirms. People who own something tend to try to protect and preserve it; tenants have a different attitude, wanting to make the most of their rights while they have them; and traders have an even shorter-term view, being interested only in the moment of transaction. The oil companies have passed through each of these stages. Now, they are effectively only marketers; and will soon face the further constraint of declining markets. It is no wonder that they are short-term in behaviour, being virtually forced to be so by the investment community, which itself is under market pressure to deliver immediate results.

The Middle East share continued to slide as production grew in other areas, such that it had fallen to sixteen percent by 1985. Prices collapsed to a low of $6/b in what became known as the "Oil Glut". It was not really a glut, because the new sources were small in comparison with the established Middle East: it was simply flush production produced at maximum rate by those with a very short view of the future. It was also partially driven by discount economics that depreciated future earnings under the rules of Discounted Cash Flow (DCF) and Rate of Return (ROR). Much of the new production came from offshore and remote areas that incurred high front end investments: the companies had every incentive under conventional economic principles to produce as fast as possible.

I remember an occasion when I was advising the Bulgarian Government at the time it was trying to attract Western oil companies to explore the country soon after the fall of the Communists. The official concerned was reading one of the applications for a concession, and turned to me with a puzzled expression "What means please DCF?" he asked. I replied "Us capitalists like to make money fast". He shrugged his shoulders and confided that such an idea was rather obscene to him. He had a point so far as exploiting a finite resource is concerned.

But then the trend changed again. It was a fundamental discontinuity as important as the shocks that preceded it, although much less spectacular: many people did not even notice it. *The share from the Middle East began to rise.* This time it is set to continue to do so, as now there are no new provinces even in sight to counter the Middle East dominance, save perhaps in Russia. It has nothing to do with politics or economics, and everything to do with the fact that the Persian Gulf area was uniquely endowed with both oil source-rocks laid down in the warm Jurassic seas and with tight seals from the salt deposited when the seas subsequently dried up.

This general trend of rising share is controlled by geological factors, but there was a political factor that triggered the fall of oil price in 1986.

According to the remarkable revelations of the book *Victory*[7] by Peter Schweizer, King Fahd dropped the price in an orchestrated effort by the United States Intelligence Services to undermine the Soviet economy, which relied on oil exports for foreign exchange. The First Gulf War was perhaps one of the consequences.

SADDAM'S STRATEGIC OIL RESERVE: A SINISTER SPECULATION

We are so used to hearing about the mistakes of governments and the ineptitude of their policies that it has become difficult to credit that the unfolding of events could reflect a deliberate policy, immaculately carried out. The image of the Intelligence Community is equally tarnished with their defectors, double agents and spooks eavesdropping by night: the stuff of a Le Carré novel.

Perhaps we are mistaken to be so disparaging. To read Peter Schweizer's book *Victory*[8] would certainly suggest so. It is a remarkable book, not only for its contents, but that it should see the light of day. How credible is it? Internally it seems to be, with its numerous references to cited interviews and now released Presidential Executive Orders. Externally too it seems plausible, judging from the comments of, for example, Hodel[9] who was President Reagan's Energy Secretary at the end of the Cold War, and the explicit confirmation by Allen, the chief foreign policy adviser.[10]

The thrust of the book is that President Reagan and Mrs Thatcher were not content to coexist with the Soviets but resolved to bring down the regime by economic means. The Reagan Administration set up an inner secret cabinet to orchestrate the endeavour through the offices of the CIA. There was little public money available for it, so it had to be paid for by covert means. Enter King Fahd of Saudi Arabia, who was not short of cash, and certainly had a motive to prevent Soviet influence spreading to the Middle East or his Kingdom, already endangered by dissident movements based in neighbouring Communist Yemen. It transpires that he cooperated in the purchase of arms to send to the, then applauded, Taliban freedom fighters in Afghanistan who began to undermine Soviet military credibility. Star Wars was then

Fig. 4-12. Peter Schweizer, author of a remarkable book
Photo: Atlantic Monthly

Fig. 4-13. Ronald Reagan, who, with Mrs Thatcher, brought down the Soviet Empire

proposed as an immensely expensive undertaking which the Soviets would feel obliged to match even if it bankrupted them, especially in terms of foreign exchange that was earned by the sale of oil. Enter King Fahd a second time in 1985 to be persuaded to open the oil valve and flood the World with oil, causing the oil price to collapse and, with it, critical Soviet earnings of foreign exchange. That is what the book says anyway. But the fall in oil price also threatened to bankrupt a large section of the American oil business, and explains why the Energy Secretary of the day, who had been working hard to try to come to the industry's rescue – so vital to America, by then impor-ting almost 50% of its needs – felt so resentful that he was not informed of what was happening.

As discussed in Chapter 3, Saudi Arabia became the cornerstone of American oil interests in the Middle East, when Chevron successfully

Fig. 4-14. King Fahd of Saudi Arabia, who helped them

leased it in 1933. It contains the Ghawar oilfield which holds almost 5% of the World's ultimate endowment. The northern end (Ain Dar) was found in 1948 but the full extent was not determined until 1955 with the drilling of a well 200 km to the south at Haradh.[11] The country owns almost a quarter of the World's remaining oil.[12] Once populated by a few tribesmen, now twenty-five million people live there, of whom many are non-Saudi nationals doing most of the work. Almost all rely directly or indirectly on oil revenue, which is a form of royal patronage. Aburish,[13] in his book, paints a disturbing picture of this strange regime with its "princely allocation" of oil for the benefit of the privileged, and their free, reserved seats in the national airline. At times, there have been more tanks in the desert than drivers: each tank having been bought on terms yielding a princely commission. The United States previously supported this helpful monarchy, and according to *Victory,* has provided various surveillance and other facilities to protect it from the growing number of dissidents at home and abroad, who felt themselves ill-served by the profligate House of Saud. It has become a curiously decadent country with heavily-veiled women flirting on hidden mobile phones despite the watchful eyes of the Wahhabi police, intent on enforcing strict Islamic behaviour, while television masts relay the often salacious material preferred by Western audiences.

But the Cold War victory carried its price: oil price. That stayed down, and King Fahd began to find it increasingly difficult to keep up the flow of royal patronage to the huge number of dependents and claimants. His problem may have worried his friends in Washington and Langley, Virginia. They needed him to survive and continue to supply America with oil at a reasonable price: not so low as to further offend the domestic oil lobby, but not so high as to burden the World economy and the US balance of payments. It is said that Schlesinger, the former US Energy Secretary, when asked about US oil policy, replied "that King Fahd should live for ever". OPEC was proving ineffectual in prorating supply, and some oil somewhere had to be taken out of the market. This leads to the conspiracy theory which bears retelling although I have no idea as to its validity.

Having so successfully ended the Cold War by covert means, perhaps attention now turned to finding a new way to prorate oil production without putting all the burden on poor old Fahd. Casting around for a candidate to perform this useful role, eyes may have fallen on President Saddam Hussein of Iraq. Here was a bellicose man who had been engaged in a long war with Iran, having been encouraged and partly armed for the purpose by the United States. He had, with justification, accused his neighbour Kuwait of cheating on OPEC quota, which had the effect of lowering oil price and thus denying Iraq its legitimate oil revenue. He also resented that Kuwait was sucking oil from his side of the Rumaila field, straddling the boundary. He was upset and wanted to teach his neighbour a lesson, but that was not the whole story as the State of Kuwait had been artificially created by Britain after the First World War out of what in a certain sense was natural Iraq territory. As he planned his invasion,

Fig. 4-15. Saddam Hussein of Iraq, here in less than
bellicose mood: was he duped?
Photo: Associated Press

presumably over a period of time, he must have tried to assess his chances of getting away with it. We do not know what soundings he made, nor what responses may have been secretly conveyed to him. He evidently did not suspect that he was being manipulated, if that was indeed the case. In any event, he satisfied himself, mobilized his army, and headed south in 1991.

It is hard to imagine that this movement passed unnoticed by the surveillance satellites that presumably keep this sensitive area under constant observation. It was therefore strange that April Glaspie, the US Ambassador, on the eve of the invasion should make a statement to the effect that border conflicts between Arab countries were of no great concern to the United States.[14] She was later called before a Senate investigating committee to explain her words, but could only offer the lame excuse that in the difficult circumstances of Iraq she, as a woman, was confined to her Embassy and did not know what was happening. Tell that to the marines![15] They meanwhile had been given another message: namely to prepare to "defend" Saudi Arabia. While all of this was going on, British arms-making machinery was being supplied to Iraq through Jordan with the knowledge of the government, as later revealed by the Matrix Churchill trial.[16] Furthermore, there were reports that Saudi Arabia had been contributing to Iraq's nuclear weapons programme.[17]

What followed is well known. A huge military build-up in Saudi Arabia was organized by the United States under a UN mandate, with loyal support from Britain and several other countries. General Schwarzkopf[18] led his troops north to liberate Kuwait under a campaign, known as Desert Storm, but not before about two billion barrels of oil (equivalent to the reserves of a major North Sea field) went up in smoke as the retreating Iraqis fired the wells. He entered Iraqi territory and was soon within a few miles of Baghdad. But, to everyone's surprise, was then ordered to halt and return, leaving the potential threat of Saddam in place.

The Gulf War was well covered by CNN; and was widely hailed as a great victory for the United States, restoring faith in its military prowess after its retreat from Viet Nam. Yellow ribbons fluttered in flag-flying America, and bumper stickers rallied the nation in support of its heroes. Politicians were seen haranguing the troops in the sand dunes. It was all good stuff and remarkable for the near absence of allied casualties: many of such that did occur were caused by road accidents or were self-inflicted in the confusion of war. We should not overlook, however, the so-called Gulf Syndrome, affecting many ex-servicemen, who were exposed to the accidental release of sarin nerve gas and the effects of depleted uranium from Allied munitions. It is a savage material that leads to long

Fig. 4-16. Two billion barrels of oil went up in smoke in Kuwait wells when Iraq fired the wells

term illness and subsequent birth defects. Iraqi mothers no longer ask if their new born child is a boy or a girl, but whether it is affected by the hideous birth defects caused by this uranium,[19] strewn across the country.

It may indeed have been another *Victory*, but not quite as generally perceived. In a few weeks it was all over. Saddam proved remarkably resilient, remaining in full control of his country. King Fahd was more firmly on his throne than ever, surrounded by a US military presence. All was well. Then the UN, led by the United States, imposed an oil export embargo on Iraq ostensibly to persuade Saddam to pay reparations and stop his alleged nuclear and chemical warfare programmes, which later proved to be largely a fiction. In this way, Iraqi oil was taken off the market; Fahd increased production, and the United States was able to import large quantities at low to moderate prices. With Iraq production removed, it became easier for OPEC to manage its quota arrangements as it was now close to capacity anyway.

At last, an effective solution to World pro-rationing had been found: Saddam would provide a huge strategic reserve for the future at no cost to anyone but Iraq and without interfering with anyone's commercial interests. It was too good to be true, but how to maintain the arrangement?

As the war ended, things returned quickly to normality. The Iraqis managed to survive the hardships despite great suffering, and seemed to support their leader. The flames at Kuwait were put out, without having done as much damage as was at first feared. But normality was the last thing that the policy needed. A perceived threat was required to justify the continuance of the embargo and the protection of Fahd. It did not prove difficult to preserve the state of tension:

  – first, there were the Marsh Arabs, a Shiite community with Iranian

links, against whom Saddam moved, triggering an Allied military response;
- next, incidents with the Kurds also prompted Allied reactions, in which a helicopter carrying UN officials was mistakenly shot-down by Allied fire;
- next came an alleged plot to assassinate President Bush Sr. in Kuwait for which some Iraqi culprits were quickly found. That triggered the firing of a missile from the Red Sea on a ministry in Baghdad;
- next, an undaunted Saddam conducted some military manoeuvres, which were quickly interpreted as a prelude to a new invasion of Kuwait, and red alerts were sounded in the Allied camp.
- next, a new face appeared on the scene. Turkey marched into northern Iraq to harass the Kurds to the consternation of the World with the sole exception of Saddam, who shared the pressure from these people with their strong separatist inclinations.

It is worth repeating the following section from *The Coming Oil Crisis* written in 1996 to recall how things seemed before the Crisis Years opened.

"How long can "Saddam's Strategic Reserve" be maintained? The need for it will not be for very long. Oil demand has been almost stagnant for many years, but is now expected to grow, driven largely by the East and the burgeoning population of the Third World. Production in most non-OPEC countries is past or close to peak. Soon the boot may be on the other foot, with calls for Saddam to release his oil. Perhaps reason will soon have to be found to show that he is not such a bad chap after all. Unlike Fahd, he is not maintained on a throne by US marines. Iran, also vilified by US policy, has no motive to give its oil away and may bury the hatchet with its former enemy and press for higher oil price.

The oil market now relies heavily on the futures market. With the fear of a return of Iraq production overhanging the market, many refiners have preferred paper to physical oil in their tanks, stocks of which have sunk to a twenty-year low. Prices began to firm in the winter of 1995/6, and there was growing popular concern in America about soaring gasoline prices. Then, surprise, surprise, the UN talked of relaxing the embargo for "humanitarian" reasons, and oil prices fell back temporarily.

In the summer of 1996, America launched a virtually unprovoked attack of 44 cruise missiles on southern Iraq. It was ostensibly a reaction to Saddam Hussein's support for the Kurdish Democratic Party, which controls the lucrative customs posts on the

Turkish border and which had invited Saddam to help them oust their rival the Kurdish Patriotic Union. The latter were successfully removed from the provincial capital of Arbil near the giant oilfield of Kirkuk, with in fact minimal intervention by Iraqi troops. Refugees were welcomed by Iran with which the KPU evidently had some ties. The missiles had virtually no effect, and the action was widely condemned by all but the United States' faithful ally, Britain. It is too soon to analyse this extraordinary move. It is variously attributed to Clinton's election campaign; a diversion to allow CIA agents to escape from Arbil; a miscalculation; or a continuation of previous policy of maintaining tension in the area to justify the continued support of King Fahd. One can imagine two other scenarios: King Fahd is under greater threat from his dissidents than is known, and the United States needs justification to maintain its grip on the country; or the United States realizes that it will soon need Iraq's oil, and that Saddam Hussein is likely to be a less generous seller than was King Fahd. Perhaps policy has changed on this realization, and now they actually do want to dispose of him. It was in any event playing with fire, and the World's oil supply. For a while, it looked as if the title of this book would be already out of date by the time it is published as oil prices rose to levels not seen for many years during the autumn of 1996. The embargo was then unexpectedly relaxed, despite the missiles, allowing Iraq to export four billion dollars' worth a year. Prices have not so far fallen markedly, confirming the underlying growing scarcity. It is too soon to evaluate the real significance of these events.

It is useful to look back at these words and note that hints of what was to follow during the *Crisis Years* were already being observed, although naturally the precise timing and nature of events could not be predicted.

The *Transition* witnessed important scientific and technological progress. Great advances were made in geology, as for example by the growing understanding of *Plate Tectonics* and their significance. Geological processes are now much better known. As a consequence, it became much easier to understand the circumstances responsible for oil accumulation, and hence to predict them.

New exploration tools were developed. The advent of the computer chip brought in digital seismic, which radically improved its resolution. Geologists could now see the structure of the oil zones with remarkable clarity. So-called 3D surveys were used to still further enhance resolution,

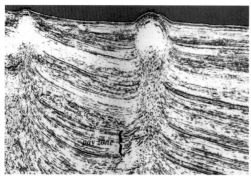

Fig. 4-17. Seismic section: a sort of X-ray of the Earth that allows geologists to study the oil zones. Here, a salt-plug in the deep Gulf of Mexico gives rise to a small trap on its flank.

and later came 4D surveys, which are permanently installed over oil fields and allow the actual movement of oil in the reservoir to be observed. The work station has brought amazing computing power to the interpretation process. In parallel with this have come great advances in well-logging so as to determine accurately where the oil and gas zones lie, along with much other useful information.

The greatest breakthrough of all in the context of resource assessment was the advances in geochemistry in the 1980s that for the first time showed where and when oil and gas were generated, as discussed in Chapter 2. Geochemistry also tellingly explains why large tracts are barren, countering the argument of those who claim much oil remains to be found in statistically under-drilled areas.[20] There are usually sound geological reasons why they are under-drilled.

Drilling and production technology advanced all round, as did offshore construction. Highly deviated wells could now reach far out from the platforms to chase thin reservoirs; multi-lateral wells, in which several branches are drilled from a single borehole, increased the extraction rate; and sub-sea completions and floating production facilities allowed smaller accumulations to be tapped. However, most of these developments were specific to the offshore and especially the smaller fields. They have much less impact on the old onshore fields which hold most of the World's reserves.

Fig. 4-18. The drilling bit, on which everything depends

But an important new frontier was conquered, namely the deepwater domain, here defined to lie in water depths above 500m which has yielded some major discoveries in the Gulf of Mexico and on the margins of the South Atlantic in very unusual geological conditions. All of this represents an amazing achievement for which the engineers can be justifiably proud. But what has been the consequence? Production rate has been increased but less has been found. The short explanation is that there is much less left to find.

Lastly, there has been a radical shift in the way the industry conducts its upstream business. In early years, the major companies did everything, but later encouraged the contracting business which grew in a competitive market, allowing the major companies to contract out most of the work related to seismic surveys, drilling and construction. When oil prices collapsed in the mid 1980s, as already described, the major companies started wholesale purges of their technical staffs, hiring many back again as, or through, contractors, but free of long-term career commitments to them. Lastly, came the so called *Alliance* in which a group of contractors cooperate closely with the oil company to undertake a project.

It is a story of the declining role of the major oil company. In a sense, it reflects the ever-diminishing technical risks, consequent upon the technological progress and knowledge. There is clearly less and less of a practical role for an entrepreneurial oil company when the precise parameters of a project can be defined in advance. The role of the major companies is being increasingly reduced to that of no more than bankers and investors. Their function in coordinating the contractors' work may also be undermined with the arrival of consultants specializing in that sphere. There is a flaw in this, however, insofar as they are also eating the seed corn of technical expertise. Whereas contractors can mop up skilled technicians who were trained in the major companies, it is much more difficult to picture the hard-pressed contractor being able to provide the same breadth of experience. So, once this generation of expertise has gone, there may not be much left to replace it.[21] Perhaps it will not matter anyway if production declines and exploration dries up as this study suggests is inevitable.

Before leaving this theme, it is worth commenting that the national companies of the major oil-producing countries, which control most of the World's reserves, have little technical incentive to open their doors to the international companies, when contractor alliances can do the job as well or better. All the companies really offer is money; and you would think that there are other ways of finding that without surrendering national control of the resource. It will in any event be much easier to find the money as oil prices rise. It can be claimed that the major oil companies offer downstream markets to the producers, but as these markets become increasingly hungry, access to them will not command a premium. There is little difficulty in persuading a starving man to eat.

For many people, there is still not a supply cloud on the horizon. They are encouraged in this view by the irresponsible, bland and near-meaningless scenarios published by institutions, such as the International Energy Agency, the US. Department of Energy, and even Shell. The man in the street still drives to work in crowded streets; air line fares have never been lower. Even in the quiet of rural areas, there is the nearly perpetual thump of distant tractors at work, burning diesel oil. The European Commission in its Green Paper on *Energy to 2020*[22] admits to growing oil demand, but does not even ask the simple and obvious question of how much there is going to be available.

In the following pages, I will examine how dangerous this complacency is. I think the World is about to face, not another shock if that means a short and sharp interruption in supply, but a fundamental discontinuity when oil production starts an irreversible decline. It is time to take stock of the resources.

NOTES

1. In fact, not quite the last national, for my friend Jeremy Gilbert, who lives nearby in Ireland, was left behind, sick in hospital, and later had to walk out (see Yergin D., 1999).
2. See, Kinzer S, 2003, All the Shah's Men: an American coup and the roots of Middle East terror.
3. Nerheim G. and B.S.Utne, 1992 give an excellent account of the Smedvig Company, which began in shipping but successfully became a drilling contractor.
4. Lloyd George, the British Premier in the First World War, liked the idea from the standpoint of a homeland for the Children of Israel in romantic biblical terms and not as new Judaic secular state, see Fromkin.
5. Yergin D., 1991.
6. Meadows D & L, J.Randers and W.W.Behrens, 1972.
7. Schweizer P, 1994 gives a remarkable account of the secret policy of the United States to bring down the Soviet Union by economic means and with the help of the oil weapon applied by King Fahd of Saudi Arabia.
8. Schweizer P., 1994.
9. Nation L., 1995 reports on the dismay of Hodel, the former US Energy Secretary, to discover that he was not privy to his government's policy. He is knowledgeable enough to express concern about the coming oil crisis.
10. Allen R.V., 1996, the Foreign Policy Adviser at the time, explicitly confirms the validity of Schweizer's book.
11. See Perrodon A., 1985 for a description of the details.
12. Referring to *Regular Conventional Oil* (as defined herein).
13. Aburish S.K., 1994.
14. According to Roberts (1995 p.315) Glaspie did have several meetings with Saddam, in which she said that the United States did not take sides in disputes but respected sovereignty. Roberts states that Saddam thought he had a "green light" to press his claim against Kuwait, but explains it as a failure of the language of diplomacy.
15. It transpires that the statement was in fact counter signed by James Baker, the Secretary of State, in Washington, so it was far from an errant remark by a distant Ambassador (see Helge Graham).
16. Scott Report., 1994, a judicial enquiry, details the British Government's connivance with the export of arms making equipment to Iraq.
17. Colvin M., 1994, reports that Saudi Arabia contributed to Iraq's nuclear programme.
18. General Schwarzkopf, who led Desert Storm but turned back at the gates of Baghdad, does not explain the reasons for this strategic retreat.
19. See a remarkable documentary entitled *From the Beginning to the End* by the Swedish film producer, Maj Wechselmann, released in October 2003.
20. See the many writings of P.R.Odell

21. Robert Arnott, an Oxford economists at the Institute of Energy conference in London in November 2004, went so far as to suggest that the disappointing results of exploration reflected the senility of the explorers, whose average age is apparently 49.
22. European Commission, 1995, issued a Green Paper on Europe's energy policy to 2020, oblivious of the resource constraints to oil production. Is it blind? Or is it politically more advantageous to react to a crisis rather than anticipate one?

Chapter 5

INSIGHT AND INFORMATION

"PUT A DIME IN THE slot, and magic fingers will console you": we did, but nothing happened save an imperceptible vibration of the bed. We had arrived in the spring of 1968 in San Francisco to stop over in a motel after a long flight across the Pacific from Australia. It was our first experience of the technological and, indeed, the commercial prowess of the New World. America then seemed an exciting place of opportunity; and we had arrived as "landed immigrants", following a well-trodden path.

In Chapter 1, I described how I had fallen in love with Colombia and had become deeply absorbed in its geology. I felt that I was on the brink of great discoveries into the nature of mountain building and continental structure. The geological revolution of Plate Tectonics was then breaking; and I felt that my Andean studies could somehow contribute[1]. So, after nine unpleasant months in Australia and Papua-New Guinea to which BP had transferred me after the golden Colombian days, I had quit to try something new. My hope was that I would somehow find my way back to Latin America. I had lined up interviews with both the State University of New York, and Amoco.

We continued to New York where we stayed with our good friends from Colombia, the Barry's: Richard had recently joined Mobil in New York to work in a new subject called Operations Research.

The interview with the University was not a success: the faculty terrorised me over lunch with questions on academic subjects which had long since passed from my mind. Surprisingly, they did offer me an Assistant Professorship, but I lacked the confidence to return to academia. I realized that I did not know how to teach freshmen, nor indeed did I know what they were. I thought I would try Amoco before accepting. When a few days later I presented myself at their office at 555 Fifth Avenue, I felt much more at home.

It was not in fact then Amoco, but the American International Oil Company, the international arm of the Standard Oil Company of Indiana, one of Rockefeller's own, based in Chicago.

I liked the office immediately. I was greatly impressed by Bill Humphrey, the Vice-President of Exploration. He was a short man with

Fig. 5-1. Dr Nestor Sander, who ran the regional evaluation

clear blue eyes and brushed back hair, who had a photographic memory. He had had much experience of Latin America, and we were soon discussing the details of the age of the *Santa Rosa Formation* in Colombia. We seemed to hit it off. I also very much liked the Regional Geologist, Nestor Sander, a silver haired Californian, who sat with his distance glasses perched on his forehead, in a darkened room surrounded by books. He told me of his early experiences in Arabia, and of the time he had spent in Paris, where he had married a French girl, Georgia. He was a civilized and knowledgeable man. There too, was the Chief Geologist, Ward O'Malley, a charming Irish American, whose father had been a well-known journalist on the New Yorker. It was a small and delightful office; and when they offered me the job of regional geologist for Latin America, I thought that I had landed on my feet.

We explored the beautiful countryside of New England with its white clapboard houses, reminiscent of Scandinavia, and found a house to rent near Greens Farms, close to Westport. I remember that it belonged to a splendid New Yorker, named Zerlene Joffe, the wife of a crocodile skin importer of all things. She confided that "she did not function under stress", and saw her psychiatrist frequently in the best traditions of the ladies of that city.

Commuting into New York on the bankrupt Newhaven Railroad was an experience in itself. "Sure glad they don't run an airline" was how one traveller succinctly put it as we walked along the track to a relief train, a common occurrence. The bar was full of Wall Street brokers, and there was a special elite carriage called the "Southport Club Car" hooked on the back for distinguished gentlemen from Southport, who played cards and drank gin-and-tonic, served by a Negro steward.

But this new life was not to be. Within a few months of our joining, the Company announced that it was going to consolidate the international company with the domestic organization in Chicago. It was the classic reaction of a head office which, green with envy, resents nothing so much as a successful affiliate. I was to experience the same phenomenon later to my own cost. So before long, we found ourselves driving through the acrid atmosphere of the steel mills of Gary, Indiana, the gateway to Chicago, not having been reassured by the numerous billboards through the prairies, advertising *chewing* tobacco.

One of the first tasks given to the new consolidated organization was to try to determine just how much oil the World had as a basis for defining its new strategy. The project was placed under the direction of Bob Blanton, a chain-smoking rather stressed man, with a clipped southern accent. I found myself trying to assess the potential of Latin America. As I look back, I realise that it was not a very sophisticated analysis.

The first step was to search out the maps and reports on each country, put out by Petroconsultants of Geneva. The maps showed each basin, together with its oilfields and exploration wells. The next step was to look up the reserve and production data, as published each year by the Oil and Gas Journal.

We developed a standard approach which involved calculating the area and rock-volume of each basin; determining how much had been found by how many wells; and then making some intelligent guess for the undiscovered potential. In those days, there were no spreadsheets or even electronic calculators, and so we did not even try the various statistical techniques that are now easily applied. Instead, it was simply pragmatic geological judgment. Unfortunately, I don't remember what the global total was, but there have not been too many surprises in Latin America, save for the deep water discoveries off Brasil. Having previously recommended the Llanos Basin of Colombia as prime territory, the subsequent giant discoveries both there and in adjoining Ecuador were broadly anticipated.

Until then, although I had been aware of the limitations of the areas in which I had worked, I had always thought that the World was a large place with plenty of scope for new discovery. Now, for the first time, I began to realize that the prospects worldwide were indeed as limited as in the places I knew: distant fields turned out to be no greener. Even in those days, when much less was known than now, many areas did not smell right for oil: and indeed they have proved to be barren. As the implications of this new insight seeped in, I began to look at life differently. I realized that the mindless consumerism, nowhere more evident than in the country where I was now living, was insupportable. I began to have a doomsday view of the future. We were, I came to realize, approaching an abyss.

It absolutely changed my attitude to exploration. I realized that the few prospects that actually did fully meet the criteria to succeed had a transcendental value; and I realized that most of the others would fail. I began subconsciously to appreciate the 20:80 rule: 20% of the people own 80% of the wealth; 20% of the patients use 80% of the State health funds; and only 20% of the prospects succeed, if that. This polarity was not understood by the Company, whose management systems tended to make everything look the same under the strait-jacket of hypothetical economic evaluation (see Chapter 10). As a consequence, it missed out on the few prime prospects, where the concession terms were usually tough, and frittered away its effort on the 80% that lacked the essential ingredients.

My career in exploration was to continue for twenty years, but this insight never left me. Circumstances, especially arising from government licensing policy, often forced me to apply for acreage in which I had little confidence as part of a sort of mad competitive game. I hardly ever expressed an objective opinion to my own management, and when I did was accused of pessimism. They were substantially blind to what they were doing, but often it did not matter because the cost was deductible from high marginal tax. I will describe this situation more fully in Chapter 12, covering the Norwegian experience.

We did not like living in Chicago. The climate was awful: icy cold in winter when winds blew off the frozen Great Lakes; and sweaty hot in summer. When we asked Mrs Muther, the real estate agent, to find us a charming house in the shade of Evanston University, she sniffed with the comment *"Very definitely integrated"*, before coming to the point with the question *"how much do you earn, young man?"* That information placed us, with few options, in Forest Avenue, Wilmette, a northern suburb. It was a pleasant old house in a tree-lined street, where the neighbours complained that there were weeds on our lawn. It was not a colourful or interesting place in which to live, although the job was excellent.

For two years, I worked in the small new ventures team, concerned mainly with Latin America, but often appraising opportunities in other parts of the World as they came in: Portugal, Korea, Svalbard, Guatemala, Peru, Borneo. I frequently found myself writing persuasive memoranda for Bill Humphrey's signature, the Vice-President of Exploration, who preferred to travel and search out opportunities and contacts, rather than feed the corporate bureaucracy. I tried always to end them on a Churchillian note with a call to arms, hoping to galvanize the hierarchy with enthusiasm: *"I urge the Members of the Committee to authorize our negotiators to set forth and open discussions on this exciting new opportunity …"*.

Eventually, I managed to persuade the Company to go into Ecuador to explore a promising area in the Amazon headwaters, and I got myself transferred there as Chief Geologist. Since I spoke Spanish, I soon found myself doing less geology and more negotiating. Everything from permission to use jungle airstrips to labour contracts, not to mention applying for new concessions, involved labyrinthine negotiation. It would be wrong to describe Ecuador as a corrupt country, but it was necessary to find ways to pay for services received: a subtle but valid distinction. I became a specialist in this murky but intriguing business: finding the intermediary who performed; and judging the style and the amount required to make the wheels turn in our direction. I remember the delightful case of the Beauty Queen. One day a creature of rare beauty presented herself in the office seeking sponsorship. I asked her why someone with her obvious qualities should need money. "To pay the judge" was the straightforward reply, leading, I am glad to say, to her success at the next contest.

Fig. 5-2. Ecuador, which enjoyed an oil boom in the late 1960s. We lived at Farsalia, at 8500 feet altitude outside Quito, in a beautiful house surrounded by eucalyptus woods and with a view of seven snow-capped volcanic peaks.

Fig. 5-3. Discussing negotiations with Juan-Carlos, the intermediary

Fig.5-4. I look more urbane than previously

It was not Ecuador that was corrupted, but myself: geology was abandoned in favour of more facile amusements in management. I was 38 years old. Mapping the Cretaceous unconformity had become a little stale, although an interest in the regional geology of the Andes was far from extinguished.

When this venture ended and a recall to Chicago beckoned, I quit and

Fig. 5-5. Farouk al-Kasim, of the
Norwegian Petroleum Directorate who
supported research into oil resources
Photo courtesy of Farouk al-Kasim

returned to London, having been invited by some friends to become manager of the Shenandoah Oil Corporation of Fort Worth, Texas, whose principal shareholders were no less than Julie Andrews and James Stewart. Now, instead of the Ecuadorean generals, I had to deal with the socialist ministers who were deciding how to allocate North Sea oil rights. I had learnt my lessons well and knew what to do, when signing up as a partner the owners of the Daily Mirror Newspaper, which supported the Labour Party. This chapter lasted several years and was followed by others that need not concern us.

I will pick up the story of when I was Executive Vice-President of a Belgian oil company in Norway in the late 1980s.

Whereas the wheels in Ecuador were greased by money, those in Norway were turned by "Research". Norway had a passion for research; and the oil companies' applications for promising concessions were awarded partly on the degree to which they had contributed to Norwegian research. It became a racket as the companies postured about the claims

Fig. 5-6. A cartoon of the late Buzz
Ivanhoe, an expert who showed us
the way

for their particular projects, few of which resulted in anything particularly useful.

When wondering what project I could find to support our claims for preferential treatment, I fell upon the idea of sponsoring a research project into the World's endowment of oil, which I thought would be of singular value to the Norwegian government in planning its strategy. I aired the idea with Farouk al-Kasim, an Iraqi who was effectively running the Norwegian Petroleum Directorate, and had an enthusiastic response. Initially, I was not able to do much myself on the project, but it began to take shape. Later, when I had become the victim of a palace revolution, and was locked up in solitary confinement in an outhouse with nothing to do while the Company plucked up courage to show me the door, I was able to dedicate myself to the study full time.

The first step in the project was to invite the late Buzz Ivanhoe to come and visit us. I had read an article by him in the Oil and Gas Journal about resource depletion, which immediately struck a chord. It was evident that he shared my underlying concern. He had done some interesting work, plotting discovery against footage drilled[2] and investigating distribution patterns. I invited him to Norway to give us the benefit of his advice and knowledge.

He was an interesting man, who had been an adviser to the Occidental Oil Company, but was now retired (see Chapter 7). He had been investigating the subject for a long time and had cooperated with the legendary M. King Hubbert, who in 1956 had correctly predicted when production in the United States would peak, developing the so-called Hubbert Curve.[3]

He alerted us to the work being done by the United States Geological Survey on the subject. We accordingly sent someone to the United States to meet the team, which was led by Chuck Masters,[4] and to obtain their material.

Next, we secured the database on historical oil reserves and production from the Oil and Gas Journal, and a listing of giant fields from Petroconsultants.

I continued this work with the Directorate after my precipitate "retirement", working with an enthusiastic team who soon saw the importance of the work. We started building the database, and began to investigate various analytical methods. I began to systematically collect references.

At an early stage of the study, we spotted the anomalous increases in reserves reported by several OPEC countries in the late 1980s, and we began to collect data on published estimates of the *Ultimate Recovery*. I wrote an article for Noroil[5] at that time, which coined the term "political reserves" for the anomalous increases. The expression caught on. I remember sitting up all night in Stavanger with my rather primitive computer, trying to calculate depletion curves, using a programme designed to determine mortgage rates.

We had enjoyed life in Norway and had really no wish to leave. But the prospect of retiring there with the high cost of living and the punitive tax regime, made the future seem fraught with hazard. Furthermore, having been at the centre of the oil business for many years, it now seemed strange to be a nobody. It takes only a few weeks to become out of touch.

For some years, we had been taking vacations in rural France, and so we thought we would move there, eventually buying a beautiful old stone house near the sleepy village of Milhac, where I may have been one of the younger inhabitants. Rural France is a magnificent place where people with smiling faces still till the land, as they have for centuries. It will be survival territory after the *Oil Crisis*.

We did not regret the move, especially since the advent of fax and the personal computer made it possible to stay in touch. I set up my office in the loft surrounded by ancient beams, heaps of papers on the floor, and an old copy machine. I shared it with a herd of spiders.

The Norwegian authorities kindly allowed me to publish the results of the study I had made with them as *The Golden Century of Oil 1950-2050 – the depletion of a resource.*[6] Although with hindsight, it seems a rather primitive analysis, it did capture most of the essentials.

To that point, my estimates of the undiscovered potential were based on my own intuitive judgment and knowledge of the underlying geology. I mistakenly had accepted that so-called *Proved Reserves* as publicly reported were good estimates, when in fact it transpired that they were often very conservatively stated, being primarily a financial term.

The next step forward came when I was invited to the Total Oil Company in Paris to discuss the subject. There, I met Jean Laherrère, a specialist, also recently retired, who had read the *Golden Century*. That meeting in turn led to an invitation from Petroconsultants to develop the subject further using their authoritative and comprehensive database. Jean Laherrère began to explain some of the statistical techniques he had developed to refine old fashioned judgment. Since I am not very mathematical, it took me some time to grasp the *parabolic fractal* and the *shifted logistic curve*, but gradually I came to understand them and see the strength of the techniques. I will cover these things in a later chapter.

As will be discussed further, the great difficulty in any study of this subject is to secure reliable data on what has been discovered so far, as a basis for extrapolating future discovery. Petroconsultants had been accumulating this information for many years, and access to its database proved invaluable.

Interest in the subject began to grow, which led me to write articles and speak at conferences, as will be explained further in Chapter 15. I don't do it very well, but the subject itself usually does command attention. Many people intuitively understand the general position without necessarily having exactly defined it in their minds. In the next few chapters, I will put some meat on the bone, explaining the conclusions of the study.

AMOCO: KING OF THE MID-WEST

Standard Oil, J.D.Rockefeller's remarkable empire, was broken up by anti-trust legislation in 1911 into thirty-seven independent companies with their own boards of directors, although Rockfeller interests did retain a substantial share-holding in most of them. One such daughter was the Standard Oil Company (Indiana), or simply, Standard of Indiana,

Fig. 5-7. Horse-drawn Standard tanker

based in Chicago. It inherited the company's premier refinery at Whiting, Indiana, seventeen miles from the centre of Chicago, just across the state line into Indiana. It was here that William M. Burton conducted successful research into new ways of refining petroleum to remove the sulphur content, and to use catalysts and other methods to optimize the production of a range of products, especially gasoline.

When it commenced its independent existence, the enterprise was strictly a refining and marketing company, having retained the brand name *Standard*, under which it sold products throughout the Mid West. It had no production of its own, securing its supply from other pipeline and production companies, notably the Prairie Oil Co., the Midwest Oil Co., and later its competitor, Sinclair. From time to time, shortages of crude caused its refineries to be partly shut down which let it into the upstream to secure its own sources of oil. In 1918, a Col. Robert W. Stewart became the Chief Executive, and set about acquiring production. He had started his career as a lawyer in Dakota, but soon rose to prominence in the business world. He is described as intelligent, shrewd, 6ft tall, 250 lbs in weight and endowed with tremendous physical strength and a commanding personality. He was in control for eleven years, during which time he acquired numerous producing properties and other companies, including Pan American. It was a company that had been built by Edward Doheny, a Californian mining prospector turned oilman,[7] having substantial rights in Mexico and Venezuela, as well as a sales arrangement with the American Oil Co of Baltimore. It was well on the way to becoming a major international oil company.

Despite these achievements in securing production, Stewart's career ended on a bad note when he was deposed by the Rockefellers who thought he had had his hand in a dubious slush fund connected with the infamous Teapot Dome scandal, in which the Secretary for the Interior was corrupted.

Fig. 5-8. Robert W. Stewart, an imaginative force in the early days who tried to build up production for the company

In 1930, Standard decided to sell Pan American's major assets in Venezuela and Mexico to Esso. It was facing the Great Depression and the threat that import levies might be imposed to protect US producers. It was also nervous that a clause in the 1917 Constitution of Mexico, providing for the expropriation of foreign assets, might be invoked, as indeed it was eight years later. However logical at the time, it was one of those transcendental bad decisions that affect the destiny of a company: with it, Standard of Indiana surrendered the chance of becoming a leader in the World of oil, which it was then close to grasping. With the loss of these foreign holdings, the Company became a strictly domestic enterprise, serving primarily the Mid West. The sale of Pan American's properties gave rise to protracted litigation with the Blausteins, who owned the American Oil Company, and relied on it for their oil supply. The dispute was not settled until 1954, when Standard acquired American and put the Blausteins on the Board. There had been earlier disputes with Esso over the use of the Standard brand name, and the Company then adopted the name, "American", from its acquisition, which was later contracted to Amoco. It was an unfortunate acronym in Latin America, sounding like *muco*, meaning mucous.

The Company built up its domestic production in the inter-war years by a long sequence of complex acquisitions, in which all sorts of residual minority interests survived, providing the lawyers with a field day. It operated upstream variously under the names of Pan American and Stanolind. By 1952, all of these endeavours had resulted in it securing production of 235,000 b/d (or 3.5% of the US total) about half of its marketing needs. It was not in fact a great deal: being less production than provided by a single North Sea field.

In 1948, it put its toe again into

Fig. 5-9. Bill Humphrey, a charismatic geologist who combed the World for prospects

international waters, taking up abortive rights in Colombia, but withdrew a year later. It was not until 1957 that it began to make a serious attempt to expand overseas, creating an arm in New York with the name of The American International Oil Co (AIOC) to which it recruited two dynamic executives, Chris Dome and Bill Humphrey.

Within the span of a few years, these two men had secured rights in Trinidad, Argentina, Venezuela, Colombia, Egypt, Indonesia, Iran, the Netherlands, United Kingdom and Norway, which forty years later still provided about sixty percent of the Company's oil supply. They were men who knew their job, and worked on their own initiative with a minimal staff, most of whom were hand picked, coming with international experience from other companies. In Iran, they broke ranks with the industry offering the government a higher take than was normal. In Britain, they teamed up with the influential British Gas Council (predecessor of British Gas), and secured a key position in the southern North Sea gas province. They understood the international environment, where they needed friends at court.

But Humphrey and Dome had too much independence and initiative for the head office in Chicago, which probably resented their success. Dome was thrown out, later being killed piloting his own stunt plane. The New York office was closed, and the international arm was integrated with a domestic organization in Chicago, later moving to Houston. The staff was expanded enormously, and a massive bureaucratic structure of committees was erected. The predictable consequence was that the Company hardly found any more oil, and generally withered away in the areas such as the North Sea where it had been a respected pioneer. Its global exploration efforts were a dismal failure apart from a few gas discoveries.

A telling article[8] by its Exploration and Production Manager explained how the Company has devised a complex economic formula to make money out of exploration, heavy with all the buzz words of "Impact", "Creativity", "Focus", "Vision", but it said it all when it revealed that it was premised on an estimate that World Ultimate recovery of conventional oil is an unrealistic four trillion barrels, more than double the number proposed here. It demonstrated a poor grasp of the true position. Unfortunately, no amount of "creativity" can find what is not there to be found.

Amoco at heart was a company of the Mid-West, where it had had a long and successful role as a marketer and refiner. Its achievements had been primarily in the refining domain where it invented several important new processes. It was also a successful post-war explorer in the United States and Canada, at one time having more acreage under lease than any other company, a policy smacking more of quantity than quality. But by then, most of the World's major fields had been found: it had missed the boat. Thanks to its marketing expertise, it remained nevertheless one of the largest corporations in the United States. Rumours circulated from time to time of a possible merger with Phillips Petroleum, but eventually it fell to BP. One must suspect that this was primarily a financial exercise, earning colossal fees for those effecting the transaction. It may be no coincidence that both the Chairman and Chief Executive of BP sit on the Board of Goldman Sachs. It would be interesting to make a retrospective audit of the transaction to see if a fair price were paid.

Amoco was a good and generous company to work for in personal terms; and one that was far more welcoming and open to foreigners than were some of its European counterparts as I was later to learn. I am grateful for the time I spent with it.

DATA BASE

In this chapter, I have summarised how this study of the World's oil endowment evolved. As much as anything, it sprang from an early very pragmatic evaluation of Colombia made for BP in 1966, followed on a world scale by another made for Amoco in 1969. As I will explain later, the great difficulty to be faced by anyone attempting such an exercise is to find valid data on how much has been discovered so far and when. To know that is clearly the first step in extrapolating what remains to be found. A cornerstone is the simple and obvious recognition that there is an *Ultimate Recovery;* and that production will one day end because it is a finite resource, formed and preserved only rarely in a few places and times in the Earth's long geological history. Once this mental barrier of there being an *Ultimate* is crossed, plausible estimates of future production can be made, even if the numbers are not very precise, and even if there are disagreements as to the details.

There are four principal sources of information:

Petroconsultants (now IHS)
This firm was set up in 1955 by Harry Wassall[9] to provide an industry news service with information on concessions, drilling, production, reserves and much besides. It began in Latin America but later provided worldwide coverage from its headquarters in Geneva. It maintained close contact with the oil companies and governments, and assembled a colossal database, which it provided to the industry. Its reports formed a ready reference, and were the first port of call when evaluating a new area. In particular, its maps, showing wells, fields and concessions, were an invaluable source of information.

In addition to data on production and reserves, it carried important information on drilling, which is especially critical to the evaluation. Its data relate to individual fields, with reserve revisions properly backdated to discovery. The reserves are classed as "median probability" (or P_{50}) reserves, meaning that the risks that the actual number will be above or below the estimate are equally matched. This largely overcomes the distortions deriving from apparent "reserve growth", discussed more fully in Chapter 6. The database did not cover North America, but otherwise was indispensable for any serious analysis of the World position.

In the late 1990s, Petroconsultants was bought out by IHS, a US database company, and much of the original expertise, built on many years of personal contact, was lost. Furthermore, the task of assembling data became ever more difficult as the number of small exploration companies proliferated and as the State companies, often having a political agenda in what they report, took a more important role.

Other Proprietary Sources
There are several other useful proprietary databanks covering the United States and Canada, and Wood MacKenzie of Aberdeen also compiles valuable world information.

Oil & Gas Journal
The Oil & Gas Journal has for many years published data on reserves and production by country. The data come from a questionnaire distributed to governments and others, and is reported as received. In recent years, much of the information reported by governments in many countries has become unreliable, partly for political reasons. The reserves are supposedly *Proved Reserves,* although in reality they are often far from that. The numbers relate to crude oil and condensate. In the 2004 compilation, as many as 76 countries implausibly reported unchanged estimates.

BP Statistical Review of World Energy
BP has published a booklet each year, which contains a wealth of energy statistics. However, it is important to recognize that in the case of oil reserves, prior to 2003 it simply reproduced the numbers from the Oil & Gas Journal, adding natural gas liquids (NGLs) for North America. Many analysts assume that these numbers have the tacit support of a prestigious company with considerable knowledge, but that is not the case, as made clear in a footnote. Since the reserve numbers are erroneous, and the concept of Reserve to Production Ratios, which are depicted in the booklet as a measure of security of supply, is grossly misleading, it is unfortunate that the Chairman in the Introduction should describe it as a "bible" intended to "improve our understanding of the World energy scene and so

make better-informed decisions". In fact, it does just the opposite. In 2003, the Company ended the long practice of reproducing Oil & Gas Journal data, but selected various additional sources, thereby assuming greater responsibility for the misleading information.

World Oil
World Oil publishes comparable reserve and production data to that of the Oil & Gas Journal. The numbers are however different in many cases, being more reliable for several OPEC countries.

United States Geological Survey
The USGS has published assessments of World reserves and undiscovered potential in successive meetings of the World Petroleum Congress.[10] It is important to recognize that the USGS has its own definitions, which need to be decoded before the data can be put to use.[11] As an arm of the United States government, the organization has to be circumspect on what it says on this sensitive issue. In 2000, it issued a new report, which was a radical departure from the earlier serious investigations, greatly exaggerating the scope for new discovery and so-called *Reserve Growth* for the period 1995 to 2025. It has been largely discredited by the fact that actual discovery has been running at less than half the indicated average value, almost ten years into the study period. In large measure, the weakness arises from the methodology which assesses alternative subjective probability rankings of the estimates and then runs so called Monte Carlo simulation to compare every combination[12]. The *Mean Probability* case is commonly quoted, but in fact only the *High Probability* case (F_{95}) deserves serious consideration. Perhaps it is only logical that the *High Probability* case turns out to have the highest probability of being right. It is furthermore important to remember that the study does not forecast actual commercial discovery, but indicates what is "available for discovery", which in the real world is a very different thing.

NOTES

1. See Stoneley R., 1969, formerly Professor of Petroleum Geology at Imperial College, who incorporated some of the results in a global study.
2. See Ivanhoe L.F., 1985: The late Buzz Ivanhoe was one of the first to investigate the size distribution of oilfields and the patterns of discovery. He remained a leading authority on the subject until his death in 2003.
3. Hubbert M..K., 1956, was the father of production forecasting based on an appreciation of the underlying resource, having correctly predicted when the United States would peak almost twenty years before it did. He was an eminent scientist who wrote on many other topics. He is immortalised in the "Hubbert Curve", the mathematical details of which he did not publish until he was 80 years old. In fact one may suspect that the original study, before the days of the personal computer, was based on little more than a common sense representation of graph paper.

4. Masters C.H., 1987, 1991, 1993, 1994, published valuable resource estimates at successive World Petroleum Congresses. He skilfully found a way to meet the political needs of his principals in the USGS, while maintaining his professional integrity. His definitions need to be decoded before the numbers, which are High Case numbers, can be used.
5. See Campbell C. J.,1989: one of his first articles on this subject.
6. Campbell C. J., 1991: a book resulting from a study for the Norwegian Petroleum Directorate. While the study has evolved greatly since then, the *Golden Century* curves do provide a depletion profile linked to the estimated ultimate recovery, not found previously elsewhere in the public domain.
7. See Knowles R.S., 1959 for a colourful description of the early oil days.
8. Schollnberger W.E. 1996, describes Amoco's economic analysis system, but lets slip that it is premised on near "infinite" resources, quoted at 4 trillion barrels. With this unrealistic premise, the system did not work.
9. See a booklet dedicated to Harry Wassall, published by Petroconsultants in 1996, which describes the entrepreneurial drive of this imaginative man, and also a Dedication at the beginning of this book.
10. Masters C.H., 1994.
11. Campbell C.J.,1995, explained the definition in earlier USGS studies: in effect the reported Ultimate of 2300 Gb reduces to about 1750 Gb if the numbers are redefined in terms of

Chapter 6

HOW MUCH OIL HAS BEEN FOUND

I N THE EARLIER chapters, I have discussed both the origin of oil in geological terms and the history of the industry. Now, I approach the heart of the matter, and ask how much has been found, how much is yet to find, and, most important of all, how much remains to produce, not only in quantity but when and where. It is, however, first of all, necessary to decide what we are talking about, for, as explained, petroleum is a slippery substance that comes in many different forms, each having its own characteristics and depletion pattern. It is essential at the outset to distinguish *Conventional* from *Non-conventional* oil and gas, but that is easier said than done because the term *Conventional* is used in such widely different senses. It is accordingly useful to identify what we may call *Regular Conventional Oil*, and define it to exclude:

> Oil from coal and "shale"
> Bitumen and derivatives
> Extra-Heavy Oil
> Heavy Oil (10-17.5° API)
> Deepwater Oil (>500m)
> Polar Oil
> Natural Gas Liquids from gasfield plants

REGULAR CONVENTIONAL OIL

As already discussed, the unfettered production of oil, as for example occurs in an uncontrolled blow-out, rises rapidly to a peak, and then declines exponentially until the reservoir is depleted. To understand this, it is necessary to think of the reservoir. The oil lies under high pressure in the pore-space between the grains making up the rock. When the well taps the reservoir, the pressure in the adjoining pores is sufficient to force the oil into the well through which it flows to erupt at the surface. The well itself comprises a pipe, normally 6 inches in diameter, which is cemented into the reservoir. Holes are made in the sides of the pipe opposite the oil-bearing zones to let the oil in. In an uncontrolled well, or *blow-out*, there is a plume of escaping oil several hundred feet high, driven upwards by the reservoir pressure far underground (see Figure 3-26). As the pressure in

BLOWOUT

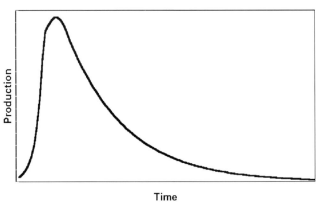

OILFIELD DEPLETION

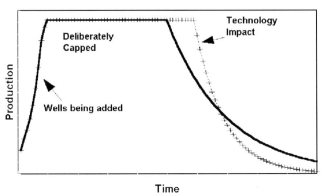

BASIN & COUNTRY

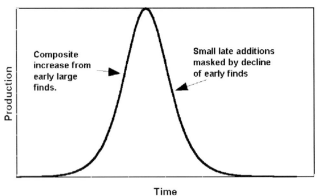

Fig. 6-1. Depletion profiles

the immediate vicinity of the wellbore is depleted, so oil farther and farther away has to be tapped, but instead of flowing through a pipe several inches in diameter it has to make its way through a network of interconnecting pores. The pore-throats, being the passages between the grains, restrict the flow, as do the capillary pressures at the interface of the oil and the rock grains, or the film of residual water that commonly surrounds them. It is obvious therefore that the flow will progressively decline due to the declining differential pressure between the formation and the wellbore, as well as the physical constrictions to the flow. It is clearly easier for light oil to flow than for heavy viscous oil. The pressure itself is due variously to the simple expansion of the oil that is compressed in the reservoir under the weight of the overlying rocks; the effects of an expanding gas-cap above it; the effects of gas coming out of solution within the oil itself; or a water drive due to the encroachment of water under a hydrostatic head from below the oil contact. To begin with, the oil flows to the surface under its own pressure, at any rate in most fields, but as the natural flow rate begins to taper off, efforts have to be made to stimulate the flow. Gas is commonly injected to help maintain the pressure, and water is pumped into the flanks of the field to physically sweep the oil towards the wells. In most modern fields, such stimulation commences early in the life of the field, when it delivers the best results. The procedures are known as gas injection and waterflood, and are collectively termed secondary recovery. Eventually the wells have to be pumped.

Almost all production to date has relied on primary and secondary recovery techniques, and it is likely that the same will be true of most future production. For the reasons explained above, the production of an oilfield has a natural depletion pattern: starting at zero, rising rapidly to a peak or managed plateau, and then declining exponentially to exhaustion. In many cases, production is deliberately choked back to optimize both recovery and the installed facilities, delivering a plateau rather than a peak. I will later turn to *Non-Conventional Oil*, including particularly heavy oil, which has a very different depletion pattern, rising slowly to a long low plateau of production. In short, *Regular Conventional Oil* production is cheap, easy and fast, whereas *Non-Conventional* production is the precise opposite.

About 16 million barrels a day of *Non-Conventional* oil[1] are already in production, but in practice it is not easy to draw a firm line between the two categories because the statistics are not always sufficiently accurate to do so. That said, it is of cardinal importance to make the distinction even if the boundary is a little fuzzy. What concerns the World most is the onset of overall decline, when shortages begin to appear. Since peak production is primarily driven by *Regular Conventional Oil*, that is the stuff we have to think about most. I will accordingly confine my attention to it in the next few chapters, and all the statistics I mention will relate to it unless otherwise specified. Later, I will come to *Non-Conventional* oil that is likely to be an important, but very different, resource in the future.

We are unfortunately not quite out of the woods yet, however, because there is another confusing element: *Natural Gas Liquids* or NGL. As already explained in Chapter 2, gas at high pressure in the reservoir commonly contains dissolved liquid hydrocarbons that condense on being brought to the surface, being known as *Condensate*. Additional liquids, comprising mainly butane and propane, can be extracted from gas by processing. These substances are in many cases fed directly into the oil pipeline and are not distinguished in the records of production. Ideally, they should all be treated separately, but here we will have to be content to identify the amounts produced from gasfields by dedicated plants for which statistics are available.[2] They will deplete in relation to gas. There are admittedly large grey areas. For example, Peru is a case in point having a large part of its liquid reserves in the giant Camisea gasfield, which, if included, would distort the assessment of its oil endowment.

CUMULATIVE PRODUCTION

Once we have overcome the difficulties of defining what to measure, the concept of *cumulative production* is a straightforward one: it is simply the total amount produced in the area under consideration as of the reference date.

Naturally, the records in early years were not particularly accurate in many places, although the amounts involved were by today's standards not large enough to be significant. It is also uncertain how much credence can be placed on the statistics in countries where the oil industry is controlled by a national company and is producing oil for the domestic market. Many countries do not report production by field, sometimes with production from several fields being metered together. Territorial changes also affect the statistics: for example, the *cumulative production* of Israel in some databases is exaggerated by the inclusions of oil from temporarily occupied Egyptian fields. The treatment of Kuwait war-loss, when about two billion barrels – an amount equal to the size of a large North Sea field – went up in smoke in the First Gulf War is another issue. Vast amounts of oil are currently being lost in Iraq as patriots blow up pipelines in defiance of the invaders. In principle, it should be treated as production insofar as it depletes the *reserves* by like amount, but it certainly was not supply, and is omitted from the statistics. There were of course other war losses in Romania, Indonesia, Burma and the Soviet Union during the Second World War, which are hidden in the pages of history. There are also leakages, especially in the Former Soviet Union, and operating losses. Furthermore, amounts used for military purposes may fall outside normal records[3]. We may note in passing that gas statistics are even more unreliable both because of the amounts being flared or reinjected, and because natural gas often contains non-flammable components such as nitrogen or carbon dioxide which may, or may not, be distinguished in the records. Lastly, is the issue of NGL as already discussed. As we will see, however, the problems of knowing how much has been produced are minuscule in comparison with those of arriving at valid numbers for *reserves*: the amount yet to produce from known fields.

While on the subject of *cumulative production*, we may note first that it eats into the *reserves*: as the one goes up, the other must come down, other things being equal. Secondly, *cumulative production* when it ends is what is termed *ultimate recovery*. Obviously, it is production which is all that really matters to us: how much was generated and how much is held irretrievably in the ground are in themselves irrelevant except to the degree that they influence the assessment of production.

Produced by Country	Production Kb/d 2004	Cumulative Production Gb
US-48	3560	173
Russia	8950	130
Saudi Arabia	8750	100
Iran	3940	57
Venezuela	1879	47
Kuwait	2050	32
Mexico	3410	32
China	3494	31
Iraq	2070	29
Libya	1550	24
Nigeria	2350	24
Produced by Region		
M. East Gulf	19362	245
N.America	4660	193
Eurasia	14308	191
L.America	7673	110
Africa	7438	83
Europe	5431	47
Far East	3735	45
M.East Other	2877	27
Other	647	4
WORLD	66131	944

Fig. 6-2 Past and Present Production (2004)

Figure 6-2 shows *production* together with *cumulative production* by region and the major countries as of the end of 2004. It is always important to stress the reference date.

It is noteworthy that the United States has produced more oil by far than any other country, and indeed more than all but two of the World's eight regions. It is not difficult to accept, on this basis alone, that the country is at an advanced stage of depletion. The importance of Russia is also evident.

The World as a whole had produced some 944 billion barrels of *Regular Conventional Oil* as of the end of 2004: that is the first of the key elements to address when considering depletion. It amounts to just over half of the ultimate recovery, as we shall come to see. That says a lot about the future.

RESERVES

Whereas oil may be readily metered at the surface when it is produced, there is no direct way by which to measure the amount of oil in the reservoir far underground. The term *Reserves* is an unsatisfactory one in many ways. In normal parlance, it means something sure and fixed, such as financial reserves or reserves of troops in a battle, whereas in the case of oil, it is no more than an estimate, and normally a poorly defined one at that.

When a prospect is evaluated prior to drilling, estimates are made of how much oil it might contain, namely what is called its *oil-in-place*. As discussed more fully below, from this starting point, a certain assumed recovery factor is applied to estimate the *recoverable reserves*, meaning by definition how much will be produced.[4] As already explained, much is held irretrievably in the reservoir. If the first borehole, termed a *New Field Wildcat* (NFW) is successful, a new estimate will be made with the new information about the reservoir obtained from the well, which in turn is normally followed by appraisal wells to delineate the field. Prior to the start of production, the estimated *Reserves* are termed *Initial* (or *Original*) *Reserves,* which are progressively reduced by production. The term *Reserves* refers to estimates of what remains to produce at the reference date:[5] namely the *Initial Reserves* less the *Cumulative Production.* To make the distinction clear, they are sometimes termed *Remaining Reserves,* although it is a tautologous usage, since all reserves by definition are remaining. When production ends, as one day it must, the *Ultimate Recovery* of the field is equivalent to the *Initial Reserves* subject to such revisions as have had to be made. These terms are confusing and often misunderstood.

I will come back in a moment to discuss the confidence attaching to the estimates of the *Reserves* of a discovery of oil or gas, but first will consider the more difficult issue of how to describe the estimates of what remains to be found in new discoveries yet to be made. They are clearly not *Reserves,* which by definition refers to something known, or rather, estimated within a defined range of uncertainty. What we mean are estimates of what may become *Reserves,* if found. There is not a good word for this category of "reserves". I evade the issue by referring to it as the "*yet-to-find*" or the "*undiscovered*".

It leads us to another difficult term, namely "*resources*". In coal mining, for example, we can picture the deposit covering a wide area, as defined perhaps by core-drilling, which could be said to hold *Resources* of a certain quantity, and to distinguish that from the amounts that are actually being exploited in mines, which could be termed *Reserves*. The procedure works less well with oil, essentially because it is fluid and a very profitable one, meaning that, with some important exceptions, most of the oil in known deposits in production qualifies almost immediately as fully exploitable *Reserves,* leaving little to fill the *Resources* category. Some classifications treat the estimates of the *yet-to-find* as *Resources,* but I find that misleading, implying that they are there to be found and produced, which is a good deal less than sure. The term *Resources* is sometimes also applied to the amounts of oil that may become available as a consequence of radically changed economic or technical circumstances, but they are very much more dubious estimates than applicable to the unexploited part of a coalfield. Another usage of *resources* makes it synonymous with *Non-Conventional* oil, which I discussed at the beginning of the chapter. On balance, I think it is better to avoid the term *Resources* altogether, except to describe the material in very general terms, as in the expression "natural resources". In any event, it is very important to avoid the misconception that the World is endowed with almost limitless *"resources"* of oil and gas that will be inexorably converted to *reserve* status by technology or investment – a view mistakenly held by many economists[6] and institutions, who should know better. They like to speak of the *Resource Pyramid* of which only the upper part has been tapped so far.

Returning to consider further the nature of the estimates of *reserves*, it is obvious that they, as all estimates, are subject to degrees of uncertainty. The *initial reserves* of a prospect are determined on the basis of seismic surveys that, so to speak, provide a scan of the structure indicating its volume. Certain assumptions are then made about the reservoir's character and thickness, as well as the extent to which the trap is full, to provide a basis for calculating the *oil-in-place.* A recovery factor is then applied to calculate the *Reserves,* also making a volumetric adjustment for expansion at surface conditions of temperature and pressure.

A wealth of new information becomes available when the first well on the prospect has been drilled. It permits a more accurate estimate of the likely *Reserves,* but the distribution of the reservoir within the prospect is still uncertain. It may thicken and change in quality in particular directions, reflecting variations in the sedimentary environment in which the reservoir rocks were laid down. The trap may be found to be divided up into separate fault-compartments, each with its own characteristics; or there may be more than one reservoir, each having its own variable distribution and characteristics.

It is therefore normal to drill several appraisal wells to delineate the field before embarking on a development, especially offshore where expensive facilities have to be installed. It is only then that sound estimates of the *Reserves* can be made.

The explorers make their best estimate the amount of the size of the prospect, but in practice are under the pressures of "selling" their idea to secure the funds from their management. But the engineers, who take over on a discovery, adopt a much more cautious line, identifying what is absolutely sure. They tend to work on a step by step basis reporting the lowest estimate delivering a satisfactory investment for each phase. They, rightly, don't build bridges with a fifty percent chance of them standing up: they want to be quite sure.

The uncertainties have led to the practice of classifying reserves as *Proved (or Proven), Probable* and *Possible,* with meanings the words imply.

A more modern method is to apply statistical probabilities to the several parameters, and to recognize a *High Case* (with a 5–10% probability), a *Low Case* (with a 90–95% probability) and *Mode, Median* or *Mean Cases.* The *Mean Case* is statistically the more correct value, but the *Median* (P_{50}) *Case,* is more commonly used, meaning generally that the risks of the actual recovery proving higher or lower than the estimate are equally matched.

Probability theory has its followers but others see it as a form of pseudo-science, given that every prospect is unique and that the assessments are subjective.[7] A hybrid definition, whereby *Proved & Probable Reserves* together are defined as having a 50% or *Mean Probability,* is a compromise in increasing use.[8] In reality we are looking for what in plain language is the *best* estimate, whatever terms we may choose to apply.

Much of the uncertainty regarding the size of the field will have been removed by the time the decision to develop it is taken, especially offshore or in remote locations where substantial investments are at stake. However, some uncertainty will always remain until the field is finally abandoned.

In practice, more and more emphasis turns to the actual performance of the wells as the depletion proceeds, as there is by then no particular need to consider the notion of how much oil might be in-place or what the recovery factor is. It becomes simply a pragmatic matter of how much may be actually extracted from the wells. Commonly, the early estimates of *oil-in-place* remain on the files, and are compared with the new estimates of what the wells will actually deliver. This often indicates an *apparent* improvement in recovery, which is commonly attributed to advances in technology or improved management, for which medals are awarded. But it is obvious that genuine revisions to reserves based on increased knowledge should be as often down as up. Modern technology is certainly far in advance of what was available fifty years ago when many of the large fields were found, and that has had an impact on reserves. All of these strictly technical uncertainties affect the estimation of reserves, but by and large the procedures are straightforward and well understood, with the appropriate degree of uncertainty being quantified in technical terms.

The reporting of reserves is an entirely different matter. In the absence of strict, enforceable, auditable and universal norms for definition and determination, there is scope for considerable latitude in the reporting of reserves, whether by companies or governments. The reporting of reserves is effectively a political act, subject to under- or over-statement depending on the circumstances and motives of the reporting entity. Some people try to exaggerate their assets when going to the bank to borrow money and understate them when meeting the tax man.

The major oil companies tended to understate *initial reserves* for a variety of motives: they may have desired to be less visible; they may have wished to smooth their asset appreciation so that the occasional discovery supports the lean intervening years; there may have been stock-market or management implications; there may have been debt considerations where reserves were collateral; there were almost certainly tax implications, especially in countries such as the United States where a depletion allowance encouraged understatement; and there was the laudable tendency of general caution, with the risks attendant upon overstating normally being greater than those of understatement. In practice, the major companies tended to treat *reserves* as a form of inventory: to be held as low as prudent management and financial needs dictated, irrespective of what was actually there.[9]

Those days are now effectively over, as Shell's recent experience confirms. One may imagine how this Company's financial department, putting out the Annual Report, followed their traditional method of adding just sufficient reserves to meet their financial needs, confident that there were always ample amounts in the unreported stock in the many ventures around the World. Only gradually did it begin to dawn on them that the stock had been almost exhausted, which was so far from past experience that the management could hardly believe it. Understandably, they sailed on reporting what they needed to do hoping for the best and assuming that the many grey areas would come to their rescue. But eventually in 2004 they had to bite the bullet and downgrade their reserves amidst much acrimony which eventually cost the Chairman his job. The other companies were basically in the same position but managed to obscure it by merger.

Smaller companies sometimes overstate reserves with their eye on the stock market. The different approaches are well illustrated by the early reports of the size of the Cusiana Field in Colombia: one partner, an "independent" company reported ten billion barrels; BP (whose shortly to be fired Chairman was in desperate need of good news) reported three billion; and Total (whose technical people like to preserve some good news with which to satisfy their management on rainy days) reported one billion.[10]

The reporting practices are much influenced by the Securities and Exchange Commission (SEC) in the United States, which moved to impose strict rules in the early days of oil in that country. It is almost unique in that the oil rights largely belong to the landowner, meaning that the ownership of the fields is highly fragmented, even to the extent of different reservoirs within the same corner of a field having different owners. The reserves were a financial asset against which money could be borrowed, and the SEC very properly moved to control what could be reported. It accepted two forms of *Proved Reserve,* namely: *Proved Producing,* meaning in plain language, estimated future production from current wells; and *Proved Undeveloped* for the estimated future production from wells yet to be drilled between the existing ones. It was a perfectly satisfactory system for the purpose it was designed in the special environment of North America.

In short, the rules frowned on exaggeration but smiled on understatement, which was regarded as commendable prudence. It meant that the reserves of the fields in question have been subject to progressive upward revision as they were drilled up, and as more and more of their reserves qualified under the strict SEC definitions. I will return to this issue in relation to the subject of "reserve growth", which is much misunderstood.

It is to be stressed that no particular conspiracy is implied by the over- or understatement of *Reserves* by companies. The management was simply exercising its sensible and normal judgment on how it should manage its inventory, which in practice was its principal asset. For most purposes and within limits, the size of the reserves and the dating of revisions did not matter much: it only becomes critical when used as a basis to predict future discovery, as in studies like this.

Governments and government companies both overstate and understate, and are much less subject to audit than are private companies. The most blatant case was the huge increases announced by certain OPEC countries in the late 1980s.

Figure 6-3 shows the reserves of these countries as reported to the Oil & Gas Journal. The first apparent anomaly was in Iraq in 1982, when an eleven billion barrel increase was announced, but in fact it was a delayed reporting of the East Baghdad Field, discovered in 1979. The next anomaly was by Kuwait in 1985 when a fifty percent increase was announced without any discovery to justify it. Iraq accused Kuwait of exaggerating to secure a higher OPEC quota, which was partly based on *Reserves:* apparently with good reason.

It transpires that Kuwait started reporting the total of what it had found, not its remaining reserves. The other countries pondered their reaction until forced to act to a further small increase by Kuwait. In 1987,

	Abu Dhabi	Dubai	Iran	Iraq	Kuwait	Neutral Zone	Saudi Arabia	Venezuela
1980	28	1.4	58	31	65	6.1	165	18
1981	29	1.4	58	30	66	6.0	162	18
1982	31	1.3	57	30	65	5.9	166	20
1983	31	1.4	55	41	64	5.7	169	22
1984	30	1.4	51	43	64	5.6	169	25
1985	31	1.4	49	45	90	5.4	167	26
1986	31	1.4	48	44	90	5.4	167	26
1987	31	1.4	49	47	92	5.3	170	25
1988	92	4.0	93	100	92	5.2	258	56
1989	92	4.0	93	100	92	5.2	258	58
1990	92	4.0	93	100	92	5.0	258	59
1991	92	4.0	93	100	95	5.0	259	59
1992	92	4.0	93	100	94	5.0	259	63
1993	92	4.0	93	100	94	5.0	259	63
1994	92	4.3	89	100	94	5.0	259	65
1995	92.	4.3	88	100	94	5.0	259	65
1996	92.	4.0	93	112	94	5.0	259	65
1997	92.	4.0	93	113	94	5.0	259	72
1998	92	4.0	90	113	94	5.0	259	73
1999	92	4.0	90	113	94	5.0	261	73
2000	92	4.0	90	113	94	5.0	261	77
2001	92	4.0	90	113	94	5.0	261	78
2002	92	4.0	90	113	94	5.0	259	78
2003	92	4.0	126	115	97	5.0	259	78
2004	92	4.0	126	115	99	5.0	259	77

Fig. 6-3. Anomalous reported reserve increases

Abu Dhabi decided to increase its report from 31 to 92 GB, exactly matching the amount reported by Kuwait. Iran went one better reporting 93 Gb (up from 49 Gb), and Iraq, not to be outdone, came in with a round 100 Gb (up from 47 Gb). Saudi Arabia faced a difficulty because it could not match Kuwait since it was already reporting more, but in 1990 decided to follow Kuwait's practice of reporting total discovered instead of remaining reserves, increasing from 170 to 258 Gb.

Venezuela for its part more than doubled its reported reserves from 25 Gb to 56 Gb in 1988. It is noteworthy that the Neutral Zone announced no such increase: it is owned jointly by Kuwait and Saudi Arabia who had no common motive.

This interpretation of the highly anomalous increases seems very plausible,[11] also explaining why the numbers have barely changed in twenty years despite production. It may have made sense from an OPEC quota standpoint to have a relatively fixed number, avoiding the need for endless re-negotiation.

In addition to this obvious case, a large and increasing number of countries report unchanged numbers for years on end, which is obviously implausible. Production eats into reserves unless matched by new discovery or revision, so it is inconceivable that the reserves should stay exactly the same. In 2004, as many as 76 producing countries, including several important producers, reported implausible unchanged numbers.[12]

The reason, apart from the OPEC practice, may be simply that the reserve estimates are not updated, or that those responsible find it politically unpalatable to announce falling reserves. It is prudent, in the absence of other information, to reduce these *reserves* by the *cumulative production* for any period of unchanged reports.

Mexico too has confessed to exaggerating its *reserves*,[13] apparently in connection with collateral for debt.[14] The Chicontepec Field is the cause of much of the confusion. It may have as much as 100 Gb of oil-in-place, but only a small proportion is currently recoverable for geological reasons.

Not all countries overstate their reserves. The most remarkable example of understatement is the United Kingdom, which reports 4.5 Gb as *Proved* for 2004, when the industry confidently expects to extract almost double that amount. It seems unlikely that there should be such a range of technical uncertainty for a shelf as well known as the UK North Sea, but understatement is said to be a British characteristic. Part of the explanation is that reserves are sometimes not reported as "proved" if the field containing them is not yet in production: still another cause of confusion.

These examples demonstrate how difficult it is to come by reliable numbers for reserves. Apart from the obvious case of the countries which had a motive to exaggerate for OPEC quota reasons, no particular conspiracy need be imputed. Much of the problem goes back to definition.

It is uncharitable to suggest that qualified engineers are so incompetent as to systematically under-estimate the size of oilfields, but by all means they are entitled to report whatever serves their management's purpose.

The upward revision of *Proved Reserves* has been widely misunderstood, being attributed to technological progress rather than being a natural consequence of initial understatement or strict definition. This raises the vexed question of the impact of technology and economics on reserves. It much depends on the environment of the field under consideration. Obviously, the scope for the application of advanced technology is greater for small fields in deep offshore waters than for mature giant fields onshore, where well tried established methods are sufficient to extract the oil. Furthermore, it is obvious that heavy oils and difficult reservoirs are more susceptible to treatment than are light oils in normal reservoirs.

The introduction of the technology of the semi-submersible rig did bring in large quantities of new reserves by opening the offshore to routine drilling. But that was probably the last major technological breakthrough affecting significant global reserves. There have been many innovations since then, including the subsea completion, the horizontal and multi-lateral well, two-phase flow in pipelines, as well as innumerable improvements in drilling and production performance all round. It is probably fair to say, however, that the main consequence has been increased production rate and thereby profit, leading to accelerated depletion rather than the addition of reserves. Such reserves as have been added by technological factors are generally small and in difficult conditions, such as in the deep offshore or in special reservoir conditions. The impact of economics is even less important. Most of the World's reserves are very profitable to produce in a price range of less than, say, $15–$20 a barrel. The huge reserves of the Middle East are producible at less than $5/b[15] (see Figure 6-5). So, not much is added by increasing prices above say $20, remembering always that we are talking about *Regular Conventional Oil*. The entry of *Non-Conventional* oil, which *is* much influenced by economics, is another issue to be covered in a later chapter. Gas, for the present, is more susceptible to economic factors, related not so much to production as to transport from remote areas. Higher prices could have a significant impact on gas reserves, allowing huge deposits in the Middle East and in places such as Nigeria and Algeria to qualify as *reserves*, once they are connected to a market. Again, we have to try to distinguish producible reserves in the ground from those being currently produced. Much could be done to improve the reporting procedures.

It is evident from this discussion that the assessment of *reserves* is an exceedingly difficult issue: not so much in a technical sense as the

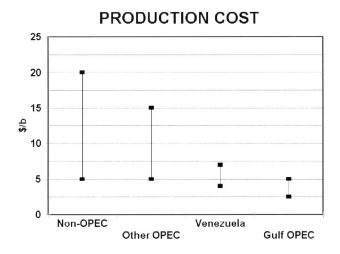

Fig. 6-4. Production costs

procedures are straightforward and well understood, but in relation to definition and reporting procedures. For many purposes, the present laxity does not particularly matter, but to have valid *reserves* is a critical element in assessing the World's endowment and how much is *yet-to-find*. It is by far the most difficult and serious obstacle to be overcome, or at least addressed, in studies of this sort.

Those with access to the industry databases are able to use the best available information on every important field.[16] But here we will have to be content with applying a factor to the average of published *"Proved Reserves"*[17] in an effort to remove the effects of anomalous reporting. In other words, we try to assess what percentage is reported, which may be more or less than actually there. While certainly not claiming to be precise, the numbers are probably correct to within reasonable limits and may be used as a point of departure, to be revised and improved as new information and insight come in.

RESERVE GROWTH AND TECHNOLOGY

Many analysts, commonly using unreliable public reserve data, give emphasis to what they term "reserve growth", which they attribute to advances in technology and improved economics. Some anticipate that *Reserve Growth* will continue in the future in response to further assumed technological progress, and solve the indicated coming oil crisis. The USGS in its influential, but flawed, study of 2000 claimed that as much as 688 Gb would be added by *Reserve Growth*, although careful reading of the text finds qualifications with the admission that they were uncertain about how to treat the subject, considering three options: either to ignore it; or to apply it to certain countries where they had some knowledge; or to assume that US experience would apply to the World. They opted for the latter option mistakenly assuming it to be a technological rather than just a reporting phenomenon, which is understandable since they lacked actual oilfield experience.

It is important therefore to assess the impact of technology in realistic terms. As already discussed, many of the apparent increases in reserves in recent years were simply in the reporting with no technological justification whatsoever. But when we remove these spurious increases, we are still left with some increase that may be genuinely attributed to technological progress.

The issue of under-reporting, backdating reserve revisions and technology is well illustrated by the Prudhoe Bay Field in Alaska. In 1969, one year after discovery the operating company internally estimated its reserves at 12.5 Gb, rising later to 15 Gb, but reported 9 Gb to meet strict SEC reporting practices. It realised that the tail-end of production would be costly in this remote area, and quite rightly was hesitant to commit to such investments at an early stage. The field reached plateau production, set by the pipeline capacity in 1980, and pressure maintenance began soon afterwards. It may be assumed that the very best technology was employed, but decline commenced in 1989 and has been relentless ever since. Extrapolating the decline shows the field will barely make the original internal estimate which made a realistic assessment based on all available technology. In other words, the revision was in the reporting and not based on any technological factor.

Most of the large North Sea fields have ended up some 30% larger than initially reported for the same reason. However, under-reporting is no

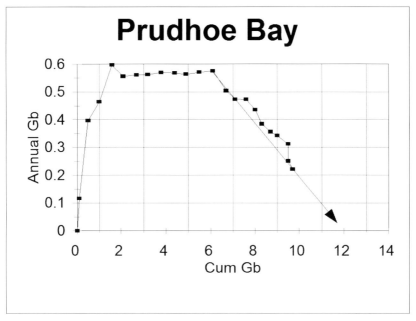

Figure 6-5 The Prudhoe Bay of Alaska illustrates that "reserve growth" is a reporting not a technological phenomenon

longer an option with the smaller more difficult fields, where it may be necessary even to take optimistic assumptions to justify the project at all, often leading to disappointment, as Norwegian experience confirms.

As mentioned above, the introduction of the semi-submersible rig was a breakthrough that opened up the offshore reserves to routine drilling. Offshore production is undoubtedly a high technology business, and so routine technological progress has had an impact. I think it has had two influences. First, it can extend plateau production on offshore facilities, which brings economic rewards but in fact accelerates depletion without adding reserves as such. Second, it can affect the tail end of depletion delaying abandonment which does indeed add reserves. I thus accept that

Reserves (2004)	Gb
Saudi Arabia	144
Russia	75
Iran	70
Iraq	62
Kuwait	55
Abu Dhabi	46
Libya	30
Kazakhstan	30
Nigeria	24
China	24
Regions	
M.East Gulf	373
Eurasia	146
Africa	80
L.America	71
Far East	23
Europe	22
M.East Other	15
Other	1
WORLD	764

Fig. 6-6. Reserves by major country and region (2004)

advances in technology can add reserves by extending the life of fields, but think that it will have a negligible impact on peak production.

The World's total of *Regular Cconventional Oil* reserves at the end of 2004 is here assessed at 764 Gb, which is 512 Gb less than as reported by the Oil and Gas Journal. The discrepancy is principally due both to the anomalous increases reported in OPEC countries in the late 1980s, as discussed above, and to the recent inclusion of some 170 Gb of Canadian *Non-Conventional* Reserves.

Figure 6-6. lists the ten largest countries and the regions.[18] It will be noted that almost half of the World's reserves lie in just five Middle East countries, designated as the Middle East Gulf Region.

GLOSSARY OF RESERVE TERMS

With such a plethora of terms, it is not surprising that much confusion surrounds the subject.

Reserves

The amounts of petroleum that are estimated to be recoverable from a known petroleum accumulation as of a stated reference date on certain stated or implied economic and technological assumptions, normally those current or foreseen for the life of the field at the reference date. The estimates may be divided into categories designated as *Proved, Probable* and *Possible* or described in terms of *Probability* (see below).

Cumulative Production

Total produced as of the reference date.

Decline Rate

The percentage reduction in production from one period to the next (month or year).

Demonstrated Reserves

USGS term[19] approximating to Proved + Probable Reserves.

Developed Reserves
Reserves in a field with installed facilities.

Depletion
The process of producing a finite amount of petroleum (or other resource).

Depletion Rate
Annual production as a percentage of the amount remaining to produce (Reserves plus Yet-to-Find, or Ultimate less Cumulative Production).

Discovered-to-date ("D-t-D") or **Discovered** or **Total Discovered**
Total discovered, as of the reference date, namely Cumulative Production plus Reserves.

High Case Reserves (=Proved + Probable + Possible)
Reserves estimated to have a low probability of occurrence (5–10%).[20]

Hypothetical Resources
USGS term meaning Undiscovered oil in a productive basin.

Identified Reserves
USGS term for High Case Reserves.

Indicated Reserves
USGS term approximating to Probable Reserves.

Inferred Reserves
USGS term approximating to Possible Reserves.

Initial Reserves (also **Original Reserves**)
The reserves as of the commencement of production with any revisions backdated as if known at that time.

Low Case Reserves (= Proved)
Reserves estimated to have a high probability of occurrence (95–90%).

Mean Probability Reserves
The statistical mean of a range of reserve probabilities.[21]

Measured Reserves
USGS term approximating to Proved Reserves.

Median Probability (P50) Reserves
Estimates in which the risks that the actual recovery will prove to be higher or lower than the estimate are equally matched, or the Median probability value

Oil-in-Place
Estimated amount of petroleum in an accumulation, of which only a percentage is producible.

Possible Reserves
Unsure reserves that fail to qualify as **Probable**.

Probable Reserves
Less sure reserves that are likely to occur but fail to qualify as **Proved**.

Proved Reserves (also Proven)
Reserves judged to have a high probability of occurrence:

approximating with Low Case reserves. Also a financial term referring to estimated future production from current and planned wells (see Proved Producing and Proved Undeveloped),

Proved & Probable Reserves

Reserves of Proved and Probable categories that together are estimated to have a *Median* or *Mean* probability of occurrence: in plain language, a *best estimate*.

Proved Producing Reserves

Reserves being produced by current wells, in commercially viable projects as accepted by the SEC

Proved Undeveloped Reserves

Proved Undeveloped Reserves, referring principally to the reserves attributable to infill locations in fields where the wells themselves have yet to be drilled, as accepted by the SEC. The term is used more widely in international operations, and is the cause of much of the current confusion, partly because of greater commercial uncertainty.

Recoverable Reserves

Tautologous synonym for Reserves but sometimes used to emphasise the distinction with oil-in-place.

Recovery Factor

Percentage of oil-in-place that is producible.

Reserve Growth

Upward revision of Proved Reserve estimates, primarily due to initial under-reporting, but commonly attributed to technological progress or economic incentive.

Resources

Notional amounts of oil and gas in Nature including those lacking reserve status irrespective of technological or economic constraints.

Remaining (Yet-to-Produce or "Y-t-P")

The Ultimate less Cumulative Production, or Reserves plus Undiscovered.

Remaining Reserves

Tautologous synonym for Reserves but sometimes used to emphasise that the number applies as the reference date as opposed to the commencement of production or discovery.

Speculative Resources

USGS term for Undiscovered oil in a non-producing basin.

Static Reserves (also sleeping or dormant)

Reserves not being produced or developed.

Static Reports

Periods during which the reported reserves are unchanged.

Undeveloped Reserves

Reserves not being currently developed, namely lacking installed

facilities.
Undiscovered (or **Yet-to-find** : "**Y-T-F**")
> Amount of oil estimated to be found in the future and attain reserve status

Ultimate or **Estimated Ultimate Recovery** or **EUR**
Cumulative production when production ends due to depletion; as applied variously to the world, a continent, a country, a basin, a field or individual well. (The term is also used confusingly for the sum of the initial reserves of fields in a basin or country, excluding the Undiscovered).

THE DISCOVERED-TO-DATE

The sum of the *Cumulative Production* and the *Reserves* gives the total discovery at any reference date. It is a very important starting point, from which to determine how much remains to be found and produced. It allows us to study the discovery pattern to see when the discoveries were made, and to investigate the size-distribution of fields.

There is a certain size distribution in Nature, which I will discuss in detail in a later chapter; and there is what might be called the natural environment of exploration, in which the larger and easier prospects tend to be investigated first, reflecting, it could be said, the well-known human attributes of greed and laziness.

The pattern of discovery has also been influenced by the impact of evolving knowledge and technology, and of course there have been many political influences. In general, however, it can be concluded that, although the actors may change, there has always been a strong motivation to find oil over the past Century both by industry and government. Oil is valuable stuff and a source of great wealth. So, in global terms the progress in discovery of oil has not been unduly hampered by artificial constraints: the discovery pattern of even the Soviet Union with its central planning was not markedly different from that of the United States. Some may argue that the Middle East has been under-explored in recent years, but there is no possibility of matching its early super-giants. The pattern of discovery is therefore substantially a natural one, notwithstanding the other influences. It is, accordingly, capable of valid extrapolation.

In earlier chapters, I described the historical evolution of the business, concentrating mainly on the upstream side. As different territories were opened up, the explorers went into action. At first, they had to learn the geology, and drill wells to provide the essential information during what can be called the lead-time. Then came the moment-of-truth when the basin either delivered, or failed to deliver, in which case it remained forever barren, having failed to possess the essential geological characteristics, especially source-rock. Already by 1908, the World's largest petroleum system, the Middle East, with almost forty percent of the World's ultimate endowment, had been found and confirmed.[22] The broad picture became fairly visible early on.

The opening of the offshore after the Second World War was another great step forward, and the same basic pattern of discovery was repeated but at a greater pace. Whereas the lead-time during the early days onshore could last for a decade, the ability to conduct inexpensive, high-quality and comprehensive seismic surveys offshore reduced it to no more than a couple of years.

When the first well in a basin is drilled nothing is known about the ultimate distribution of its fields, but when the last well has been drilled, everything will be known. It is like watching a photographic print gradually take shape in the developing tray. We have, to-day, reached the point at which the image has appeared, not in the finest detail, but in more than broad outline. This picture shows that now there are virtually no new major provinces left to find,[23] and that efforts will have to concentrate both on ever smaller and obscure prospects in established basins, and on trying to increase the recovery from what has already been discovered. The law of diminishing returns applies very much to the discovery of oil.

The pattern of discovery is much affected by the way in which areas are defined. Ideally, a clear-cut geological natural domain with common characteristics, termed a *Petroleum System,* should be considered, but in practice the boundaries of geological provinces are often fuzzy. Sometimes, more than one system is superimposed. National frontiers exert an influence in that different countries provide greater or lesser incentives to

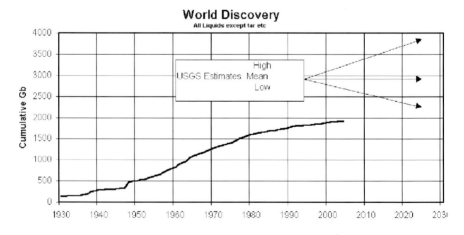

Fig. 6-7. World cumulative discovery

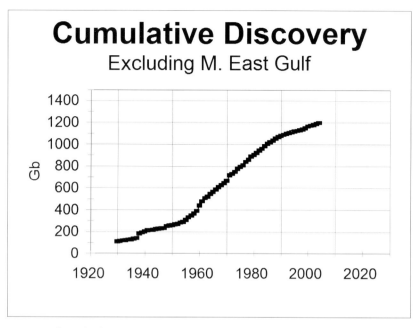

Fig. 6-8. Cumulative discovery, excluding the Middle East Gulf Region

explore, which are far from constant over time. Furthermore, most of the statistics in the public domain relate to countries, not geological basins.

Discovery is decidedly cyclic. When a new productive trend is found, one discovery follows another with the larger fields coming in first to give a clearly identifiable peak. Some countries or basins have only one such trend, and their profile has a single peak. In other cases, there may be several cycles, perhaps reflecting a move offshore, each giving a peak; although taken overall the larger cycles, which tend also to come early, mask the smaller ones.

In terms of distribution, we may consider the number of fields and their size; and plot both parameters over time, noting the inflections, when the rate of rising cumulative discovery begins to decline. To do that requires access to the industry data bases, but published information on giant fields (namely those with initial reserves of in excess of 500 Mb) gives a useful indication.

Figure 4-10 showed the discovery of giant fields, and Figure 6-7 shows the World's cumulative discovery by region, illustrating the mid 1960s inflexion as the rate of discovery began to fall off. I will later come to the issue of the key countries of the Middle East Gulf, namely richly endowed Abu Dhabi, Iran, Iraq, Kuwait, Neutral Zone and Saudi Arabia. Figure 6-8 shows discovery in the Non-Middle East Gulf countries. As will be discussed in Chapter 8, their aggregate production will have an important

Total Discovered Gb	
Saudi Arabia	245
Russia	205
US-48	198
Iran	127
Iraq	81
Kuwait	87
Venezuela	82
Abu Dhabi	65
China	55
Libya	54
REGIONS	
ME Gulf	618
Eurasia	370
N.America	226
L.America	181
Africa	163
Europe	68
Far East	67
M.East Other	42
Other	5
WORLD	1708

Fig. 6-9. Total discovered (2004)

impact on oil price. Their discovery inflexion in the 1970s presages a peak or plateau, which is a key element in the transition to decline and the related unfolding world crisis.

Figure 6-9 lists discovery for the ten largest countries, and the regions. As of the end of 2004, the world as a whole had discovered about 1.7 trillion barrels. Peak discovery occurred in the 1960s. Ninety percent of current production comes from fields more than twenty years old; and seventy percent from fields more than thirty years old. These are very telling numbers.

Understanding how much has been discovered so far is a critical first step in any assessment of the World's ultimate endowment. It is not an easy task, given the unreliable nature of much of the data in the public domain. In the next chapter, I will come to discuss how much is yet to find. The *Discovered* plus the *Undiscovered* gives the *Cumulative Production* when it ends, or in other words, the *Ultimate* recovery. It is a vital concept in depleting a finite resource.

ALAIN PERRODON AND THE PETROLEUM SYSTEM

Fig. 6-10. Alain Perrodon, who developed the concept of the Petroleum System

Q Recently, I have been involved in editing the English versions of several French geological reports, and was impressed by the marked differences between French and Anglo-Saxon geological thinking. It is not just a matter of translation. It seems to me that the Anglo-Saxons give emphasis to disciplined almost sterile description, whereas the French concentrate on trying to understand the broad dynamics of geology, integrating structural and stratigraphic factors. One could say that it is almost a more philosophical approach. I don't know if I put this clearly, but do you see what I mean?

A *Whether or not there is a philosophical reason, certainly we take a more* *global view of geological phenomena that seeks a coherent integration of* *the dynamic evolution of the Earth's crust through time.*

Q The contribution of French scientists to petroleum geology has been impressive. Professor Tissot and his colleagues made the critical geochemical break-through with hydrous pyrolysis which allowed us for the first time to really understand where the oil came from, and when. This was a critical step in the work that you did to develop the concept of the Petroleum System. Could you explain this historical evolution? And what makes the French contribution so special.

A *Petroleum geologists of my generation, especially European geologists,* *have lived through quite an exceptional adventure. We started our* *professional lives just after World War II without any petroleum experience,* *but we were swept ahead by a tremendous current of petroleum history. We* *engaged in the conquest of vast sedimentary basins, unexplored or sometimes* *deliberately ignored by eminent experts of those times. In Algeria, and more* *specifically in the Sahara, for instance, we found our first theatre of* *operations. We undertook, often with more enthusiasm than technical means,* *the systematic exploration of territories without knowing in advance if they* *would contain promising basins. Despite many failures, the global results went* *far beyond our expectations.*

Q Tell us more about how your career unfolded specifically?

A *I had a marvellous apprenticeship of field geology, wellsite work and* *subsurface studies in northern Algeria and the Sahara, lasting for seven* *years. I then had the opportunity to take part in the first Saharan discoveries:* *the giant oil field of Hassi Messaoud and the giant gas field of Hassi R'Mel.* *Three years later, I entered the Bureau of Oil Research, which coordinated the* *research undertaken by the French national oil companies and the predecessor* *of the Elf Group, where I spent the rest of my career.*

In 1956, having submitted a thesis on the Chelif Basin in Algeria, I started *teaching a course of petroleum geology at the School of Geology in Nancy. I* *became responsible for the Elf Group's geological services and research.*

Q What were the milestones in developing the concept of the Petroleum System?

A *There is no doubt that the concept of the Petroleum System sprang from
   the ideas provoked by the theory of Plate Tectonics. It progressively drew
attention to a dynamic dimension in geological phenomena that demanded a
logical integration. Since hydrocarbons are intimately linked with sediments,
it became natural to include them in the broader scheme of things. Geological
thinking moved progressively from the analysis of the structural framework to
dynamic studies of sediments and their fluids, particularly hydrocarbons. The
notion of "petroleum system" replaced that of "petroleum habitat". These
achievements are of course in large degree due to the results of technical
progress, but they also stem from an extraordinary blossoming of new ideas
and from a major revolution in Earth Sciences.*

Q It seems to me that you have found a role that is at the same time
   industrial and academic. You have published widely and received the
recognition you amply deserve, including the award by the American
Association of Petroleum Geologists. I think you like to see things with a
historical perspective. But history does not necessarily only refer to the
past. What do you think about the future of the oil industry?

A *Increased knowledge means we can better evaluate. We now understand
   how oil is formed and trapped. We can confidently declare that large tracts
are barren of potential. It means that frontier exploration is dying out, even if
it is not already dead in many areas. Carried as we are on the powerful wave
of technical progress we must not, however, underestimate the fragility of our
hypotheses and theories. As the late J.A.Masters said at a AAPG Conference,
"Technology has its role and I do not mean to diminish it – but the world is
changed by dreamers".*

Q It seems to me that there are very far reaching consequences for
   Europe and France itself, if oil production is about to peak as seems
inevitable from what we know of the reserves and potential. I sometimes
think that we are living on the brink of an abyss, when you realize how
dependent we have become on cheap oil-based energy.

A *France, more than anywhere, has long realized its dependence on
   imported energy. It is for that reason that we decided to develop our
nuclear capability. Many people fear a repetition of Chernobyl or radiation
leaks, but I think that an efficiently managed nuclear industry in state hands
can be a solution with acceptable risks. Although oil will become more
expensive, it is not about to run out; and there are still possibilities to bring in
new gas production. I do not therefore see exactly an energy crisis, but I do
think that the energy issue is an increasingly important one. I think it will
indeed lead to fundamental changes in the way the world lives. Here in France,*

we are privileged by having a large rural countryside which will help sustain us, perhaps bringing us back to a more traditional style of life, which personally I would welcome.

NOTES

1. See Stosur G., 1996.
2. See US Department of Energy.
3. See ASPO Newsletter No 46, 2004. Also the Defense Energy Support Center Annual Fact Book 2001, which says that the military consumed 5.6 billion gallons of fuel that year.
4. It is important to distinguish "will" from "can": "will" is to be preferred as the more pragmatic case taking into account economic and technological factors.
5. They are sometimes termed *Remaining Reserves* to distinguish them from *Initial Reserves*, although strictly speaking this is a tautologous usage.
6. See the writings of Adelman and Odell, the high priests of this heresy. Also Statoil mistakenly states "Total hydrocarbon resources with the potential to produce liquid fuels are so large that they can be considered infinite for the purposes of this analysis".
7. Laherrère J.H., 1995, has discussed in several papers the issue of reserve definition, being a firm proponent of the "probability" system. As he points out, there is now standard agreement for equating the deterministic terms of *Proved, Probable* and *Possible* with their probabilistic equivalents, a subject also well covered by Beardall (1996).
8. Sometimes it is said that the reserves have a *better than* 50% chance of occurrence or production. The Probability equivalent of Proved and Possible reserves also differs from one classification to another.
9. See Barry R.A. 1993, for an excellent objective discussion of the manner in which oil reserves are calculated and reported by the industry.
10. Petrie P., 1992, describes the Llanos discoveries: see also Cazier E.C., 1995.
11. Barkeshli, 1996, an official with the Ministry of Petroleum in Tehran, has now confirmed that these increases were reported for what he calls the "quota wars".
12. Abu Dhabi, Afghanistan, Albania, Angola, Austria, Azerbaijan, Bahrein, Bangladesh, Barbados, Belarus, Benin, Bolivia, Brunei, Bulgaria, Cameroon, Chile, China, Congo, Croatia, Cuba, Czech Republic, Dubai, Ecuador, Egypt, Equatorial Guinea, Ethiopia, Gabon, Georgia, Ghana, Guatemala, India, Indonesia, Iran, Iraq, Ireland, Italy, Ivory Coast, Japan, Jordan, Kazakhstan, Krygizstan, Lithuania, Madagascar, Malaysia, Morocco, Mozambique, Myanmar, Namibea, Neutral Zone, Netherlands, Oman, Qatar, Papua-NG, Philippines, Poland, Qatar, Ras al Khaimah, Romania, Russia, Rwanda, Serbia, Saudi Arabia, Sharjah, Slovakia, Somalia, South Africa, Spain, Sudan, Surinam, Syria, Taiwan, Tajikstan, Tanzania, Thailand, Trinidad, Tunisia, Turkey, Turkmenistan, Ukraine, Uzbekistan, Viet Nam, Yemen and Zaire.
13. Los Angeles Times, 1991, carried a report by an ex-Pemex executive, stating that the reserves had been exaggerated, mainly by inclusion of non-conventional reserves in the Chicontepec, which according to Macgregor, 1996, has as much as 100 Gb of oil-in-place. There are serious geological difficulties, meaning that only a small fraction is currently producible.
14. Duncan R., 1996, explains how the loans granted to Mexico in the wake of the collapse of the peso have oil reserves as collateral.
15. The issue of cost is however difficult because it involves capital costs, accounting practices and tax issues which differ from country to country.
16. Campbell C.J., and J.H. Laherrère, 1995, give the most comprehensive analysis of the world's endowment, based on Petroconsultants' material in a report summarized and reviewed by Mabro R. (1996).
17. Campbell C.J., 1996, converts published "proved reserves" into "median probability reserves" by use of a factor.
18. See Appendix for a full listing.
19. The USGS terms are not used by the oil industry.
20. Strictly speaking the probabilistic values cannot be related to the deterministic values of *Proved,*

Probable and *Possible,* but for practical purposes the correlation shown here is close enough.

21. The Mean or Expected value in a log-normal distribution equates with ((3 x Proven)+ (2 x Probable) + Possible)/3.

22. This is perhaps a generalisation, for several sub-systems can be recognized.

23. Jennings J.S. (1996), former Chairman of Shell, makes this very clear between the lines if not exactly on them.

Chapter 7

THE ULTIMATE AND THE UNDISCOVERED

WE NOW COME TO a semi-philosophical concept: the idea of an *Ultimate* recovery. How can anyone contemplate the ultimate of anything: it is like thinking about infinity. We naturally shy away from any such idea, which is foreign to our way of life, our attitudes and experience. Yet, all of us do have knowledge of depleting a finite resource: our own life-span. We are not immortal. Most of us do not however dwell on the thought of our deathbed, or the depletion of our lives: we comfort ourselves with the hope of advancing medical technology and defying the average. The hard-bitten insurance companies approach the subject with more reality when they use actuarial tables to say that my life expectancy has shrunk from when I wrote *The Coming Oil Crisis* to only four more years now. If I dwelt on that I would not be at this screen.

So it is with oil. It is a decidedly finite resource, with its own life-span, about which we care not to think. In this chapter, I will discuss how to estimate that life-span. As we shall see, oil is in fact at its middle age, or the midpoint of depletion.

"Are we running out of oil?" is a question that is often asked. The short answer is, "yes: inevitably we are, but not for a long time". What matters more is the less often asked question: "When does decline set in?" The peak of production in itself is no earth-shattering event, when what really affects us and our survivors is the long downward slope that follows.

Shortage means scarcity and higher prices. As I have said, peak production more or less corresponds with the midpoint of depletion, and to calculate that we need to have an idea about the *Ultimate* life-span.

We need to model future supply on these considerations. It is true that we cannot exactly imagine the production of the last barrel of oil: staring down the last wellbore from which nothing emerges. In some regards, it is a case of *reductio ad absurdum*. Nevertheless, it is necessary to develop an idea of the practical limits: even if we cannot know the exact number, nor quite grasp the idea of an *Ultimate*. Already, many oilfields have had to be abandoned when their production fell to almost nothing, and they became uneconomic. To introduce the idea of economic limits, however, brings its own pitfalls. What is the economic limit? Old stripper wells,[1] whose production is down to a few barrels a day, are shut-in in the United States when oil price falls below a certain critical level. It matters to the owners of the wells, but in global terms, it is not really very relevant because the amount these dying wells deliver is very small in a world context. The difference between very little and nothing is not important in practical terms, whatever the philosophical distinction.

What matters to us is the discontinuity between rising supply, as we have known for 150 years, and the onset of decline which will inevitably characterize the rest of this Century. It heralds the opening of a gulf between supply and demand with far-reaching consequences. Demand is itself a somewhat flexible concept. A starving man may demand a scrap to keep him alive; while a teenager may demand another bottle of coke to maintain his idea of happiness. Both reflect "demand" in economic terms, but one is a good deal more pressing than the other.

It is evident, mathematically, that adding ten years to the *Ultimate* life-span, advances the midpoint by only five years. So, even an inexact estimate of the *Ultimate* gives a good approximation of the midpoint date. In principle, *Ultimate* corresponds with *Cumulative Production* when it ends, but it is more practical to define it as the *Cumulative Production* at some date in the far future, such as 2100, so as to avoid worrying about the irrelevant tail end, which delivers next to nothing. The difference has a negligible impact on the peak of production which is the most important issue. Making again an analogy with human life-span, think of our spending pattern: it means little if we have an expensive christening or a jolly wake, most of what we spend will still be incurred in middle age.

With these reservations and qualifications, we can therefore try to estimate a practical *Ultimate*. It is the sum of today's *Cumulative Production*, the *Reserves* and the amount *Yet-to-Find*. It should be a rounded fixed number that is subject to only periodic revisions based on a thorough new global assessment. One could consider several cases with higher and lower assumptions to test the sensitivities, but I think it is better to bite the bullet with the best estimate one can make, and stick with that until there is good reason to revise it. Revise it, we certainly should, when new evidence so justifies, recognizing that it is not cast in stone.

For practical reasons, I include in the *Ultimate* estimate a buffer of some extra oil, described as "unforeseen" to provide a rounded total. In the mechanics of the model, it is convenient to derive the *Yet-to-Find* (or *Undiscovered*) by subtracting the *Discovered* from the *Ultimate*. But, of course, the *Yet-to-Find* is assessed in arriving at the estimate of the *Ultimate*.

METHODS FOR DETERMINING THE ULTIMATE

There are qualitative and quantitative methods for determining the *Ultimate*. Both are built, directly or indirectly, on extrapolating past discovery in relation to the underlying geology. We can use our heads and deliver old-fashioned judgment, or we can devise statistical methods that generate more abstract numbers. In fact, we need a combination. In a perfect world, the statistical methods would no doubt provide the more accurate estimate, but it is not a perfect world with many uncertainties about the validity of the input and the definition of what we try to measure. Accordingly, it is well to use common sense and judgment in finding the most reasonable solution. It is not an exact science.

THE INTUITIVE

In Chapter 5, I described making an intuitive assessment of the *Yet-to-Find* of Latin America. I knew the Continent well, and had an intuitive feel for the characteristics of its many basins. I could make reasonable guesses of the potential. Subconsciously, I was thinking about complex geological relationships, and was asking myself how large the tracts were; how much work had been done; and what the results had been. I was saying to myself "looks to me as if we could hope to find about half as much again" – or some reasoning like that. For many basins, I said "smells wrong: the few wells that they have tried, didn't work. I don't think this place has what it takes – there's no good source". For example, I intuitively wrote off large tracts along the Pacific Coast of Colombia, sensing that they were deficient in source-rock and endowed with poor reservoirs due to the volcanic content in the sediments. Nothing has happened since to change that assessment. Most places with oil exhibit some signs of it in seepages or early wells: and by 1969, when I did the study, most ideas had been tried. The old adage of "where there is oil, there is more" is a good one, but it carries a corollary: "if you find it at all, you find it soon." It is true that I did miss the deep water basins off Brasil. Our knowledge of the offshore was then limited, and we tended mistakenly to extrapolate onshore conditions in the absence of any other information. I also remember, for example, turning down a prospect around the Natuna Island in the South China Sea because non-prospective basement rocks[2] occurred on the adjoining lands, and I didn't have the imagination to anticipate the unexpected trough in between. It was a bit like that in the early days of the North Sea. The prolific Viking Graben cuts across the older non-prospective Caledonian trend of Norway and Scotland. Now, forty years on, almost every offshore region has been covered by at least some seismic surveys and a few test wells, outlining the regional geology. It is no longer necessary to extrapolate from the adjoining lands. It is, accordingly, inconceivable that any sizable new offshore basin of interest has been missed.

The intuitive method is not a bad one, if handled objectively by knowledgeable geologists with enough global experience to have a good sense of judgment and proportion. Objectivity is however now the difficulty. In corporate bureaucracies, they no longer consult specialists, but hold committees, often of people having no particular insight on the subject. The quest is for consensus, the hallmark of a committee, giving a politically acceptable answer. Often, the management has its own ideas of what it desires from an evaluation, which is then duly massaged into shape. The explorers have come to behave more like salesmen than doctors. So, it has become very dangerous to rely on studies of this sort made by oil companies. Government agencies, such as the US Geological Survey or the IEA, also have their own politics to worry about, and besides they tend to have a more academic view. In fact, the USGS considers only theoretical geological prospects, declining to soil its hands with economic or technical constraints. Figure 6-7 showed how far the estimates lie from the trend of past discovery. It has been under threat of being abolished by a cost-saving government. No doubt, it is under pressure to offer an optimistic assessment in the pragmatic world of politics where no one is paid to deliver an uncomfortable message. Indeed, in mediaeval times they often killed the messenger of bad tidings.

My own intuitive study of 1989, as published in a book the *Golden Century of Oil,* proposed an *Ultimate* of 1.65 trillion barrels, only 200 Gb less than the 1.85 trillion now indicated with the benefit of vastly improved data and analytical techniques. The difference is within accepted range of error.

THE PUBLISHED RECORD

The record of published estimates of the *Ultimate* is a useful approach, largely reflecting the evolving intuitive judgment of explorers, based on their experience at the time of the estimate. In 1942, when most oil production was onshore, and the United States dominated world production, 600 billion barrels seemed a good number. But the opening of the offshore and the discovery of giant fields, especially in the Middle East, led to progressive upward revision to a peak of 3.5 trillion barrels in 1969. Then, the estimates began to fall with the realisation that giant discovery had peaked, and that the offshore was far from universally prospective. The consensus has dropped to just below two trillion barrels on a falling trend. It is of course important to check carefully into the definitions being used in such evaluations, and to see if the anomalous reserve increases of the late 1980s have been properly recognized. Figure 7-1 shows how the 2000 USGS Study lies far from the consensus of earlier estimates, including those by that organisation itself.

The creaming curve is so named because the better prospects are tested

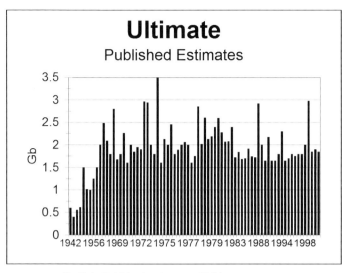

Fig. 7-1. Published estimates of Ultimate recovery

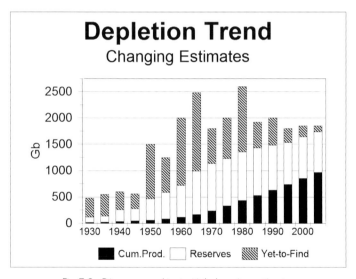

Fig. 7-2. Discovery and potential: changing estimates

first and taken off the top of the list, like skimming cream off milk. The curves plot cumulative discovery either against cumulative *wildcats* or over time. Such curves can be made only where there is accurate data on discovery and *wildcat* drilling, with reserve revisions being properly backdated to the discovery of the fields to which they relate. In practice, it calls for access to the industry data bases.

The plot over time is distorted by political events, such as wars, government licensing rounds or oil price shocks, while that against

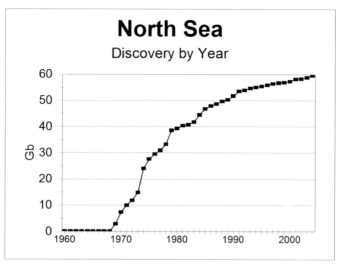

Fig. 7-3. Discovery plot of the North Sea: discovery over time

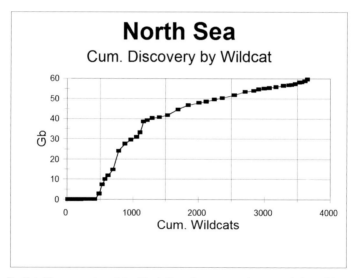

Fig. 7-4. Discovery plot of the North Sea: discovery against cumulative wildcats

wildcats gives a more natural distribution of discovery, but even here there are distortions, such as where wells are drilled on poor prospects to fulfill concession commitments. In the simplest case, in a basin with only one major play, such as the northern North Sea, the plot is hyperbolic, because the larger fields are found first, being too large and obvious to miss. In other cases, where the country or basin contains several different plays, there will be a series of hyperbolas, stacked one upon another, with the overall trend probably also being hyperbolic, insofar as the larger provinces tend to be found first.

The hyperbola can be constructed by taking the relationships between the number of wells drilled at the first discovery; the cumulative discovery at the present number of wells and the discovery when half the present number of wells had been drilled, and then extrapolating to asymptote. Alternatively, computer programmes with regression analysis can be used. The asymptote corresponds with the *Ultimate*, subject to a cut-off for discoveries too small to be economic under any realistic price scenario, and, more generally, the time-span implied, taking into account current and projected drilling rates. Again, we cannot aim at extreme accuracy and have to be content with notions of how much oil will have been found with, say, double the present number of wells, or after, say, thirty years of exploration drilling on current trends. It comes to much the same within the accuracy of the numbers at our disposal. We must remember that exploration in a province is likely to end when production ends, if not before, largely for tax reasons. So, not all the oil in the theoretical distribution will in fact be found. It is useful to plot creaming curves both against wildcats and over time to compare the results and obtain a hint of any defects or distortions in the input. For example, the listing of wells does not normally identify the objective of the well. In areas where there are both oil and gas systems, some of the wells were aimed at gas and should therefore be excluded from the oil analysis. Alternatively, we could convert the gas into oil equivalence and look at combined values, but there are pitfalls here too. There are many points of detail to resolve. It is a case of using common sense to balance all of the factors, always remembering that the input data are often unreliable.

In addition to the simple curve, we can also plot discovery by size-classes. In many cases, the plot of giant fields has been flat for many years, suggesting that none is left to find, whereas the discovery of the smaller classes of field-size may still be rising. We can plot average size; and the number of discoveries to obtain yet more indications of the pattern.

THE LATE L.F. ("BUZZ") IVANHOE: ONE OF THE FIRST TO EXPRESS CONCERN
(This interview from *The Coming Oil Crisis* is included in memory of one of the pioneers, who died in 2003)

Q Buzz, I well remember reading one of your articles in the Oil and Gas Journal in the 1980s. Here, I thought, was someone who understands the global situation, and could really explain what is going on. I got in touch and invited you over to speak at a meeting of exploration managers in Norway. Later, you helped organize a study for the Norwegian

Petroleum Directorate. I know about your subsequent work, but how did you become interested in the first place?

A *I grew up in Brasil, but had my secondary education in the United States taking a mining degree before going on to do an M.Sc. in geology at Stanford. I had a varied early experience as a mining engineer in Ecuador and running seismic crews in Venezuela and Canada, before joining Chevron.*

Later, I worked for a small company in California which really taught me the oil business at the sharp end. In 1957, I became an international consultant working in Turkey, Israel, Libya, Peru and Australia for a number of clients, including Occidental. The latter retained me as a full time consultant from 1973, following the First Oil Shock.

Q It was a shock, and the scramble was on to find new sources of oil. Everyone was becoming concerned about security of supply. That must have been your baptism on this subject.

A *Yes, there was much concern. My experience gave me an intuitive sense of value, so that I was able to rapidly appraise worldwide exploration opportunities. It was a case of sifting through all the data and trying to find the better areas on which to concentrate. In August 1974, the US government directed the Geological Survey to hold a seminar at Stanford University to review their methodology. It soon became evident that its evaluation was giving much higher numbers than the oil industry experts thought reasonable. The Survey was assuming that the future would be as rosy as the past, whereas we realized that the average size of discovery was falling. When I reported the outcome to the Company, it asked me to make an in-house study of the world's petroleum basins as a basis for its exploration strategy. I believe in visual presentations, and I assembled my data into ten large maps covering the world. I worked on my own, and the Company was very generous in allowing me to publish the general conclusions.*

Q Once you had the basic framework of where the basins were, what was the next step?

A *I set about refining the maps to highlight the more prospective tracts within the basins, and began to quantify their "oiliness". In 1979, I published a paper contesting a Financial Times report "Window on Oil" by Bernardo Grossling who was the USGS expert on global assessment. He had calculated the area of the world's basins and assumed that they would ultimately be as rich as Texas. I planimetered the prospective tracts on my maps, and came up with half the area claimed by Grossling. I pointed out that Texas was not a good analogy and that many basins were much less prolific. In*

1983, I was asked to try to put a number on each nation's"oiliness" potential. I did this by resurrecting an earlier method called the Discovery Index, namely the reserves added annually per foot of exploratory drilling. It involved evaluating the validity of the reserves claimed: the reports were often unreliable for political and other reasons. One needed experience and judgment to smoke out reasonable numbers. The Company again cooperated in allowing me to publish the final results in 1984. My estimate of Ultimate recovery, incidentally, was 1.7 trillion barrels

QThat is pretty close to my latest estimate of 1.8 trillion. How did it compare with other contemporary estimates?

A*It was less than the 2.0 trillion published by King Hubbert in the June 1974 issue of the National Geographic Society magazine, which concerned me. So I arranged to meet him at his home in Washington. He was then 81 years old, but as sharp as mustard. It did not take long to realize that here was a man of superior intellect. He explained that he was interested in the life-span of oil, recognizing that production had to start at zero, rise to a peak before ending at zero. He predicted in 1956 that production in the US Lower 48 states would peak around 1970 and decline thereafter, which it did. He said that he did not predict global ultimate production as such but simply used the 2.0 trillion Ultimate as a model upon which to build a global depletion curve that the general public could understand*

QI think the great contribution of Hubbert was to stress that oil is a finite resource and that peak production will more or less come at the midpoint of depletion. Whether we build the depletion profile on an Ultimate of 2.0 trillion or 1.7 trillion will only shift the date of the peak a few years one way or the other. Given that these numbers are within a realistic range, it means that world oil production will peak in a few years' time and after that will come shortage. It seems an immensely important issue, considering that oil provides about forty percent of the world's traded energy. Do you think that governments are sufficiently aware of the situation?

A*Regrettably no. I do not think that the US public or its politicians have any realization of the constraints to global oil supply. It is like telling a 40-year old jogger that he has cancer and less than ten years to live: he simply does not want to hear the message. Democracies thrive on Good News. One is branded as a crank or doomsday merchant if one points out the reality of the position for which there is ample evidence. Economists, who have never had the practical experience of actually looking for oil, project past trends and misunderstand reserve-to-production ratios to claim that there is lots of oil for* *the next Century.*

I feel strongly that everything possible must be done to publicise the position. There is not much time to prepare. I have accordingly spent the last year organizing the "M.King Hubbert Center for Petroleum Supply Studies". I chose the name to commemorate the remarkable achievements of someone who can rightly be regarded as the "father" of this subject and was ahead of his time. It was inaugurated at the Petroleum Engineering Department of the Colorado School of Mines on October 8th 1996. The plan is to issue a quarterly newsletter to inform newspapers, magazines, politicians and the public at large.

Americans have always been largely isolated from the world's problems, and have had no recent experience of the wars and strife that affect so much of the world. "It can't happen here" is their instinctive reaction. A tankful of gasoline is regarded as their unalienable right, but as shortages grow after world peak, they will have to compete with the rest of the world for precious supplies. I am doing my best to alert people to what is coming. I think that there is much that could usefully be done, given the understanding and political will. It is a very grave situation that we face. It is madness to ignore it.

PARABOLIC FRACTAL

There have been many attempts to plot the distribution of oil fields by size. Log-normal plots have been the most commonly used;[3] and the USGS has one called the *Shifted Pareto*. Jean Laherrère[4] has investigated all of these approaches, and has come up with a law of distribution, which he terms the *parabolic fractal*. It states that a parabola describes the distribution of objects in a natural domain when size is plotted against rank on log-log scales. It relies on a law of *self-similarity*, whereby a complete segment of the distribution describes the whole. It sounds complicated, but is really quite simple, although it is not easy to know why this relationship holds true. As an example, we can plot the size of the larger towns in the United States (based on physical city limits rather than administrative units, which are not natural domains) against their rank: New York, Washington, Chicago, San Francisco etc., to determine the parameters of the parabola. It can be extrapolated to the smallest unit, which could be two people in a tent. Such a plot gives the population of the United States to within a few percent of the latest census. Laherrère quotes several other examples, including: spoken languages; the size of species; galactic distances, all of which confirm the validity of the law. Probably, it has something to do with there being no such thing as a straight line in the Universe, everything being eventually parabolic. Throw a stone into the air and it will follow a parabolic trajectory, obeying these same fundamental forces. Entering the log-log domain, which gives high numbers, probably brings this into the scale of the Universe. Einstein would have understood but I am afraid I don't, beyond seeing that it works. It is especially useful when applied to oilfields, where the largest are normally found first. It works best in a clear-cut natural domain – a single *Petroleum System* with a common source-rock. The Niger Delta or the Viking Graben in the North Sea are good examples. It can also work to a degree for very large populations that come to form a sort of super domain of its own: such as the continents or the giant fields of the World.

The difference between the parabolic plot and what has been

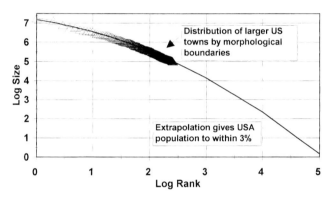

Fig.7-5. Parabolic Fractal of US population

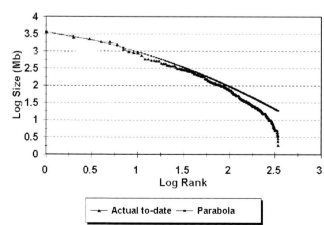

Fig 7-6. Parabolic fractal of the distribution of oil fields in the North Sea

discovered gives the *Yet-to-Find*, subject to a cutoff for small and uneconomic fields.

PRODUCTION EXTRAPOLATION

Since the *Ultimate* is defined as *Cumulative Production* when it ends (or at some distant date), we could determine the *Ultimate*, if we could find a way to extrapolate the production curve to zero. M. King Hubbert,[5] to whom I have already referred, was a distinguished American scientist, who was the first to try this method to predict future oil supply, using a *logistic* curve, or simply stated, a bell-shaped curve. It had in fact been used much earlier in population studies. King Hubbert correctly predicted in 1956 when the United States (48 States) would peak, some fifteen years before it did. He was much reviled at the time, but has since been amply vindicated.

His formulae[6] are a bit complicated for non-mathematicians, but in principle, it is quite simple to apply the formula from three input parameters: date of peak, peak production rate and Ultimate. I suspect that he began his study on very pragmatic reasoning using graph paper in the days before computers. He realised that the unfettered production of any finite resource starts at zero, rises to a peak, and ends at zero. He made a guess of what the *Ultimate* might be and drew the peak at the midpoint of depletion, namely when half the *Ultimate* had been produced. The simplest result is a triangle, but plotting the actual production to-date showed a curved plot as production increased and then declined. Hubbert realised that once the rate of increase began to slow, it meant that about one-quarter of the *Ultimate* had been produced. An S-shaped, or sine-curve, was developing. Once this inflection was spotted, the plot could be extrapolated over the top and down the other side, giving the classic bell-shaped curve. So, once about one-quarter of the *Ultimate* had been produced, the inflection in the curve could predict what the *Ultimate* would be.

The reason behind it is of course that the production curve mirrors an earlier discovery curve: it being axiomatic that oil has to be found before it can be produced.

It is a particularly useful tool because it relies only on production data which are generally of good quality and in the public domain. There is no need to worry about the problems of reserve definition and inaccurate reporting. Production could do it all. It works well in a place like the United States where production has been unfettered, save for the period of pro-rationing by the Texas Railroad Commission. It works less well for individual countries with a small population of fields, or for OPEC countries where production around midpoint has been capped for quota

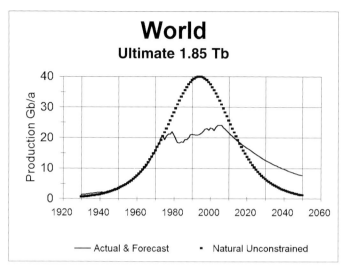

Fig. 7-7. The bell-curve models the natural unfettered exploitation of the resource. Actual production may be constrained by prorationing or other reasons.

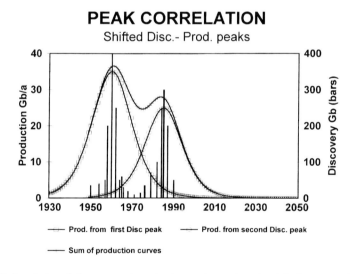

Fig. 7-8. Correlation of discovery and production peaks, combined with bell-curve modelling is another method.

reasons, thus distorting the curve, but even so it can give a hint.

Considered in greater detail, many countries exhibit multiple discovery peaks as different geological trends were developed, each of which being later mirrored in a corresponding production peak. Jean Laherrère[7] has investigated this relationship, and has found that it is possible to correlate

Ultimate				Yet-to-Find			
Major Country		**Region**		**Major Country**		**Region**	
S. Arabia	260	M. East Gulf	669	Iraq	19	M.East Gulf	51
Russia	220	Eurasia	370	S. Arabia	15	Eurasia	33
US-48	200	N.America	226	Russia	15	Africa	15
Iran	135	L.America	192	Kazakhstan	8.3	L. America	11
Iraq	100	Africa	163	Iran	7.9	Europe	7
Kuwait	90	Europe	75	Venezuela	5.7	Far East	7
Venezuela	88	Far East	74	Kuwait	2.7	Other	5
Abu Dhabi	70	M.East Other	43	China	4.6	N.America	3
China	60	Other	10	Abu Dhabi	4.5	M.East Other	2
Mexico	56	Non-M.East	1181	Norway	3.2	Non-M.East	108
Libya	50	**WORLD**	1850	Libya	5.5	**WORLD**	142
Nigeria	48			Mexico	2.7		
Kazakhstan	45			Algeria	2.6		
Norway	33			Azerbaijan	2.5		
UK	31			UK	2.4		
Indonesa	31			US-48	2.3		

Fig. 7-9. Distribution of the Ultimate and Yet-to-Find

discovery and production peaks with a certain time-shift, individual to each country. He then constructs bell-curves for each such peak and sums them to yield an estimate of the *Ultimate*.[8] It is an elegant approach, which is simple to use and well matches actual production in mature countries.

Duncan[9] has developed a phase diagram, as a variant of the Hubbert method, called the D-Model, that plots annual production against *Cumulative Production*, using translated coordinates. It is based solely on production statistics, and seems to give good results, although laborious to produce. Mathematical skills, which I lack, are needed to understand it.

These are the methods that can be used. Having been through them all, and having checked the reasonableness against drilling and discovery rate, I conclude that a good number for the *Ultimate* is 1.85 trillion barrels.[10] I stress that this relates only to *Regular Conventional Oil*.[11] My extrapolation shows that of this less than 300 Gb remain to be produced after 2050. Rounding it off within the accuracy of the model, one could as well say that this 1.85 Gb would be *Cumulative Production* by, say, 2100. It avoids the mental anguish of having to think about the circumstances of depleting the last few barrels. It will do well enough to drive the depletion model, which I will come to in the next chapter.

But first, we need to consider the distribution of the *Ultimate*, as depicted in Figure 7-9

Only three countries hold more than 200 billion barrels each, and only twenty-six hold more than 10 billions barrels each (less than six months' World demand). Almost forty percent of the World's endowment is in just one region: the Middle East Gulf.

THE YET-TO-FIND (OR UNDISCOVERED)

The *Yet-to-Find* (or *Undiscovered*) is calculated in the model by subtracting the *Discovered* from the *Ultimate*. It works out to be 142 billion barrels, distributed as shown in Figure 7-9.

The Middle East Gulf Region is assessed to have the greatest potential at just over 50 billion barrels. Again, most of the giant fields in the region have been found, but the rich endowment of source-rocks and the effective seals mean that there is probably much more to find, albeit in ever smaller

Fig. 7-10. Map of the Persian Gulf area: the wide distribution of oilfields implies that most of it has been thoroughly explored.

fields, hidden in the complex fold-belts of Iran and Iraq.

Relatively few exploration wells have been drilled in the Middle East, compared with, say, the United States. It is sometimes claimed that this is reason to expect that colossal amounts of more oil could still be found in the Middle East if exploration were stepped up. But it may not be so because the Middle East petroleum habitat is a concentrated one for geological reasons. Gentle structures, many with their salt cores, grew over long periods of geological time, so that they were able to effectively drain the catchment areas around them. In fact, seal is as important as source in the Middle East. Half of Saudi Arabia's oil lies in just two fields: Ghawar and Safaniya. What remains to be found is smaller by far. Figure 7-10 shows how closely spaced the major oilfields are.

Russia is held to have the next highest potential with some 15 Gb, excluding the Arctic, here treated as *Non-Conventional Oil*. To give a sense of proportion, that is equivalent to about one-third of the total found in the North Sea, the largest new province found since the Second World War. Although the larger fields and basins have already been found under the systematic exploration of the Soviets,[12] there may well be much left to find in deep basins beneath a salt seal, and in generally smaller fields. Some of the more remote areas are under-explored. It is of course difficult to assess the potential on the information available: it may have more than here thought likely, but distant fields are not always greener. The political situation and the difficult fiscal and contractual environment also make it uncertain how much will actually be discovered at least during the next critical decade or so. Indeed, the Russian Government appears to be slowing the pace of development to make the resource last longer. It is as well to take a cautious number for the purposes of modelling peak production. Detailed reviews of country by country potential have been published,[13] and my latest assessment is summarized in Appendix 2.

Twenty-eight billion barrels are attributed to "unforeseen" discoveries in new or currently producing countries so as to deliver a rounded total. It could occur in stratigraphic traps along the Atlantic margin, or possibly in some rather unlikely areas, such as offshore Namibia or the Falklands. There may be a few surprises in complex fold-belts or remote and difficult areas. But mainly it will come from ever smaller fields in mature basins. Some of the "foreseen" in certain places may not materialise and may have to be reallocated to the "unforeseen" elsewhere.

The offshore is relatively easy to explore. Seismic surveys can be conducted rapidly and cheaply. The results are also of superior technical quality in the absence of topographic distortions, meaning that wells are generally located on firmer evidence. Modern exploration rigs can be towed in without the expense of road building in difficult terrain. It means that most of the major offshore finds have been made, save in the few areas having significant ultra-deep water potential.[14]   Attention may therefore return to the onshore, especially to difficult areas or those that are, or have been until recently, closed for political reasons, but the size of whatever secrets remain is unlikely to be large.

According to industry sources, the World has been finding an average of seven billion barrels a year over the past twenty years, on a long-term downward trend. Exploration drilling peaked in 1981 when over 11,000 *wildcats* were drilled, but has now declined to one-quarter of that number for the simple reason that there are fewer viable prospects left to test. Technological advances have also made it easier to know what prospects are viable. At this rate, significant exploration will be over within about twenty-five years, even if the tail drags on a little longer. Probably, almost all the important new discoveries will have been made within the next decade or so in areas outside the Former Soviet Union and the Middle East Gulf, where politico-economic constraints may delay and curtail effective exploration. Less than 100 billion barrels are likely to be fully accessible to the international industry.

Fig.7-11 Wells have been drilled in all prospective basins and under difficult conditions

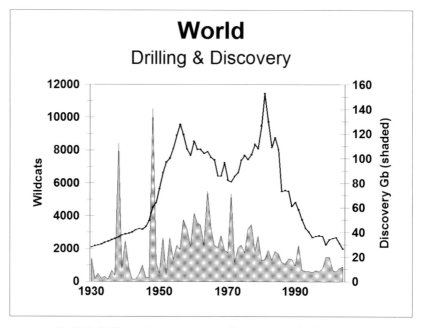

Fig. 7-12. Drilling and discovery rates: effort is not matched by reward.

Some 475,000 wildcats have been drilled out of an estimated ultimate number of about 500,000. But these are misleading numbers because 400,000 of them were drilled in North America under the special circumstances of that region. Many were drilled long ago without the advantages of modern methods. The same amount of oil could be found with far fewer wells today.

The high oil prices of the 1980s prompted a surge in drilling, but the results were disappointing. The present oil price shock may prompt another boom, but the results are not likely to be any better.

It is important to remember that, in many countries, exploration expense has been deductible against taxable income, so that it already enjoys a considerable hidden subsidy. The subsidies, which are higher at times of high oil price, delivering more taxable income against which costs could be set, in effect, meant that dubious, high risk prospects were drilled. Governments, however, may change the tax rules, as has happened in the United Kingdom, when they found that the deduction no longer delivered sufficient revenue producing projects. The subsidy is a form of national investment that has to be justified by results.

On the other hand, much of the potential for exploration lies in countries where State companies have exclusive or dominant positions. They enjoy no tax incentives for exploration, and spend real money for which there are many other competing claims in the national budget.

Profitable niches in exploration will no doubt continue to be available for enterprising independent companies and perhaps contractors, but it is difficult to anticipate a contribution to World supplies much greater than that indicated, especially when the time and drilling rate constraints are taken into account. It is uncertain what role the major companies will have: they are ceasing to behave like integrated companies, but rather as holding companies with increasingly independent affiliates. I will discuss later the advantages and disadvantages of this development. I am already struck, reading the *World Oil*[15] review of worldwide operations, by how much is in the hands of small and relatively unknown companies.

JEAN LAHERRÈRE AND THE PARABOLIC FRACTAL

Fig. 7-13. Jean Laherrère, who discovered the Parabolic Fractal, which helps predict oil discovery

QJean: I believe a motor accident in France affected your career causing you to change from geology to geophysics.

A*Yes, I was injured in 1955, and in those days a geologist needed strong legs for field work in remote places, so my boss thought it a good idea to move to geophysics. I had a good grounding in mathematics, and enjoyed the subject, having graduated at the Ecole Polytechnique*

QHow did your career unfold?

A*I had joined Total, the French oil company, previously known as CFP, Compagnie Française des Pétroles, which had its origins as a founder shareholder in the Iraq Petroleum Company after the First World War. France's strategic needs after the last war called for a more dynamic search for oil as an operator. We concentrated on Algeria, which was then French territory, but was regarded by the US companies as having very limited potential.*

QWhat led you to Hassi Massaoud, which was found in 1956, but is still the largest field in Africa?

A It was found by the now little used refraction seismic technique, which along with aeromagnetics, was the only method to map beneath the thick salt sequence which overlies it. Even modern reflection surveys barely penetrate. My first paper in 1959 was on the anisotropy of seismic waves to calibrate refraction events. The survey needed an enormous explosive charge to record refraction arrivals 15 km distant (see Figure 4-3). Incidentally, we used four tons of fuel and fertilizer to make the change: the same as they used in the Oklahoma bombing. My second paper in 1961 was on synthetic seismographs.

Q After the Sahara, you visited and worked in many places around the world. At that time it seemed as if there was still plenty of oil to find. When and how did you begin to question this assumption?

A It dawned on me slowly as my career unfolded and as I began to have responsibilities for exploration throughout the world. I spent time in Australia and Canada, but in 1972 returned to the Head Office in Paris. I had a wide experience, being at different times in charge of negotiations, making Total's first deals in South America, basin studies, and research. I also served as Deputy Exploration Manager. In addition, as President of the French Oil Industry Commission, I supervised the preparation of manuals on exploration techniques. With the passage of time, I began to see relationships such as how the larger fields tended to be found first, as was the case with Hassi Messaoud. During my career I watched most basins become mature exploration areas, and I realised that very few new provinces were being found.

Q You discovered a law of distribution which you call the parabolic fractal. Was this something that resulted from working on the computer? Or did the inspiration come, so to speak, on the back of an envelope?

A I am a visual man: to remember anything, I have to see it. I have always looked for graphic representation, and dream of finding a good and simple representation of the laws of Nature. Being retired, I now have the time to study these things. I am in fact writing a book on natural distribution and inequality. Life is a race, with equality at the starting line but not at the finishing gate.

It has long been a habit of mine to plot relationships and numbers. I plotted oilfields in the order of their size, and then used log paper to bring out the relative sizes. At first, I used a straight line method (as first used by Pareto and Mandelbrot) but it did not fit, especially for the smaller fields. I realized that the plot was curved, and as I thought about it, I realized that many distributions in Nature are curved. The horizon, for example, appears a straight-line but in fact is part of the Earth's circumference. It dawned on me *that the simplest curve is the parabola, and I found that the best fit for almost all distributions is parabolic (galaxies, urban agglomerations, languages etc). I then tried to understand more of the theory, delving into fractals, chaos theory, and the law of self-similarity, whereby a complete segment of a distribution describes the whole: one branch mimics the whole tree. Perfect symmetry is linear from one infinite to the other, but in Nature neither perfection nor infinity exists. Since the self-similarity is not perfect in Nature, the fractal is parabolic. I have written a paper for the Académie des Sciences on the subject.*

Q It seems to be a remarkable tool to predict future discovery.

A *In certain circumstances, where the data are valid and where there is a clearly defined natural domain, it can give excellent results, but it is not a tool to be used blindly. I believe in approaching each case from as many vantage points as possible, realising that in practice the data are often weak and that there are extraneous influences at work. But in general I think that we have reached the point at which we can make a sound assessment of the world's future production of conventional oil, using a combination of these methods*

Q What do you mean by conventional?

A *I mean oil that is producible at a price less than about $25/band over a certain period of time. A large number of small fields will be left behind, when production ends. The law of diminishing returns explains that it will take an impossible number of wells to find them all. The world has large amounts of heavy oil that is not competitive against ordinary oil such as has been produced so far. Some of it will become viable in the future, but we have to bear in mind the time frame as well as the volume. Heavy oil is slow to produce and will have a very different depletion pattern. There is a big difference between a field with wells flowing at 10 000 b/d and one with wells producing at less than 100 b/d. The amount of work involved is several magnitudes greater. The critical issue is not so much the size of the resource but the number of wells is will take to produce the heavy oil.*

Q Do you think that there will be a gradual transition from conventional to non-conventional production.

A *No, I anticipate a crisis because we live in a short-term world in which governments are unable to plan ahead. Conventional production is set to peak within a few years on the basis of my studies. About half the World's* remaining oil lies in just five Middle East countries, whose control of supply must inevitably increase with all that that implies. Even if a crash programme to develop Non-Conventional oil were put in place immediately, it would be years before it could make a global impact. Nevertheless, there are few large fields yet to find, and the companies may be able to invest in helping the national companies produce difficult fields.

Q Can you explain the evidence for the coming peak?

A It is fairly obvious. Before you can produce oil, you have to find it. Countries often have several cycles of discovery as different geological trends are opened up. You can correlate them with production cycles after a time-lag, at least in countries producing at capacity. Peak discovery occurred in the 1960s and is about to be reflected in production, which in world terms will peak around 2000. It is not quite as simple as this because of the uneven distribution, with a few swing countries having an exceptional endowment.

Q It is an immensely important message which deserves more attention. What can be done?

A We now have the analytical techniques to determine what the situation is. What we need is better information on reserves and stricter definitions. Reporting reserves is a political act ! We like to appear poor in front of the tax inspector; and rich when we meet the banker. The national companies, which own most of the reserves, should be urged to cooperate in providing the essential information. It is not so much a technical problem as a political one, in which the buyers as well as the sellers can exert their influence. They need to know the security of their supply. So long as OPEC quotas are based on reserves, the published numbers will be questionable. More than fifty percent of the countries, whose reserves are listed in the Oil and Gas Journal, report unchanged numbers from one year to the next, which is obviously implausible.

Q One reads many economic evaluations that say it is just a matter of investment and technology. Indeed, most reputable institutions appear to think that there is not a cloud on the horizon. Are they mistaken?

A I am sorry to say that they are. One of the reasons is that they have to rely on unreliable published data in which reserve revisions are not properly backdated to the discoveries they relate to giving a totally false impression of the trends. Another is that they misuse Reserve to Production ratios, failing to understand that all oilfields decline after middle age. Economists do not seem *to quite grasp the implications of depleting a finite resource. Many people confuse reserves and resources. For example, the Russian classification was not based on realistic recovery factors: they took the maximum theoretical case without regard to economic or technological constraints.*

Q Could you provide an example of one of your plots that demonstrates your concerns well.

A *This is a plot of the remaining reserves over time for the World outside North America. The first curve shows reserves with revision taken on a current basis (from the API Basic Petroleum Data Book). The second curve shows the reserves with the revisions backdated as if known at the date of discovery. The first curve shows a plateau after the marked spurious increase in the late 1980s, which was related to OPEC quota. The second curve shows the actual decline, despite all the technology, we hear so much about, and despite unprecedented drilling levels.*

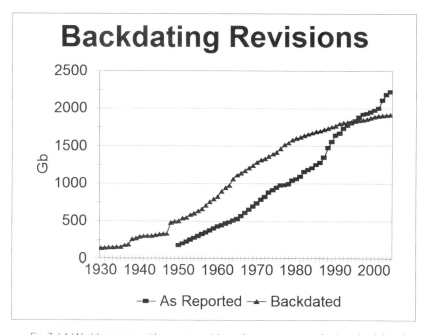

Fig. 7-14. World reserves with reserve revision taken on a current basis or backdated

NOTES

1. A term for old wells producing a few barrels a day as are found in several nearly exhausted US fields.

2. A term meaning rocks devoid of prospects, such as the igneous or metamorphic rocks of the ancient shields.

3. Although the log-normal distribution tends to work well for the parameters used in estimating the reserves in a single field, it tends to under-estimate the reserves of a basin because it gives equal weight to large and small fields, when in nature there are many more small fields yet-to-find than large ones.

4. See Laherrère J.H., 1996.

5. See Hubbert, M.K., 1956, 1982.

6. Hubbert did not release his formulae until he was 80 years old in his 1982 paper. Indeed, he probably started with a simple graphic approach and only later used a formula to draw the bell-curve.

7. See Campbell C.J. and J.H.Laherrère, 1995.

8. In the same way as sound can be defined as the sum of a limited number of harmonics (Fourier analysis), so oil production can be modelled as a small number of symmetrical cycles, each reflecting the practical discovery and production of oil from a separate geological trend or "play".

9. See Duncan R.C., 1996.

10. My first estimate in 1989 was 1.65 Gb, but I did not then appreciate the industry practice of understating the initial reserves of a discovery. I according increased it to 1.75 Gb to take that into account. At the time of *The Coming Oil Crisis*, I had increased it to 1800 Gb. The current version is based on 1850 Gb. The changes in part reflect evolving definitions including particularly the treatment of Condensate, which is now included with oil.

11. In principle, this is *Regular Conventional Oil* only, but it probably includes relatively small amounts of *Non-Conventional Oil* not distinguished in the statistics. It certainly excludes North American NGLs, which are recorded separately.

12. See Tull S., 1997, for a sanguine assessment of the FSU.

13. See Campbell C.J. & J.H.Laherrère, 1995, and Campbell C.J., 1991, provide such studies: the former based on Petroconsultants data, which incorporate Russian official statistics. Masters of the USGS has also evaluated the situation over a period of years, although care must be taken to decode his definitions.

14. Say in water depths of more than 200m.

15. See *World Oil*, 51st Annual International Outlook August, 1996.

Chapter 8

FUTURE PRODUCTION – CRISIS AND BEYOND

RE-READING CHAPTER 8 IN *The Coming Oil Crisis* recalls a viewpoint and assessment that seemed reasonable in the last decade of the last Century. It was indeed in a certain sense the last decade of the old order. Tensions and difficulties were anticipated, but they were constrained by the relative stability that existed and the mind-sets accompanying it. The situation has deteriorated since then in many important respects, but the changes are so recent that it is difficult to grasp their full import. Even so, it can be said that the old order came to an end on September 11th 2001, when a new war was declared, changing the political and geo-political landscape in unimaginable ways. It accordingly becomes necessary to re-write the Chapter on future production in its entirety. It is not an easy thing to do because we are still too close to what may be described as the Crisis Years to see them clearly, or to be able to divine how they will evolve. An additional difficulty is that we simply do not know what to believe as more and more evidence appears to show that the official record is far from the truth. Accordingly, we can do no more than offer hopefully intelligent guesses as to how future historians will come to re-write the record. This speculation poses a difficulty for a book of this sort as it might be held to undermine the confidence in the other conclusions on which some professional qualifications can be claimed.

We cannot avoid facing up to these events because the forecast of oil production has to reflect in some manner the wider constraints. We enter here the realm of what has been described in derisory terms as Conspiracy Theory, as if any alternative view is by definition in some way extreme and unreliable. Indeed, it may be that the Establishment, if we can call it that, fosters far-out Conspiracy Theories with the motive of discrediting all but its desired rendering of events.

Be that as it may, we have no alternative but to address geopolitics, and make some attempt to unravel the conflicting versions of events. In particular, we have to try to fathom the foreign policy of the United States, with all of its hidden agendas, if we are to build a plausible forecast of future oil production. We cannot escape the consequences of the

Fig 8-1   The events of 9/11 changed
                the World

distribution of remaining oil, as ordained by Nature, and the political responses that it imposes on the countries that produce and consume oil. Every country is affected, and some begin to wake up and admit to their policies, as the following report of the British Foreign Secretary's words confirm.[1] The telling line is that referring to "Country Action Plans", hinting strongly of foreign intervention for energy supply. It is worth quoting in full because it really is an astoundingly frank statement in marked contrast with the customary double talk.

"30-10-04. Foreign Secretary Jack Straw warned that Britain's growing need for energy over the next decades has to be seen in a "changing context" due to declining production from the North Sea. "By 2020, we will probably be importing three-quarters of our primary energy needs – and we will need to adapt to that," he warned when launching his government's first-ever International Energy Strategy.

Straw's warning comes after the British Foreign Office identified energy security as being one of eight international priorities last December. The concern is that the country is no longer self-sufficient with oil and gas supplies from the UK's sector of the North Sea running out fast. The situation was underlined in July, when Britain recorded its first deficit in oil trade since 1991. The worry over gas was exemplified by the closure of the North Sea's Frigg field on October 26 after one time supplying up to a third of the UK's domestic gas needs.

The UK's oil production has been in decline since production peaked at 2.8 mm bpd in 1999. Although the current output of 2.1 mm bpd is in line with the average of the past 20 years, it is predicted it could run dry within the next decade. According to the UK Offshore Operators' Association, the country will cease to be self-sufficient in 2007, production will drop to 1 mm bpd by 2010 and virtually end altogether five years later. Of even greater concern is the situation of natural gas, where the UK is rapidly moving from a position from being a net exporter to a net importer. By 2010, it is expected to be importing around 50 % of its gas. Like the rest of the EU, the dependency is expected to rise to 70 % by 2020.

In June, a parliamentary report expressed alarm about the delay in building up an infrastructure for imported supplies. It questioned whether the UK gas market had the ability to cover demand if there was severe weather over the next two or three winters. "Transco (responsible for the national grid) has the physical capacity to transport a high-surge demand for gas, but pipelines bringing gas into the UK have little spare capacity," chairman of the EU sub-committee for Internal Market, Lord Woolmer warned. This was despite concluding that globally there were ample supplies of gas from diverse sources, which would be available to the EU and UK up to 2025 and probably beyond.

As part of the UK's International Energy Strategy, Straw announced that he would be tasking British ambassadors "in priority posts overseas" to take personal charge of implementing and delivering its objectives. "We will be developing with them individual Country Action Plans on energy and climate change. And we will be enhancing our posts' capacity on energy issues and making better use of our network of energy attaches," he said, underlining the importance being attached to the security of supplies.

The energy time bomb also has important implications for Britain's economy. Since North Sea production started in the late 1970s, oil has lubricated the economy with billions in tax revenues. Even during the decline since 2000, tax revenues have been worth some $900 mm, the equivalent to reducing income tax by some 10%.

The bonanza earned by record increase of 65% in oil prices this year could be one of the last and will leave the Treasury with a big hole to fill in the country's budget. The plight is not helped with Britain's huge trade deficit, which grew last year to a record $ 85 bn. Britain's oil industry, which still directly employs 260,000 people, is already being consigned to the history books by the rapid withdrawal of international oil majors. Companies such as BP and Shell have been retreating from the North Sea for the last couple of years and plunging their investments elsewhere. Yet the offshore industry remains optimistic that many smaller specialist operators can continue to extract oil from mature fields.

It is estimated that 30 bn barrels of oil have been pumped from Britain's oil reserves and that a further 30 bn remain to be exploited. But continued extraction will depend on costly technological developments such as horizontal drilling and water or gas injection that require oil prices to remain high. Like Houston in

Yet-to-Produce	Gb	Midpoint of Depletion
Saudi Arabia	160	2013
Russia	90	1996
Iran	78	2007
Iraq	71	2005
Kuwait	58	2020
Abu Dhabi	51	2025
Venezuela	40	1999
Kazakhstan	38	2036
China	29	2003
US-48	27	1971
Libya	26	2005
Nigeria	24	2004
Mexico	24	2000
Algeria	15	2006
Norway	14.5	2002
Azerbaijan	14.2	2014
REGIONS		
M.East Gulf	424	2016
Eurasia	179	2003
N.America	33	1973
L.America	82	1999
Africa	81	2004
Europe	29	2000
East	29	1999
M.East Other	16	1999
WORLD	906	2003

Fig.8-2 The Yet-to-Produce

the US, Aberdeen is the UK's oil capital but few have doubts that its days are numbered."

Clare Short[2] describes the nature of the British Government in which second rate Ministers are controlled by the Prime Minister's office, giving more emphasis to image than substance. It is easy to understand how they may occasionally sing from the wrong hymn sheet by mistake before a new central spin on the issue has been devised.

AN HISTORICAL BACKDROP

So here goes: for what it is worth, which may not be much, this is what I think happened. But before coming to recent events, it is necessary to set the scene by looking back in history, for all events and postures have their antecedents. As we delve into these roots, we find ourselves addressing the very fabric of the financial structure of the modern world: its principal driving force.

Some 300 million people occupied the Planet at the time of Christ. Primitive, yet sustainable, agriculture allowed their number to no more than double over the next seventeen centuries. During this epoch, most people lived hand-to-mouth existences in rural circumstances. In the absence of international trade in food, they were very much victims of local circumstances, facing famine if the harvest failed for climatic or other reasons.

The Stone and Bronze Ages had ended as people turned to iron and steel for stronger tools and weapons. At first, they used firewood for smelting, but later turned to coal as a more concentrated fuel, which incidentally helped spare the forests, some of which were at risk of being decimated. In Britain, the coal was at first collected from beaches and outcropping seams, but as these sources were depleted, people started mining for it. The mines were progressively deepened until they reached the water tables, which led to flooding. That in turn prompted the development of steam-driven pumps, which evolved into steam engines. They later powered locomotives to transport coal and other goods on railways. These events ushered in the Industrial Revolution led by Britain. It saw the rapid development of industry, transport, trade, and agriculture, allowing the world population to double to about one billion by the middle of the 19th Century. This epoch also saw the rapid growth of capital, with debt and usury forcing industrialists to seek expansion through new markets on which to sell their products and repay their debts. That meant growing trade and transport: sail began to give way to steam, fuelled by coal. In parallel came new banking structures whereby banks started to lend money in excess of what they had on deposit and charge interest on it, in effect creating money out of thin air. The loans triggered

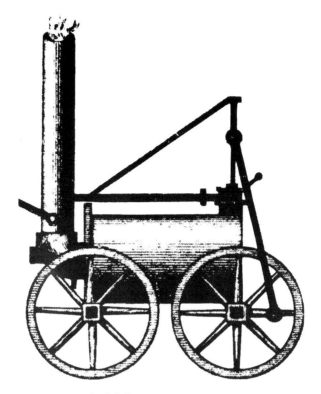

Fig 8-3. A primitive steam engine

more debt, but the system worked because the resulting economic growth, fuelled by the new cheap energy, provided the collateral. It is important to stress that, while money facilitated the process, it was the underlying energy supply that sustained the system. In view of the central but unseen role of finance, we should perhaps better rename it the *Financial-Industrial Revolution*, with finance being perhaps the more important element.

Then, as described in Chapter 3, an enterprising German engineer by the name of Nicholas Otto saw the advantages of inserting the fuel directly into the cylinder of a steam engine, creating the so-called Internal Combustion Engine, which was much more efficient. At first, it relied on benzene, distilled from coal, but later turned to petroleum, refined from crude oil. This new energy source made possible a huge expansion of the *Financial-Industrial Revolution*, allowing the population to expand six-fold, exactly in parallel with oil.

An important element in the financial part of the Revolution was the role of global trading currencies. It seems that, in the web of international transactions, great benefit accrued to the nation controlling the principal currency used. The British Empire at its peak was a magnificent construction spanning the world, but it arose not so much from military supremacy as on the back of British traders and the use of the pound sterling as the World's premier currency. Sterling debt and sterling-denominated trade expanded throughout the World, paying a hidden tribute to the City of London, with the contracts being policed by the gunboat complete with its Marine band on the quarterdeck, inspiring confidence. Confidence was what it was all about because the system required a confidence that expansion tomorrow would repay the debt of today.

Germany was overtaking Britain as an industrial power during the latter years of the 19th Century, but found itself seriously disadvantaged because the Deutschmark was insignificant in World trade compared with the pound sterling. There were of course many other inter-related factors at work, but probably it was industrial and financial control that lay at the heart of the conflict that erupted into two world wars, with the second being essentially a continuation of the first. It is significant that the first battle in the First World War was fought in Alsace-Lorraine, which holds Europe's largest iron deposit, essential to Germany's expanding industry.

At this point in the narrative, we really have to enter sensitive territory, but it can't be helped. The fact is that for various historical reasons, which need not concern us here, the financial community was dominated by Jewish families. The Jews themselves had been subject to degrees of oppression in various countries, partly because of their disinclination to integrate with their host societies for religious reasons and partly because of their role in business and finance. In those distant, simpler and mainly rural, days, finance and banking were, with a certain reason, widely perceived to be in some way sinister and underhand: the term usury acquiring a pejorative meaning. The Zionist movement was a response to this oppression with its call for a national homeland in Palestine, a territory from which Jews had been expelled by the Romans in AD 135.

In 1916, in the middle of the First World War, when the British, French and German armies faced each other in stalemate under appalling conditions on the killing fields of Flanders, Chaim Weizmann, the head of British Naval Research and a Zionist activist,[3] succeeded in persuading the British Government to promise to provide the homeland in Palestine in return for Zionist influence to bring the United States into the war on Britain's side. The so-called Balfour Declaration, making such provision, was announced a year later, when the United States did enter the war on Britain's side, confirming the connection. British banks, many controlled by Jewish families, had already established themselves in New York, and were no doubt in a position to influence political decisions.

The year 1917 also saw the Bolshevik Revolution in Russia which ushered in the Communist era. Perhaps the old aristocratic Tzarist regime had run its course in any event, but it is significant that the new leaders saw their mission as one to end the capitalist system in order to spare the people from perceived exploitation. The Royal Family and the Aristocracy were readily identifiable targets but in reality they were no more than figureheads for the underlying *Financial-Industrial System*. It proved difficult to meet the new ideals, and once in power, the regime evidently became extreme and corrupt in various ways, perhaps finding that in practice it was difficult to avoid the sweaty grip of the money-lender.

The entry of fresh American blood turned the tables in the war leading to an Allied victory in 1918, which was followed two years later by the Versailles Treaty that set the peace terms. The primary result was the imposition of onerous reparations on Germany, which were not only deeply resented, given that the country had agreed to end hostilities with its forces still on foreign soil, but were unenforceable in economic terms. A secondary, but in the longer term more important, outcome was that the treaty effectively defined the carve-up of the Middle East between the Allies, with the creation of new states including Iraq, Kuwait and, later, Saudi Arabia, also opening the door to American oil companies in an area which had previously been an Anglo-French sphere of interest.

The hardship and resentment arising from the reparations provided a fertile soil on which the Nazi movement grew, seeking to re-establish German self-esteem and power. A new wave of anti-Semitism erupted, as the capitulation of 1918 was attributed to Jewish activists with American links, notwithstanding that military defeat stared the country in the face.

But, it was events across the Atlantic that really brought the Nazis to power. The United States, having been spared the ravages of war, emerged triumphant and enthusiastic for the post-war epoch of the Roaring Twenties. The economy began to boom in the new euphoria as banks pumped more and more money into the system with an unbridled faith in the future. Investment trusts and holding companies began to speculate on the stock market, pushing up prices. In earlier years, companies had had loyal shareholders whose primary benefit flowed from regular dividends, but now shares were increasingly bought and sold by speculators gambling on their changing values. This form of income was basically subsidised by

Tony Bennett Remembers the Depression

Tony Bennett, the veteran American crooner, states in an interview[4] that he is "totally against the age of super-greed in which we're living. It's a matter of shame, the growing gulf between the have's and have-not's to-day. I grew up in the Depression, a tragic time in our history, when ordinary people were starving. Remember that, I figure. If I have enough, why should I want more?"

being less taxed than that deriving from dividends, yet again emphasising the role of Finance in the conduct of affairs.

A classic speculative bubble burst on the trading floors of Wall Street in September 1929 leading to a crash on Black Monday, the 29th of October. Within three years, one-third of the US workforce was out of work creating desperate conditions across the country, which have left a searing memory in the national psyche to this day.

The crash reverberated around the World due to the interwoven nature of the Financial System, affecting nowhere more seriously than war-torn Germany. Rampant inflation destroyed everyone's savings. Mutual recriminations flashed across the Atlantic, with the issue of war reparations being held partly responsible. The Austrian Central Bank was close to insolvency, and the Reichsbank in Germany was forced to default. In the face of these difficulties, the new Nazi movement was widely seen as offering the best hope for restoring stability, even attracting support from pragmatic Jewish banks.

In America, the Government was forced to intervene in the economy introducing almost socialist policies under the so-called New Deal, while in Britain, Meynard Keynes proposed new economic principles whereby currencies would no longer have to be backed by gold. He maintained that the economy could be stimulated by new liquidity created simply by printing money. His proposals were adopted and, risky as they seem, were perhaps justified as a response to the Great Depression.

Hitler came to power on January 30th 1933, and after stabilising conditions at home, turned attention to foreign policy, aiming primarily to expand eastwards, where several German communities had long been established. He wanted to rival the British Empire and see the Deutschmark prosper as a result.

The British Government found itself in a quandary: on the one hand, it welcomed moves against the Communists and their socialist fellow-travellers at home who posed a threat to the *Financial-Industrial Order*, yet feared the resurgent militarism of Germany, still smarting from the grievous losses endured in the First War.

At first, it tried appeasement and accommodation, but finally decided to declare war in 1939 in response to the unprovoked German invasion of Poland. The British Empire soon rallied

Fig 8-4 Adolf Hitler

Fig. 8-5 The financial centre of Wall Street

to the support of their mother country, but the United States hesitated. It might indeed have seen advantage in the demise of the British Empire, a trading and financial competitor, and could have welcomed a strong Germany to counter the Communist Regime of Russia. But in the event, developments in the Far East tipped the balance. Fearing Japanese commercial expansion, the United States had been endeavouring to settle the long standing Sino-Japanese War, which was moving in Japan's favour, and in July 1940 put pressure on Japan by embargoing the export of scrap iron and oil products to that country, essential to its war effort. The latter reacted by signing an alliance with Germany and Italy, before invading the French territories of Indochina, a year later, as a staging point for an attack on Sumatra in the Dutch East Indies (now Indonesia) to secure its oil. That in turn led it to make a surprise pre-emptive strike on the US fleet in Pearl Harbour on December 7th, 1941. The US Government was aware of the impending attack from radio intercepts, but curiously took no action to alert the fleet that suffered heavy damage and some 3000 casualties. As a result, this event rallied the nation, such that the Government had popular support when it declared war on Japan, which led Germany and Italy to retaliate by declaring war on the United States under their alliance with Japan. This chain of events brought the United States into the Second World War, which indirectly stimulated the domestic economy drawing a final curtain on the Great Depression.

After many campaigns from North Africa to the Caucasus, the war finally drew to a close in the early summer of 1945, as Anglo-American forces advanced across western Germany, and the Soviet Army stood at the gates of Berlin. Hitler committed suicide, and Germany sued for peace.

It is worth mentioning in passing that oil was a critical factor in the war, with US production giving the Allies a huge advantage. A museum at Falaise in France reveals that the countryside, a few months after the battle, which was the turning point in the Normandy campaign, was blackened by swarms of flies attracted by the carcases of horses on which the German army was forced to rely for transport.

The war in the Pacific ended soon afterwards when the United States dropped atomic bombs on Hiroshima and Nagasaki, vaporising some 100,000 people although Japan was actively suing for peace. Stewart Udall, President Kennedy's Secretary of State, explains how the military had been keen to test these new weapons, but had been restrained by President Roosevelt, who saw the wider implications. When Truman took command, on the death of Roosevelt, he generally sanctioned the continuation of existing programmes, including the atomic bomb, without having had time to fully grasp his new responsibilities.[5]

The war effort had drained Britain, which emerged with heavy US debts and a devastated economy, being soon obliged to give up its once splendid Empire, spanning the World. War-torn continental Europe was in ashes,

Fig 8-6 The atomic bomb

and the Soviet Union, which had suffered more grievously than any of the combatants, was not about to risk retreating from the territories that it had occupied, including eastern Germany. In effect, the United States emerged supreme, with a vibrant economy, having been spared the direct ravages of war. A strong dollar replaced the pound sterling as the premier world currency, bringing with it the enormous hidden tribute that such control conveys.

Before leaving this chapter of history, it is necessary, in the light of subsequent events, to mention in passing that a Zionist activist of Polish origins, by the name of Menachem Begin,[6] had established a terrorist group in Palestine in 1942 aiming to oust the British, who had been administering the country under a League of Nations mandate since the First World War. In 1947, it blew up the King David Hotel in Jerusalem, killing several British officers. It was the final straw for a war-weary Britain, which pulled out, dumping responsibility for the place onto the United Nations. That paved the way for the unilateral declaration of the State of Israel a year later, with far reaching consequences for the oil-rich Middle East and recent events. Many of the indigenous Palestinians were deprived of their lands and forced into refugee camps.

Before the Second World War had ended, the United States and its allies began to consider the financial structure for the post-war world, convening the Bretton Woods Conference, at which exchange rates were set, being again linked to the price of gold. The International Monetary Fund and other institutions were created to manage the new *Financial-Industrial System*, now led by the United States. With the coming of peace, it moved to rehabilitate Europe for fear that difficult economic conditions might encourage the spread of Communism, undermining its hegemony. Communism indeed did already have a certain following, especially in France and Italy, deriving mainly from the wartime Resistance Movements that tended to be led by Communists. Some 13 billion dollars were distributed in aid and debt, which was welcomed by the recipients although it indirectly strengthened the dollar, bringing Europe firmly under US influence.

A new conflict soon manifested itself with the coming of the so-called Cold War, as the former allies now glared at each other across the cease-fire line. Berlin emerged as an isolated divided city, with one half, separated from the other by a high wall, and being supplied by air from the West. In hindsight, it is hard to grasp what this conflict was really about. The West was led to believe itself to be threatened in military terms, while the Soviets for their part found themselves ringed by military bases containing nuclear weapons aimed at them. The Soviets were depicted as bent on world domination although the only notable convert was the island of Cuba. In hindsight, we may be tempted to conclude that both sides had commercial motivations, cloaked in military threats. The Soviets had faith in their central planning, which prompted them to give such encouragement as they could to that system, such that the rouble would grow in strength. The United States, for its part, had faith in its *Financial-Industrial System*, seeing the principle of free-market competition as a desirable ideal, despite the great inequalities it imposed. Even its poor, many living in ethnic ghettos, were persuaded to share this general view in the belief that they were not barred from privilege if luck should shine on them. A side effect of the Cold War was the continuation of a State–funded arms industry of enormous scale that delivered high returns to financiers, investors and executives.

It is worth mentioning in passing that the United States, like other countries, has a long established elite of privileged families that in many intertwined ways control the government, the finance, the investment community, big business and even academia. For example, it transpires that the present Secretary of State, Condoleeza Rice was a star pupil at a college established by John Foster Dulles, a previous Secretary of State, with the curious objective of training black priests for the Presbyterian ministry. She was taught by no less than the father of Madeleine Albright, another previous Secretary. The Bush family for its part came from a banking dynasty. The old principle of keeping it in the family seems to have been widely practised.

Europe was somewhat more ambivalent to the Cold War, experiencing the rapid growth of popular socialism and active trade union movements, supporting State intervention in the economy, which in effect represented a milder version of Communist ideology. The European Union emerged out of a cooperative system to bring order rather than competition to the iron and steel industries. Many politicians had sympathy for the Soviet system, including Britain's Prime Minister, Harold Wilson, who, it is said, even had rather direct ties.[7]

Fig. 8-7 Harold Wilson

The only hot spots of the Cold War occurred in the Far East with the Korean and Vietnam wars in the 1950s and 1960s, resulting from the division of those countries to isolate their Communist factions. On August 2nd 1964, a US warship was attacked by a patrol boat in the Gulf of Tonkin, in an incident that was exaggerated by President Johnson to secure support for the declaration of a new war. It dragged on inconclusively for many years at a cost of some 47,000 American lives, before President Nixon announced a somewhat ignominious withdrawal in the face of rising domestic opposition.

Possibly one of the most significant outcomes of these wars was that their cost forced the United States to abandon the Gold Standard, allowing the dollar to be traded freely on commercial markets, which in turn meant that it could print as many dollars as the market would stand. That in turn demanded the imposition of confidence in the system: if necessary, by military means.

The Cold War came to an end around 1990 when Boris Yeltsin replaced Mikhail Gorbachev, a moderate new leader of the Soviet Union, who was bent both on modernising the cumbersome administration of his country and improving relations with the West. As already described in Chapter 4, the collapse of the Soviets was engineered by the British and US Governments.

THE HIDDEN PRESSURES

These few lines have hardly done justice to such an important chapter of history, yet they may have served to identify certain key threads in human and political behaviour. Almost every event seems to have had an economic and financial subtext. Not far from the industrial slums were industrialists endeavouring to make money fast, backed by financiers manipulating money flows, complete with derivatives in terms of debt, usury, credit and confidence, all of which was underpinned by an underlying faith in ever-onward growth. Debt and confidence are two sides of the same coin.

The World Wars were fought between countries under themes of heroic loyalty and national fervour, but not far below the surface lay more pedestrian financial and commercial pressures. The scale of suffering caused by the system was colossal. Equally remarkable was the willingness of perfectly decent people to inflict such suffering, often themselves giving their lives with misplaced loyalty. As mentioned above, 47,000 American soldiers were persuaded to do so fighting a somewhat unspecified enemy in a distant Eastern country that by no stretch of the imagination could have posed the slightest military threat to their homeland. Control of the press to fashion public opinion must have played an important role, only now being eroded by the internet that allows uncensored commentary to flow around the world. But whatever the explanation, the developments over this epoch were made possible by an abundant flow of cheap oil-based energy. It fuelled the factories and the tanks, the airliners and the bombers, and delivered Midas-like wealth to the governments, individuals and companies that found themselves in control of the System. One important step in this direction was the structure of the corporation that had acquired the legal status of a citizen without any of his human responsibilities. The tax treatment of corporations brought many hidden advantages and distortions. The cheap oil-based energy, which made all this possible, in one sense seemed to deliver enormous benefits yet may eventually be found to have been a curse, replacing a gentler life-style by one of unsustainable excess, of which there are universal examples. For instance, the British Government has recently announced a state-funded scheme to supply the obese with life-style counsellors, while simultaneously outlawing the ancient practice of fox-hunting. It also established field hospitals in Welsh cities to deal with injuries resulting from excessive drinking at Christmas parties. Pigs that once snuffled acorns beneath oak trees now spend their lives lashed to steel frames facing the final indignity of frequent artificial inseminations as farmers try to meet an insatiable supermarket demand for pork and bacon.

It is all a bit odd and not quite real. Television and radio have played an important part, fashioning a form of synthetic life-style and mind-set, divorcing people from their actual conditions and possibilities. Here again, we can see a form of Anglo-Saxon hegemony pumping commercial music, if that is the word for it, around the world. It is now almost impossible to go into a public place without being subjected to the relentless throb of amplified sound. The once jolly folk song has been replaced by a strange wailing and, in some way threatening, lament, set to a relentless Afro-beat. The subconscious impact of this somewhat aggressive noise, to which many people are exposed during their waking if not sleeping hours must be enormous. The youth is particularly targeted, being given to attending what amount to orgies of drink, extreme noise and flashing lights in their disco-culture. It is commonplace to see young people, dressed in sawn off jeans and pulled down woollen hats, slumped in public places with earphones droning into their barely conscious ears.

THE CRISIS YEARS

These somewhat disjointed remarks and observations set the scene for addressing the Crisis Years that will surely follow the End of the First Half of the Age of Oil. It lasted 150 years and made possible the pattern of events described above, acting as the very bloodstream of the modern world: its economy, its finance, its mindset and its political agenda. Accordingly, there is every reason to expect that the transition to oil decline will be a time of great tension. It is not so much that the decline of oil itself will bring everything to a physical standstill, but rather that the perception of looming long-term decline will undermine the foundations of the *Financial-Industrial System* that depended on perpetual growth for its survival.

The ending of the Cold War left the United States in a quandary. It had built up an enormous military capability that became palpably and grossly in excess of what by any stretch of the imagination was needed to defend the homeland.

The dollar reigned supreme, delivering a handsome tribute in hidden fees. The Third World had been successfully impoverished. The procedure for doing so was for the leading financial institutions to identify a vulnerable country and then move against it on the currency markets, which made its situation even worse. Next, offers of rescue through dollar loans were made, being subject to liberalised policies. In this way, foreign companies were able to move in to use near-slave labour to produce goods and profit for export, while the country itself was burdened by interest payments on the foreign debt, policed by the IMF and World Bank. Ecuador, for example, finds itself forced to dedicate its entire oil revenue to service foreign debt, leaving the campesino as badly, if not worse, off than he was before. The system was encouraged by the local elite, and the governments they controlled, because they themselves became intermediaries in the long chain of beneficiaries between the producer in the jungle and the final consumer in the shopping mall. None of this implies evil intent on anyone's part, but rather represents an underlying ignorance of the workings of the *Financial-Industrial System* that had delivered so many apparent benefits to the winning countries that others sought to emulate, without realising that it was their suffering which made the system work.

But there was a cost to the winner too, and that was the transfer of the actual US manufacturing base overseas which led to the demise of the older industries and a shift in occupations whereby steelworkers became hairdressers. This was exacerbated by a flood of new immigrants, especially from Mexico. Today, 34 million US residents are foreign-born, being admitted to hold down wages. Earlier immigrants from Europe faced permanent physical separation from their homeland, such that they were forced to build for themselves a new identity in the country of their adoption, but now improved communications allow the Mexican to remain Mexican at heart, sending a weekly contribution to Mamita in Trujillo. They see themselves as being in the United States simply to try to make a quick buck. In short, a large sub-class has developed at, or below, the poverty line, while the elite rely ever more on their financial control of world trade. The country has been not only living on capital, but on borrowed and paper capital at that.

Looking at the oil connection in greater detail, it transpires that when Saudi Arabia, for example, sold oil to consumers in the United States, both parties had accounts with major New York banks, so that the transaction became effectively a book transaction. King Fahd simply could not eat enough dates to consume the proceeds of the oil sales. The two accounts in the vaults in New York provided the banks with the collateral to extend domestic credit: that being even in excess of the amounts on deposit. It transpires that the import of physical oil has been exactly matched by an expansion of domestic credit, meaning that the country basically secured its oil for free. Foreign central banks built up dollar reserves, consistent with the workings of the system. China even artificially held down the exchange rate to facilitate its exports markets.

It sounds like a dubious and unstable system. What would happen if the Chinese were to unload their massive dollar holdings on the market, perhaps when their energy shortages force them to turn from export markets to home food production? The value of the dollar, along with the US economy, might collapse overnight.

In any event, this was the edifice that has effectively run the World over the past Century.

THE GRAND PLAN

One can be forgiven for seeing it as something of a house of cards built on nothing more than faith in the continuation of a System, itself supported by a wide spectrum of vested interests. It may indeed have been precisely the issue of confidence that occupied the minds of the analysts in various Washington think-tanks as they tried to define the country's post-Cold War foreign policy. A grand plan for American economic hegemony and support for the dollar was needed to support the System. They soon saw the central role of oil in the equation, recognising that the country's growing dependence on imports was not about to abate because domestic production had long been in decline, mirroring an even more serious earlier decline in discovery. They realised that the country would become ever more dependent on the Middle East, including Iraq, where it had already fought a war, as described in Chapter 4.

The *Project for the New American Century* was formulated in Washington, and a letter pressing for military action in the Middle East was sent to President Clinton in 1998, being signed by Elliot Abrams, Richard L. Armitage, Richard Perle, Donald Rumsfeld, Paul Wolfowitz and Robert B. Zoellick. A further document, *Rebuilding America's Defences*, was drawn up by the same group in September 2000, together with Dick Cheney and Jeb Bush, the new President's brother.[8] It tellingly contains the following extract

> "while the unresolved conflict with Iraq provides the immediate justification, the need for a substantial American force presence in the Gulf transcends the issue of the regime of Saddam Hussein"

It could hardly be stated more clearly, with the word *justification* being particularly telling. The use of the term *Defences* in the title is also revealing, since the country was not in any way threatened in military terms. The strength of the dollar and its control of the economy may rather have been the concerns. Human rights or the status of women in Muslim society would hardly have been issues upon which to propose military

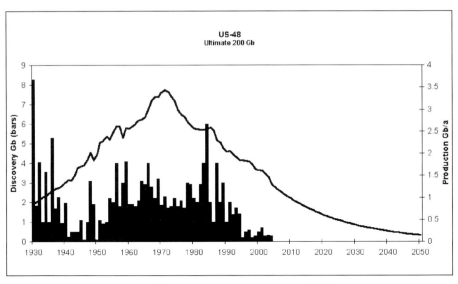

Fig 8-8 US oil discovery and production (US-48)
Note accurate information on discovery is not available

intervention. There were much more obvious objectives: namely control of oil and support for Israel with which several of the analysts had close links, even to the extent of dual citizenship. The influence of the Israeli sympathisers in America is clearly colossal,[9] as amply demonstrated by the fact that US Aid to the tune of 100 billion dollars has been given to the country for no obvious reason other than this link. Certainly, other tribal re-settlements around the World have failed to attract similar support.

It seems that the Vice-President, Dick Cheney, was given authority by the new administration to implement a decisive policy, based on these long years of planning. The change of administration itself may have been significant for the incoming members, being new to their jobs, probably found it expedient to simply extend past policies, not having had time to develop new ones of their own. They may also have lacked experience or knowledge of the wider world, giving their advisers excessive new influence.[10]

In Chapter 1, I described what I called the *Task Force Mentality* which I encountered when I went to work for Texaco in Colombia. It could be epitomised as "Action without Thought". Later, I came upon a variant, which could be called the *Committee System* when I worked in the Standard Oil Building in Chicago. There were committees for everything that delivered bland analyses and recommendations, but lacked executive authority. The Executives themselves were dynamic people but not gifted with knowledge or insight of the conditions in the foreign countries whose governments owned the oil rights. Their function had reduced to little

Fig. 8-9 Dick Cheney, Vice-President

more than managing the budget as best they could. The managers on the ground were either old retainers, put out to pasture, receiving detailed instructions on every minor step, or more dynamic figures endowed with the *Task Force* mentality. The operations themselves were normally conducted very efficiently, but overall planning and strategy were inept. Little attempt to integrate or understand the foreign communities, societies or cultures was made.

It is easy to picture how the Vice-President may have taken command within such a general administration and mindset. He was evidently cast a *Task Force* role, and resolved to take decisive action. In digesting the reports and presentations from the sundry committees, he was soon able to synthesise two clear and interwoven themes: control of foreign oil and support for Israel. He probably determined that a military solution was to be preferred, matching his own dynamic mindset. Having little sympathy for the subtleties of foreign diplomacy, he may have asked himself what good would come from trying to talk to all those fellows in their robes and fancy headdresses, who barely spoke English.

His first action was to strengthen and expand the chain of military bases around the World, many of which, including those in the Persian Gulf, had been in place since the Cold War. The key new ones were located around the Caspian and more recently on the islands of Sao Tome and Principe, off oil-rich Nigeria and Angola. It proved easy to persuade the countries in question to accept such bases with offers of loans, military assistance and training, with no doubt special inducements for the individuals in command. A base was established in Kosovo on the proposed route for a new pipeline bringing oil from the Caspian region to the Adriatic,[11] and troops were also positioned along the new pipeline from Chad to the Cameroons.[12]

It also probably did not take long to figure out that control of foreign oil demanded control the Middle East and possibly the Caspian if that should develop as hopes at the time suggested. However, a glance at the atlas showed the latter to be a difficult landlocked place, meaning that getting the oil out would call for special attention. A Balkan export route had already been secured through the Kosovo War, but it was evident that a closer presence was desirable. A further study of the atlas identified Afghanistan as of key strategic importance, especially as it had already played a role in undermining the Soviet regime, as discussed in Chapter 4.

The next challenge was to determine how to secure domestic political support for a new foreign war, as it was recognised that the general public would have little stomach for it unless galvanised into action by some extreme event. It was soon evident that a strong pretext to move was needed, indeed a very strong one to reach everyone's TV screen. A further advantage was identified in that people perceiving themselves to be threatened would be disposed to rally to their government with a sense of loyalty, which in turn would allow it to strengthen its grip on the country and improve its chances of re-election, as indeed has proved to be the case.

THE DOUBLE SIMULATION

So, we may conclude that it was decided to put in hand a bold plan of action. It took courage to do so, but they were evidently up to the occasion, deciding to implement it on September 11th 2001. The details of the operation remain obscure, but the many curious features of the event can hardly be denied or easily explained. They include:

1. The normal defences being shut down that day for a simulated hijacking.[13]
2. The rapid identification as hijackers of a group of Egyptians and Saudis, who had been given minimal flying training at a school in Florida, being supervised by Intelligence minders in their apartment building.
3. Four airliners were reported as being hijacked, having exceptionally low passenger lists.
4. Two of the airliners were filmed striking prominent buildings in New York, which exploded in what struck some analysts as controlled demolitions, the steel from the sites being later exported to China as scrap, preventing forensic analysis. The death toll was held to a minimum by timing the incident to occur before most people arrived at work, some being alerted at the last minute by the Omega messaging service not to go their offices that day. Some senior executives also found themselves attending a charity event at an air base.
5. The Pentagon was depicted as another target. An explosion occurred leaving a small hole at ground level without trace of a crashed airliner.[14]

6. The passport of one of the alleged hijackers was found in the New York rubble, despite the strength of the explosion.
7. An Israeli film crew was in position on the roof of an adjoining building to film the event.
8. The manoeuvres undertaken by the aircraft would test the skills of an experienced pilot, being far beyond those having no more than a brief training in light aircraft, suggesting that the aircraft may have been flown by remote control.
9. Within seconds of the event, a sinister figure in an Afghan cave had been identified as the ring leader of a global organisation, now named al-Qaeda, threatening the United States. He looked the part in his beard and outlandish robes, making excellent TV imagery. He was easily controlled having been previously on the CIA payroll.[15] Various videos and messages from him declaring a holy Muslim war were broadcast. Knowing full well where he was, may have made it easy to plan an unsuccessful search.[16]
10. Finally, the Vice-President took that day to be out of sight, evidently having taken command from some control bunker, while the President found himself reading to children at a school in Texas, evincing no surprise when an aide burst in to inform him of the incident.

The operation was pulled off immaculately despite a few difficulties that were experienced when the intelligence services both at home and abroad got wind of what was afoot, leading to many subsequent claims that the Government had failed to take proper note of the reported threats.

For good measure, a brief anthrax scare followed to bring home to every individual the fear that they were personally threatened. A universal sense of fear was a critical part of the strategy.

Before long, the B52s had been armed and sent into action. Images of the new sinister enemy in the form of Afghan tribesmen with their robes, beards and head-dressers, astride donkeys with a musket across their backs, were soon broadcast around the World. Within a few weeks, it was all over. The Taliban Government fell to be replaced by a puppet regime, led by Hamid Karzai. He was a western-oriented man, who had previously been a consultant to Union Oil of California. The action now was depicted as having a moral objective. Afghan women appeared before the cameras to explain how they had been oppressed by the previous regime, which had denied them education or the opportunity to find careers as dentists, teachers or shop-assistants.

But a setback to the grand strategy came when the Kashagan prospect off Kazakhstan, once billed as rivalling Saudi Arabia, was finally drilled with disappointing results. It soon became apparent that the Caspian would not in fact lessen dependence on Middle East oil to any significant degree. It was a setback but a small one, for Stage Two of the Grand Plan involving an attack on the Middle East itself had been the primary mission all along. A direct initial attack on the Middle East, with its obvious oil links, would have been widely opposed both at home and abroad, and so it was expedient to lead into it through a skirmish in remote and irrelevant Afghanistan. That campaign served its purpose by putting the country on a war footing, to which the people were now conditioned. Even so, some

Fig 8-10 Osama bin-Laden

further pretexts were needed. Iraq was accordingly accused of having threatening weapons despite evidence to the contrary from the UN Inspectors.[17] It was not even necessary to pretend that the country was in any way linked to the events of September 11th as Saddam was already established a villain in the popular mind after the First Gulf War, itself being the result of an earlier strategy related to oil price as discussed in Chapter 4.

The grand plan, if that is what it was, might have made eminent good sense in an abstract way from the distance of Washington where academic strategists spent their days moving chess pieces around the global scene. The man in command, who appeared to have a vision or knowledge of history and geopolitics that barely reached West Texas, may have readily accepted, having been further encouraged by the notion that he was in some way divinely inspired. He may also have been influenced by other lobbies from the arms industry wanting new business, from the financiers and investment people wanting to deflect attention from the basic weakness of the stockmarket and the dollar, and of course by sundry Israeli lobbyists. With a shrug of the shoulder, he may have said "if that's what it takes, folks, let's do it, but try to keep the causalities down". The successful efforts to limit casualties tends to confirm that the action was not the work of those for whom the only good American was a dead one.

Before long, the Operation "Shock and Awe", as it was code named by public relations experts, was underway, laying a carpet of flame across Baghdad. Military commanders don't blanch at what is euphemistically called collateral damage.

The venture was greatly helped by the near-incredible support given by the British Prime Minister. The explanation seems to be that his government gives high priority to imagery, spin, and the facile word, of which he is indeed an accomplished practitioner, but lacks concern for the deeper issues of substance at stake. It is furthermore Presidential in style no longer relying by cabinet discussion: a process facilitated by the appointment of low calibre Ministers.[18] He had ample control of his own party in Parliament, but was helped by his Conservative Opposition, who, with its imperial memories, may have seen themselves re-living the heroic Anglo-American invasion of the Normandy beaches, not wishing to appear lily-livered. The truth, it seems, was that he gave a casual support to the President, subject to UN sanction and other qualifications which were later ignored.[19] In any event, it is a sad reflection of the state of democracy in the Mother of Parliaments that he remained in power after the pretext for war was unmasked as palpable fraud. In earlier years, it would have meant resignation, if not disgrace, given that it led to the death of some 100,000 innocent people and destroyed a sovereign State, not to mention its priceless art and archaeological treasures. The scale of devastation is hard to grasp. Even such mundane things as the records of marriage, birth, and property rights went up in flames. The bombing of power plants took out the sewage works causing cholera and endless suffering for which the overstretched medical services were ill-equipped to deal. Whatever the motives for the invasion, they certainly did not include any feeling for humanity or the people who lived there. France, Germany and Russia, however, were not so easily duped and took a principled position of opposition.

So much for what seems to be a plausible explanation for the sequence of events that triggered the Crisis Years, which now dawn.[20] We may never know exactly what transpired on September 11th, and it does not much matter anyway, as it cannot be undone.

> For more insight, it is worth referring to the work of Mike Ruppert, a former Los Angeles policeman, who has assembled an impressive 673 page dossier.[21]

LOOKING INTO THE FUTURE

We may suppose that the sequence of events described above will trigger a looming discontinuity of historic proportions. A discontinuity it is, so forecasting the future on past trends and experience is absolutely the wrong thing to do. There is no place for a "business as usual" scenario. The decline of oil is imminent, real and unavoidable, but, as already stressed, that is only part of the problem. By all means, alternative energies can be tapped and life-styles changed, and the decline in oil production itself is no more than a gradual one at less than 3% a year. Surely, one might think, the necessary adjustments could be made with careful planning to allow life to go on albeit in modified form. The problem is that the decline of this essential energy supply undermines the very foundations of the *Financial-Industrial System*, which depends for its very existence on now unattainable economic growth. That is what causes the collapse of the world we have known.

THE COLLAPSE

I was invited to help, in a minor way, make a TV programme which the BBC broadcast on December 8th 2004. It was a fictional documentary covering the consequences of an explosion at the Ras Tanura oil export terminal in Saudi Arabia. It depicted the fate of a new swinging bank that had specialised in derivatives, and the role of a particular trader who had inserted a contract escape clause triggered if oil prices passed $50 a barrel. When prices actually passed this level while the film was being made, the producers had to lift the threshold to $100 a barrel. In the film, the bank fails, causing a meltdown of the world financial markets, accompanied by scenes of riots in the streets as a Second Great Depression falls upon the world. It may prove to have been prophetic.

Taking a longer view, it is easy to see why a general financial collapse is inevitable. In short, to explain it yet again, the banks have been lending money in excess of what they had on deposit. The interest charged represented new money, created out of thin air, which itself gave rise to further debt, with the essential collateral being created by the resulting economic growth. Furthermore, other hidden money flows arose from the control of trading currency: previously the pound sterling and now the dollar, which in effect become the principal artefact and benefit of empire. Lastly, it is evident that the system creates ever greater disparity of wealth, which is an invitation to tension. Even I, as a humble pensioner, probably effectively kill one or two Africans every day, as an indirect result of my minimal demands for goods and services. In short, one man's wealth has to be another's poverty in a world of finite resources.

This essential relationship has manifested itself many times in the past when the limits of the available resources to communities, factions or tribes were breached, leading to revolutions, industrial strikes and violence of various types and in various degrees. But now, the very limits of the Planet are being breached, whether we speak of wheat, water, or oil and gas. Accordingly, it takes no great feat of imagination to conclude that the past system has now run its course and must collapse.

In fact, it looks as if the collapse has already begun, as usual, striking the weakest first. Africa is in turmoil, whether we speak of Dhofar and Sudan, civil war in Uganda, the Ivory Coast or Nigeria, or rape and tribal warfare in the Congo. Kidnapped children are recruited to fight in the armies of warlords. Famine and Aids stalk the continent. Not long ago, the Hutus and the Tutsis in Rwanda were engaged in a genocidal conflict, costing the lives of millions.

These are described as *Developing Countries,* although that is hardly their fate. The most extreme cases are those with rich mineral deposits, whether we speak of diamonds in Sierra Leone or oil in Angola. Their revenues not only distort the established sustainable basis of life, but lead to intense competition between different factions seeking to get their hands on the booty. The population of Nigeria in 1900 was about 16 million, but has risen to 125 million largely on the back of oil revenues with which to import food. As oil production declines to exhaustion this Century, the World population will likewise have to return to something close to pre-oil levels. The nature of the cull does not bear thinking about.

Nigeria is at one end of the spectrum. The United States is at the other. As described above, it has already exported much of its manufacturing base, and is grossly in debt being run by shrewd banking operatives speculating on financial markets. Once the bread-basket of the world, it has become a net food importer. It is no longer possible to speak of investment as such because there is hardly anything tangible left to invest in that is capable of real growth in the absence of the cheap oil-based energy that made it possible in the past. The speculators have a mind-set built on past experience, but are shrewd enough to begin to look ahead. Several even call me for insight into oil depletion. For example, no less than J.P. Morgan, the prominent New York bank, had me make a

Fig 8-11. Ras Tanura, the prime Saudi loading terminal

presentation through a phone link from Ballydehob in the west of Ireland to 160 of its clients around the World, such is the miracle of modern communications. So, they evidently begin to wake up to the oil depletion situation.

But what can they actually do? Not much. But we can imagine the following train of events. At first, they will try to take defensive positions for themselves and their privileged clients: building up holdings in, for example, gold, Canadian tar-sand operations, commodities and property while still placing the institutional money under their control into conventional stocks. In doing this, they face a dilemma. On the one hand, they want to try to protect their own fortunes, yet fear making moves that themselves would trigger a general meltdown.

Presumably they will start selling short or moving into hedge funds that supposedly can make money from declining markets. This process can only run so far before it leads to a crash on the market. The dollar has depreciated greatly over the past year signalling that foreign speculators are already nervous. The central banks are covering their positions in other currencies, especially the Euro. The high price of oil strangely exacerbates the situation, because it leads to a massive increase in the un-earned revenue flowing to the Middle East countries, which is promptly recycled into the world markets, increasing the instability. The traditional government response of lowering interest rate will fail as fewer and fewer people will want to borrow or lend as soon as they perceive it to be a collapsing market. Already, record numbers of people in Britain are going bankrupt. It is impossible to say when the crash will come, but it cannot be far away.

Europe is in a similar, but perhaps less extreme, situation. Britain with its close banking and other ties with the United States is probably the most vulnerable. France, by contrast, will be less affected because it is a large rural country, substantially powered by nuclear energy, and it has a fairly robust leadership that is mistrustful of the Anglo-Saxon Financial System and its political reflections. The new growing prosperity of oil-rich Algeria may cause some of the migrants from that country to return home, lessening urban pressures.

Norway at the northern fringe of Europe, with its sparse population of only four million people, controls the bulk of Europe's remaining oil and gas, putting it into a strange but vulnerable position. At the time of the last oil shock, the German Minister was pressing Norway to open its doors with the words "Norway, do not forget your history", summoning up images of the jackboot on the frozen streets. In November 2004, the German Chancellor was again in Norway pressing for closer oil and gas co-operation with his country. Its position is discussed further in Chapter 12.

China is in a desperate situation, having enjoyed a capitalist boom at five seconds to midnight. It is already facing an energy crisis head on, as its oil production passes peak heading into steep decline. It finds itself increasingly unable to feed its 1.3 billion inhabitants as the aquifers deplete and the deserts encroach. Before long it may be forced to unload its dollar holdings, which could indeed be the trigger for the general collapse.

By contrast, the smiling Indians may well survive better, despite their excessive numbers, having become accustomed to live on little : their per capita oil consumption is less than a barrel of oil a year compared with twenty-seven in the United States.

Russia is probably best set to weather the storm. The country flirted briefly with capitalism on the fall of the Soviets, but is now reverting to an earlier form of central government, locking up some of the worst of its kleptocratic oil barons and forcing others to emigrate. It has substantial energy resources, which it will likely husband for its own use. Its manufacturing base is not advanced, and consequently has less far to fall.

Other countries will face their predicaments in different ways with varying degrees of success or failure. Urban conditions everywhere will be dreadful. As the situation becomes more evident, oil producing countries will move to stop exports to conserve their own resources for their own use, putting further strains on world supply. Of particular interest are Mexico and Venezuela, two major producers on the door steps of the United States. President Chavez of Venezuela, a tough-minded paratrooper with a popular mandate, is already taking a strongly independent role. If it is spared a US invasion, Venezuela is likely to emerge as one of the few winners, having ample resources to meet the needs of its 26 million inhabitants, which is not an excessive number for such a richly endowed country. Mexico is in a more difficult position with a long common border with the United States. Already, the latter is pressing for the so-called NAFTA-Plus Treaty, which would effectively destroy Mexican sovereignty, giving the United States access to the possibly rich oilfields that likely await discovery in the Mexican sector of the deep Gulf of Mexico. On the other hand, the collapse of the American economy may prompt a massive wave of returning migrants, exacerbating the difficulties at home, especially in the desperate conditions of smog-bound Mexico City, already home to 16 million people.

The one country that emerges sublime is Cuba, where force of circumstance has imposed a self-sustainable life-pattern on that green Caribbean island. The exiles in Florida may soon be looking for whatever flimsy boat they can find for the return voyage.

It is hard to imagine the path of this economic breakdown. No doubt it will be accompanied by much violence, especially in urban conditions where crime is already rampant despite the affluence. The mental adjustment of all these people to accept the bread line will not be easily achieved:

indeed there may not even be a bread line.

Another obvious consequence will be the disintegration of countries, many being somewhat artificial constructions. There already seem to be gentle movements in this direction. Nevada is now issuing its silver dollar; the Welsh have their own assembly; and the Scots are moving to substantial independence on the primary issues affecting their daily lives. The European Union, in one of its enlightened acts, proclaimed in the Treaty of Maastricht that "no decision shall be taken at any level higher than it need be". In times of hardship, people tend to club together behind their walls for mutual help and protection. As transport becomes ever more difficult and expensive, and as incomes fall, it will no longer be possible to ship Caesar salad from California to Toronto, or unload the container ship from China. People will be forced to rely on their own resources and produce, and will likely come to enjoy doing so. The window box may make a come-back supplying a crop of tomatoes rather than geraniums.

FUTURE PRODUCTION

So, what will actually happen to oil price and production? If indeed someone does blow up the Ras Tanura jetty in Saudi Arabia, the melt-down would be short and sharp as so graphically illustrated in the BBC film. But if the jetty remains intact, we may rather face a more phased crisis with recurring increasingly severe vicious circles. Oil price shocks will trigger recessions, and dampen price, leading to brief economic recoveries before falling capacity limits are again breached, re-imposing a new price shock. It is naturally impossible to predict what may happen in detail nor to capture every eventuality. Each country faces a unique set of conditions and challenges. The best we can hope to achieve is a fairly middle-of-the-road scenario that minimises the inevitable real world departures.

The IEA market report of October 2004 sets the scene in its classic bland style by suggesting that the *growth* in oil demand will fall from 240,000 kb/d in 2004 to no more than 70,000 kb/d in 2005. A more plausible reality is that recession will cause demand to decline in absolute terms, temporarily reducing pressure on price in a market that over-reacts to small imbalances between supply and demand. If there is a financial melt-down, then the fall in demand could be quite dramatic, such that oil prices might collapse to, say, $10 a barrel. This would be received with resounding cheers from the club of flat-earth economists who would again proclaim that warnings about peak oil and depletion were entirely groundless. They would likely attribute the recession to inadequate market liberalisation or the politics of Mr Putin.

A short decline in the price of oil would have a negligible impact on the actual production of oil. Existing facilities would continue to produce at the maximum rate possible driven by the economic imperatives of sunk

Fig. 8-12 Fidel Castro

cost and discounted cash flow. The fields that rely on high prices are generally small ones, having a negligible global impact. The OPEC countries would face a desperate loss of revenue, on which they utterly depend, forcing them to produce at the maximum rate possible. The World might indeed enjoy a mini-glut.

An alternative scenario might contemplate a situation in which the economic recessions were less severe, with sustained demand yielding oil prices in the order of $50 – $100 a barrel and flat world production. But the high prices themselves would give rise to massive financial transfers to the Middle East, which would soon further destabilise an already fragile financial system after a time lag.

But gradually, people everywhere will begin to perceive that the world they had known had ended, and that their many aspirations had become no longer remotely realisable. It would lead to tensions and conflicts in many areas and especially in the cities, forcing governments into ever more dictatorial roles. The United States is already widely seen to be moving in this direction with indications that it is about to re-introduce the draft for military service, most of which will be needed to police its own cities. Britain too now moves to reintroduce identity cards, last seen in the Second World War.

The cart-horse has a long gestation, and adjustments take time. Yet, people, when pressed, do tend to rise to the occasion showing remarkable fortitude and resilience. So perhaps after all, we are justified in adopting the fairly optimistic global depletion model as described below.

PRODUCTION FORECAST

The production of *Regular Conventional Oil* in countries that have already passed peak and their depletion midpoints (when half the total has been extracted) will continue to decline at its current Depletion Rate (annual production as a percentage of what is left). Production in countries, apart from the Middle East, not yet at midpoint will be flat to midpoint, or as local circumstances suggest otherwise. Since most are within a few years of midpoint, the alternative assumptions have little impact on the general position (See the Appendices for the details by country and regions).

The countries of the Middle East are in a very special position, as is obvious. It is becoming evident that Kuwait and Saudi Arabia have been reporting *Original* rather than *Remaining Reserves* since the late 1980s, with Abu Dhabi, Iran and Iraq, simply matching Kuwait within narrow limits. They have not been deducting production, which explains why the numbers have barely changed since the anomalous increases in the late 1980s. Yet, they still do have substantial genuine reserves, amounting to almost half of what is left in the World as a whole, as well as exceptionally low depletion rates. Most of their efforts will have to be dedicated to offsetting the natural decline of their ageing giant fields, but on balance, it might be reasonable, if verging on the optimistic, to expect production to be about flat beyond the normal midpoint, such that decline does not set in until the Depletion Rate has risen to a more normal level of, say, 3%.

We may suppose that Iraq will fragment into its several component factions, and that a degree of stability will return on the departure of the invaders. Russian, Chinese and French companies may be welcomed by the diverse factions and return production to more or less pre-war levels. President Putin has moved to forgive Iraq its massive debt, which the shattered country is far from being able to meet, in return for re-establishing the rights for promising oil fields that had been negotiated with the previous sovereign government. The Cossacks will likely be greeted with garlands by a grateful people.

There are of course other scenarios whereby both Iran and Saudi Arabia are reduced to ashes by fresh invasions or internal conflicts, insurrections and revolts. One can imagine that Israel might provide a new pretext for US intervention, offering a bridgehead from which to advance through Jordan to take the Saudi fields. The fate of Israel otherwise will hang in the balance as the world economic deterioration will adversely impact the massive flow of subsidy from the United States that it has enjoyed so far, as well as its own considerable direct overseas investments. It is already subject to reverse immigration as disillusioned settlers look for better places to live and bring up their families. Possibly, the long-suffering Palestinians will be able at last to return to their ancestral olive groves from the over-crowded refugee camps, now housing some four million people in desperate conditions.

Production in Russia is likely to rise only marginally for the next few years before reaching a second peak, as a more authoritarian government does its best to conserve the resources for the future.

Turning to the *Non-Conventional* oils, we may suppose that current deepwater developments will be maintained under the watchful eyes of the US Navy, and that production will rise to a natural peak of about 7.5 Mb/d by 2015. The Mexican sector of the deepwater Gulf of Mexico may be an important new source of oil, which in one way or another will be under US

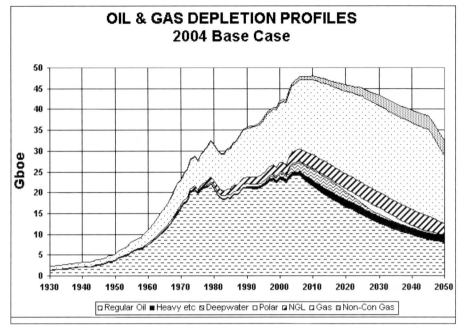

Fig. 8-13. Production Forecast

control.   Production of the heavy oil and bitumen of Canada and
Venezuela will be stepped up within limits, being found to be barely
economic in times of economic recession, and also in the case of Canada
seriously constrained by the depletion of the stranded gas used to fuel the
processing plants.

AFTER THE CRISIS

By definition, crises cannot last for long, and in due course the World will
adjust to the new reality imposed upon it by Nature. Although the
Transition will be a time of great tension, with wars, famines, pestilences
and a general break-down of the established order, there will probably be
survivors. By around, say, 2020, the new order will be beginning to be put
in place albeit at different rates and in different ways. We may begin to
speculate about the shape it will take.

There will still of course be plenty of oil for, according to the forecast,
the production of all liquids will have fallen to around 65 Mb/d by 2020,
which is approximately what it was in 1990. And world gas production
may have risen from 90 to 120 Tcf a year unless constrained by
government policy. So, indeed the hydrocarbon energy supply will still be
fairly copious, with the fundamental difference being that, by 2020,
everyone will understand that it is in relentless and immutable decline.
The extreme tensions and indeed savagery that was experienced during

the Crisis Years were entirely man-made, emotional and psychological responses, which do tend to typify times of change and uncertainty.

So, what might the World look like then? Some places will clearly emerge in better shape than others. Countries that are physical islands clearly have an advantage because they are naturally isolated and self-reliant. What for example will be the fate of Ireland, where I live? It is worth examining its particularly blessed condition.

In fact, Ireland's history has an extraordinarily relevant message to convey. By the middle of the 19th Century, its population had risen to about eight million on the strength of the potato, which was an efficient source of human energy and nourishment. But then a blight, originating in North America, decimated the crop, leading to famine and emigration, which reduced the population to half its previous number. This was Ireland's crisis with the potato almost exactly replicating the World's current dependence on a single energy crop, crude oil. The Irish crisis brought great tensions and suffering, and erupted into terrorism and civil war before a new benign nationhood was achieved in 1922.

The new government faced a huge challenge in running the country. It, significantly, gave prime attention to energy supply, building an impressive hydro-electric plant on the Shannon River, and establishing State administrations to support rural electrification and public transport. Progress was delayed during the Second World War when conditions were particularly harsh. There were many old bachelors around in those days as people did not get married until they knew they could support their families, but there was still a high level of emigration to Britain and the United States to absorb surplus population.

Eventually by the 1980s, prosperity came to Ireland, partly induced by tax holidays granted by the government to new companies that established themselves. Many did, and for a while Ireland enjoyed the reputation of being Europe's silicon valley. Furthermore, part of Dublin was declared to be a financial centre outside the domestic tax regime, bringing new wealth and prosperity to the island. Dynamic entrepreneurs made an appearance. Immense wealth began to flow to Ireland. Once charming Dublin became a throbbing metropolis with choked streets, while the *glitterati* built mansions on the sea front, as property values soared. Housewives became wage-earners, commuting long distances. The country became known as the Celtic Tiger as it mimicked the economic prosperity of the so-called Asian Tigers of Taiwan and Korea. But even before the last Century had closed, there were signs that the bubble was on the point of bursting. Wages had risen, and the international companies that had benefited from a well educated, English-speaking work force started moving on to relocate in eastern Europe and India. Dell was one of the computer companies that came to Ireland, but now, if you call them with a query, the voice that answers will have the delightful but unmistakable accent of Bengal. New immigrants began to flock to Ireland. Recently, leaving my hotel room in Dublin, I encountered three maids waiting to make the bed : one was from Moldova; the other two from Mongolia. Mandarin is Dublin's second language.

Ireland tried to find oil and gas off its coasts, falling on two modest gas fields, which at first helped provide the fuel for the exploding demand for electricity. But it was not enough, and a pipeline was built to tap into Britain's North Sea gas supply. This was being almost given away under the free market policies of Mrs Thatcher when competition drove down the prices. It was an epoch of hyperactivity, as companies were forced under market pressures to deplete the resource as fast as they knew how. Great advances in technology accelerated the process still further. The Celtic Tiger could well afford it.

So, the real shock in Ireland comes when Britain becomes a net importer of gas in 2006 and is unwilling to re-export to Ireland, save at an excessive price. The lights and all those computers will start quite literally to go out. That combined with the market collapse of 2005 will prove to be the final death-throw of the Celtic Tiger. The results will be devastating: within the span of no more than a few years the slums of Dublin reappear along the banks of the River Liffey, while the *glitterati* will be forced to build higher fences and hire guards with big dogs to protect themselves. It will begin to resemble Bogotá, but as in that capital, behind the hardship remains a smiling courageous face.

Fig. 8-14. An Irish cart

Then the principal political party, Fianna Fail, meaning Soldiers of Destiny in Irish, with its deep roots from the independence movement will re-exert its authority to introduce a national plan for survival, backed by widespread popular support. The people have a natural cooperative spirit, bred of their historical hardships and are ready to help each other. The first move of the new policy is to close the frontier to build a fortress on the island. The residual British enclave in northern Ireland also becomes an independent state when Britain is no longer able to subsidise it, leading the different factions soon resolve their differences, when they no longer have vested interests in maintaining the conflict.

Rural Ireland will revert to the regional life it had known less than a century before when farmers traded cattle in village markets. Christmas trees will no longer be grown for export to Germany, but a new and successful programme of self-sufficiency will blossom. There will more time, and the pubs stocked with home-brewed stout will remain full. People will meet in each other's houses for songs and to dance. Glimmerings of such an environment already exist where I live in Ballydehob.

In terms of energy supply, the country will also be very successful. Solar panels will be erected on south-facing roofs, and people will get used to showering only after a period of sunshine. Heat-pumps will be installed to capture ambient heat. The end of affluence will remove the mountain of waste that presently clogs the system. But the great breakthrough comes when the Government develops a simple mechanism to capture wave energy. It comprises no more than two large, inter-locking cylinders, set on the seafloor, with the upper inverted one being filled with air. The waves, above, cause the pressure to fluctuate, which in turn makes the upper cylinder bob up and down. It is equipped with a linear electricity generator, providing a near perpetual almost free supply.[22] A series of them will be installed around the coasts meeting the country's greatly reduced electricity demand. In addition, new meters will be installed in all industrial, commercial and residential property that penalise excessive use and encourage efficiency through variable utility charges. The cart-horse will be re-introduced for ploughing, and the bicycle for commuting. The railway system will be re-vitalised, proving adequate for the much diminished level of transport and travel. The cheap-fare airlines go bankrupt.

In short, the country survives. Marriage as an immutable institution is rediscovered but women will be content with small families. So, the country will be able to enter the 22nd Century with a happy sustainable population of about three million, living contentedly on their green fertile island, washed and nourished by the Atlantic waves and winds.

By no means all countries will be as fortunate as Ireland. Some face

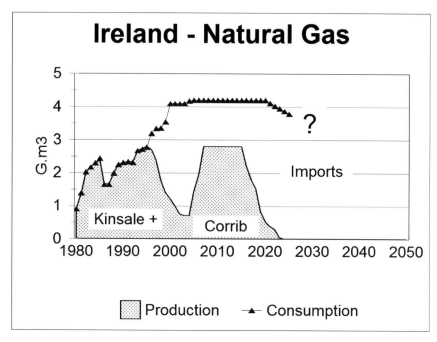

Fig 8-15. Ireland's Gas Supply

much more difficult natural obstacles, while others are exposed to deep-seated ethnic tensions and conflicts, inherited from earlier days when their natural frontiers were re-drawn by imperial powers.

Surprising as it may seem, things will not be that bad in the United States either. It will have suffered extremely during the Crisis Years, as urban life disintegrates and the very fabric of society, law and order fails. But out of the ashes emerges a new self-reliance, as people rediscover the pioneering spirit of their great grandfathers. The Federal Government, and the worthless political system behind it, will fall, having become utterly discredited in a series of abortive foreign wars undertaken in a vain attempt to secure foreign oil supply. California may be the first to secede under a new State Government, composed of Mexican immigrants with ample knowledge and experience of surviving under difficult times. At first, they toy with the idea of re-unification with Mexico, to which the territory previously belonged, but give that up when they see the appalling conditions in Mexico itself. Immigration comes to a standstill allowing the population to decline naturally to a level that is readily supportable in the vast territories of North America. The collapse of communications is accompanied by the re-development of small rural communities, each with its Sheriff, who deals with any miscreants in customary fashion, as illustrated by John Wayne in traditional films. Nature plays her part too, as the forests grow again, spared the chain saw from the paper mills, and the rivers support more fish as their content of chemical effluent declines. Tensions and conflicts occasionally erupt between communities and as frontiers are drawn between newly independent States, but by and large the well-known national attributes of enterprise and initiative give an acceptable and indeed agreeable life to the survivors.

Sub-Saharan Africa continues through the 21st Century little changed from how it began, as tribal warfare takes an increasing toll. The worst affected countries are Nigeria and Angola, where the curse of oil had been most severely felt. A collapse in birth-rate and longevity, with the help of Aids, malaria and malnutrition, brings the population down from its present excessive 733 million to a sustainable 200 million. Despite the dreadful conditions and appalling suffering of the transition which is prolonged, the survivors greet the 22nd Century with equanimity and confidence, displaying their well known optimistic character.

The Far East, including China and India, also faces a long transition, but adapt despite their grossly excessive populations, as they had been relative newcomers to the age of affluence. Famines sweep the lands as they had since time immemorial, with a late monsoon spelling catastrophe. But the survivors, perhaps now numbering about one billion, a quarter of their previous maximum, also find a new equilibrium within the conditions Nature had ordained them to live.

It is not necessary to cover all the regions, peoples and individual countries. The transitions are variously long or short, difficult or easy, extreme or terrifying, but eventually a new world is born.

A historian standing beneath the Martyr's Memorial in the university town of Oxford at the end of the 21st Century will still be preparing his lecture notes. Last term, he had covered the Rise and Fall of the remarkable Roman Empire that once spanned the civilised world. This term, he is addressing the much briefer Age of Oil. As he walks back to his college rooms, which have stood since the 16th Century, he gazes along the quiet tranquil street, observing a group of smiling students who laugh and enjoy the morning air, leaning on their bicycles. As he enters his college gate, his eye is caught by an announcement pinned to the Notice Board. It is for an educational tour to the old motor works at Cowley, where an automobile is on exhibit. He makes a mental note to tell his students to take a look at that extraordinary invention that had so briefly changed the world with such unimaginable consequences. In their Final Examination, he will ask them to explain how the fuel that drove it declined to exhaustion.

NOTES

1. Article by Hamad Chapman from Alexander's Gas and Oil Connections.
2. See, the very telling and convincing book *The Honourable Deception?* by Clare Short who was a member of the British Cabinet at the time.
3. Born in Belarus in 1874, he later became the first President of Israel (1949-52).
4. Seliger M, 2004, *Easy Does It*: The Times Magazine 04:12;04.
5. See, Stewart Udall's book *The Myths of August*.
6. The following quotation leaves no doubt as to his attitude: "Our race is the Master Race. We are divine gods on this planet. We are as different from the inferior races as they are from insects. In fact, compared to our race, other races are beasts and animals, cattle at best. Other races are considered as human excrement. Our destiny is to rule over the inferior races. Our earthly kingdom will be ruled by our leader with a rod of iron. The masses will lick our feet and serve us as our slaves".
7. See, Jerry Dan, 2003, *Ultimate Deception*; ISBN 0-9539951-1-9.
8. See, Clare Short's book *Honourable Deception?*
9. A key work on this subject is "Zionism & American Jews" (1981) by Dr Alfred Lilienthal, who explains the role of the powerful Jewish lobby in Washington.
10. A parallel was the already mentioned decision to use the Atomic Bomb (see, Udall, *The Myths of August*).
11. It is significantly named Bondsteel, perhaps a reference to steel pipes. The route having been secured, the pipeline, with a 750 000 b/d capacity, was authorised in late 2004 by the transit countries, with the construction being committed to the US AMBO Corporation. One may also wonder if the contentious election in the Ukraine, where one side was backed by massive US funding, did not also reflect the quest for a secure export route for Caspian oil.
12. There was also the curious affair of an abortive coup in oil-rich Equatorial Guinea, which was organised by old Etonian mercenaries based in South Africa, allegedly having links even with the son of a former British Prime Minister. Evidently Britain, America's ever-faithful ally, did not wish to be entirely left out.
13. This was reported by John Fulton, the official at the Defense Intelligence Agency (DIA) at a meeting of the Homeland Security Institute in Chicago on September 6th 2002, being covered by the New York Times and in an Associated Press report by John Lumpton of September 21st. See also M.Ruppert's book *Crossing the Rubicon*.
14. A particularly convincing reconstruction has been made on http://www.pentagonstrike.co.uk/flash.htm#Main
15. He had indeed been visited by CIA agents while in hospital in Dubai, being treated for a kidney complaint.
16. Some reports suggest that he is living happily in Freetown, Sierra Leone.
17. It seems that the Iraqi Government originally submitted an 11,800 page report listing all chemical, biological and nuclear weapons, including those supplied by the United States along with training facilities. The report was however intercepted by US officials who reduced it to 8000 pages removing the references to US supplies. The changes to the report made it seem suspect adding justification for the invasion. (See http://www.harrybrowne.org/articles/HusseinWasRight.htm).
18. See again, the telling exposé by Clare Short.
19. See, Clare Short.
20. If this account sounds extreme, then read Stephen Kinzer's *All the Shah's Men*, which documents how the British and American governments toppled the Government of Iran and recovered its oil to be shared between them.
21. See M.C. Ruppert's *Crossing the Rubicon*.
22. A prototype designed by Fred Gardner of Teamwork Technology is already in operation off Portugal, where it supplies sufficient electricity to meet the needs of 700 households.

Chapter 9

NATURAL GAS, GAS LIQUIDS AND NON-CONVENTIONAL OIL

T HE REALISATION THAT Regular Conventional Oil production is almost at peak and set to decline comes as something of a body-blow. The possible consequences of this discontinuity in the way the world lives cannot be known, and may be hard to accept: they sound extreme and alarmist. Surely, we ask, things are not as bad as they seem; surely some happy solution will be found. Perhaps so, but anyway, why not sit up and think about it?

I won't belabour the point further. It is time to step back from the main theme, and fill in some of the other supporting data and ideas, before returning to a synthesis in the final chapter.

As discussed in Chapter 2, petroleum is a family of hydrocarbons in gaseous, liquid and solid states. Taken together, and ignoring the size of individual accumulations, the total resource is enormous. Economists, who fail to understand the distinctions and differing depletion patterns of these substances, are misled into stating:

> "*Total hydrocarbon resources with the potential to produce liquid fuels are so large that they can be considered infinite for the purposes of analysis*".[2]

For them, the resource is to be treated as near infinite, with production being just a matter of supply, demand, profit, investment and technology.[3] In fact, the several species of hydrocarbons have very different characteristics and depletion patterns, which need to be taken into account when forecasting supply.

In the previous chapters, I have considered what I term *Regular Conventional Oil* covering the categories which have provided most of our hydrocarbon fuel to-date, and which will dominate all supply far into the future. Now, it is time to turn to the other hydrocarbons which will become more important after peak. They are subject to very different depletion patterns that control their production rates. There cannot accordingly be a seamless transition from one to the others, least of all if no plans are made to adjust to the decline and eventual end of *Regular Conventional Oil*. This is not a particularly inspiring chapter which could easily be skipped, but I have to cover it to close off this escape route for those searching for an easy way out of the predicament the World faces.

CONVENTIONAL NATURAL GAS

Natural gases fall into two categories: the combustible gases, comprising methane (CH_4), ethane (C_2H_6), propane (C_3H_8), butane (C_4H_{10}) and hydrogen (H_2); and the non-combustible gases, including nitrogen, carbon dioxide and hydrogen sulphide. Methane is by far the most abundant, making up more than eighty percent. Typically, seventy-five percent of a gas accumulation is made up of combustible gases.

Conventional natural gas comes from two sources: first, from specific source-rocks which are rich in organic material derived from plants; and second from normal oils which have been cracked to gas on being heated excessively on deep burial. Many basins contain both oil and gas fields, reflecting the admixture of source-rock types and variations in the depth of burial. But some basins, such as for example the Southern North Sea, contain only gas – in this case because it comes from deeply buried coal deposits, which have been subject to natural coking. Many fields produce both oil and gas, the gas having separated as a gas-cap above the oil in the reservoirs.

There are two main types of combustible gas, termed respectively *dry gas* and *wet gas*. *Dry gas* consists mainly of methane, although in some fields it may be mixed with nitrogen, carbon dioxide, sulphur dioxide (sour gas), as well as, rarely, helium. The composition of the gas reflects particular source-rock characteristics, as well as the thermal and bacterial conditions to which it has been exposed. *Wet gas* contains higher hydrocarbons, including butane and pentane, and is generally associated with oil accumulations.

In general, gas prone source-rocks are much more widespread in Nature than are oil source-rocks, and they are especially abundant in young geological sequences. But on the other hand, gas is more mobile than oil, due to its smaller molecular size, meaning that much has been lost over geological time. Accordingly, gas requires a much stronger seal than does oil to hold it in the reservoir. Salt is the most effective seal, and many gasfields depend upon a salt cover. Permafrost conditions in polar environments are also effective. These two factors – of wider source but the need for better seal – counter-balance one another, so that the total endowment of conventional gas is in fact less than oil. It is worth noting in passing that some gas reservoirs at shallow depth are still being charged from the source-rocks, and are accordingly productive despite weak seals.

Tcf	Produced	Reserves	Discovered	Yet-to-Find
Russia	508	1624	2132	500
USA	1028	189	1217	35
Iran	60	980	1040	100
Qatar	15	101	1024	20
Saudi Arabia	56	355	411	50
Venezuela	41	176	217	10
Nigeria	31	168	199	10
Canada	140	59	199	15
Algeria	66	127	192	15
Turkmenistan	64	93	157	20
Abu Dhabi	25	118	143	5
Iraq	13	95	108	20
Other	1153	2915	3161	1000
WORLD	3200	7000	10200	1800

Fig. 9-1 Endowment of Conventional Natural Gas

Figure 9-1 gives some tentative estimates of the *Cumulative Production, Reserves* and the *Undiscovered* potential of conventional natural gas by major country.[4]

Figure 9-2 illustrates the trend of published estimates of the *Ultimate* gas recovery.

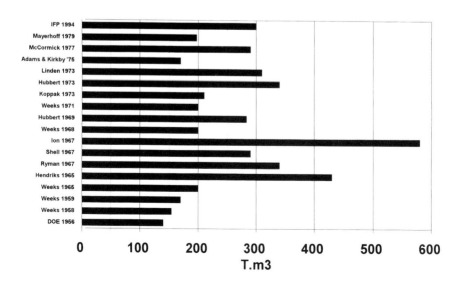

Fig. 9-2. Trend of published estimates of the Ultimate recovery of gas

Gas is at an earlier stage of depletion than is oil, mainly because of the high cost of transporting it from remote areas. Much was flared from oilfields in earlier years, and some is re-injected into reservoirs to enhance the production of oil. Gas is still being extensively flared in Nigeria, despite the penalties imposed by the government. It is instructive to look at a night-time satellite picture of the Earth, which shows the flares of gas fields over many parts of the world. The statistics on gas reserves and production are even less reliable than is the case for oil.

Gas flows more easily through the reservoir than does oil, which affects both the recovery factor and the depletion pattern. As much as eighty percent of the gas is normally recoverable, compared with less than about forty percent in most oil fields. Peak production is usually capped to optimise the facilities, or meet the terms of the supply contract, which generally prefers a long constant supply to a high peak. The production profile is therefore characterized by a rapid rise to a long plateau, followed by an abrupt fall. The terminal decline is commonly controlled by compressor capacity: compressors have to be installed when the natural flow drops. Seasonal fluctuations in demand determine the detailed production schedule.

It is clear therefore that the details of the production profile are primarily driven by economic factors although naturally subject to the overall resource constraints.

The production of gas associated with oilfields is however closely linked with the oil. Gas is commonly re-injected to maintain pressure, enhancing the production of oil, until that falls to a low level. At a certain point, late in the field's life, the operator may decide to draw down the gas cap, effectively converting the field from an oilfield into a gasfield. The Brent Field in the North Sea is a well known example.[5]

It raises the issue of how to define oil- and gas-fields, which in turn touches on the conversion factors by which to equate gas with oil.[6] In terms of value, one barrel of oil has been worth as much as a rounded 10,000 cubic feet, but that naturally depends on the respective prices which may well be subject to much fluctuation in the future. It is accordingly better to use a calorific equivalence is about 1 barrel of oil being equal to a rounded 6000 cubic feet. Fields with more ultimate gas than oil should be classed as gasfields.

Whereas it was possible, in the case of oil, to model the production profile around fairly well understood resource considerations on a country by country basis, it is not possible to do so for gas because a large part of the supply is driven by confidential supply contracts. That said, we may still suppose that the midpoint of the plateau of production will more or less equate with the midpoint of depletion.

Figure 9-3 depicts a very generalized global production and depletion profile for conventional gas. It assumes that production will rise at about 1% a year to 2025 and then level off in a plateau at 125 Tcf/a until 2045, matching the capacity of the infrastructure, and then decline steeply. In calorific terms, gas will overtake *Conventional Oil* production by around 2015, making it a particularly important fuel during the crisis and post-crisis years. Prices too are set to rise, perhaps greatly.

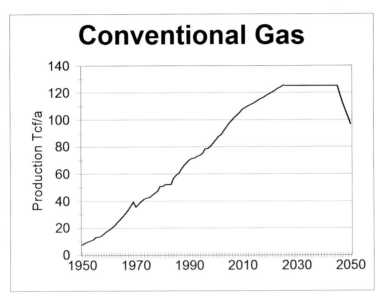

Fig. 9-3. Conventional gas production profile

Fig. 9-4. The 12,000 cu.m LPG carrier "Vestri" that operated in the 1970s for the Smedvig Company of Norway

Natural gas is normally transported by pipeline, but it may also be liquefied by lowering its temperature below −250°C, and then transported in special insulated and refrigerated tankers, although as much as 11 percent of the energy is lost in the conversion. Accidents have been few but the dangers inherent in transporting large quantities of gas in this way can be imagined. Even so, an increasing number of gas terminals are likely to be built, especially in the United States, to capture currently stranded gas in remote areas.

NON-CONVENTIONAL GAS

As discussed, the boundary between *conventional* and *non-conventional* gas is even less easily drawn than in the case of oil. By and large, however, it can be said that the production of the categories here classed as *non-conventional* gas will not become important in world terms until far into the future.

The resources of *non-conventional* gas may be very large. Some are already in production, where they are located close to market, as in the United States or in the Po Valley of Italy. Exactly how large they are is hard to estimate. I identify the following categories:

Biogenic gas
On deposition, organic material in the sediments was subjected to microbial action at more or less surface temperatures. The larger molecules were broken down with the release of methane and carbon dioxide, giving a reducing environment, which was incidentally important to the preservation of oil-bearing kerogens.

The methane, so released, filled the unconsolidated sediments with which it is in communication, although much was lost to the surface. Permafrost can act as a seal for such gas at shallow depth, and some of the Siberian gas fields contain biogenic gas. Permafrost may itself have released large quantities of methane to the atmosphere during periods of global warming in the past. In other cases, as in some of the gas fields in the Po Valley of Italy, shallow reservoirs can be continuously charged from the source, compensating for the loss of gas through weak seals.

Many basins, such as the North Sea, the Gulf of Mexico or the Niger Delta, contain disseminated deposits of biogenic gas at shallow depth, but individual accumulations are generally small, and would be rapidly depleted if exploited. Such deposits form a hazard in drilling: the release of large quantities of gas can so reduce the density of sea-water that the drilling rig above the escape may lose buoyancy and sink, a situation made even worse if the gas ignites.

The potential resource of biogenic gas is huge and almost impossible to quantify, but little will be capable of practical extraction for a very long time, if ever.

Dissolved gas
Gas is soluble in water under geo-pressures, and deep aquifers may contain large amounts of dissolved methane, depending on the salinity of the water, and the pressure. A pilot production project was undertaken in the United States but proved uneconomic because of the difficulties in disposing of the brines.

Coal-bed methane
Coal deposits contain large amounts of methane. When combined with air, it forms an explosive mixture, which has been the cause of many coal mine accidents. The methane is physically absorbed on the internal surfaces of the coal, with only about ten percent occupying the pore-space. Initially, coal-bed methane was extracted for safety reasons, but it is now viable in its own right.

The production technique is fairly laborious. The pressure in the coal-bed is reduced by pumping out the formation water in the joints, which causes the methane to desorb from the coal matrix into the pore-space. Three alternative stimulation techniques are then applied: hollowing out a cavity; hydraulic fracturing or directional drilling.

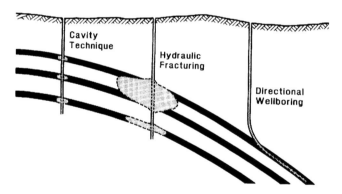

Fig. 9-5. Producing coalbed methane

Production is most advanced in the United States providing almost 2 Tcf per year, which is about ten percent of total production. Reserves stand at about 60 Tcf.[7] The removal of tax incentives in the USA in 1992 prompted several American companies to try to develop coal-bed methane overseas, especially in Europe, but so far progress has been delayed by a number of legal and environmental issues. The long term prospects seem, however, excellent, especially in countries like China, India and Australia with substantial coal deposits, and limited resources of other hydrocarbons. There are however environmental hazards from the disposal of the contaminated water used in the extraction process.

Gas in tight reservoirs

The smaller molecular size of gas means that it can accumulate in reservoirs lacking sufficient porosity and permeability to hold oil. There are, for example, large amounts of gas at depth in such circumstances in the Alberta Basin of Canada. If, however, the combined pressure and temperature conditions are too high, most such gas will be in solution, and the amounts are then constrained by the volume of the aquifer. Over-pressured shales also contain substantial amounts of gas, but it is not exploitable.

Hydrates

Methane occurs in hydrates, which are ice-like solids found in Arctic regions and in oceanic depths. They have also long been known as hazard in normal oil and gas operations, clogging pipelines. Hopes of exploiting such deposits appear to be doomed because, being a solid, the gas is unable to migrate and accumulate in commercial volumes. Reports[8] that the Messoyakha Field in Siberia produced gas from hydrates are erroneous: the hydrates referred to were nothing more than the normal deposits that form in pipelines in cold conditions. Huge amounts of misplaced research are dedicated to hydrates. The size of the resource is also most uncertain. A particular offshore seismic reflector was previously equated with the occurrence of hydrates but on closer investigation was found to be an artefact, possibly caused by thermal interfaces in the sea above. Where drilled, the deposits have been found to patchy and irregular, with the larger occurrences being related to seepages of normal gas on the seafloor. It is well said that they are the fuel of the future and likely to remain so.

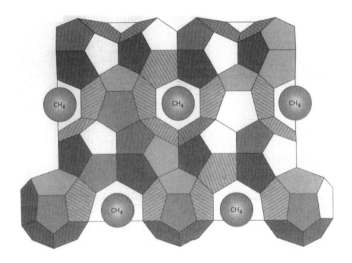

Fig. 9-6. Methane bound among water molecules in hydrates

NATURAL GAS LIQUID (NGL)

Liquid hydrocarbons, dissolved in gas, condense at surface conditions of temperature and pressure. The process of condensation gives rise to cooling, and it is curious to see an oil well in the tropics coated in ice. The product is a light clear liquid, similar in appearance to a refined product. Indeed, it can be used as an inferior motor fuel. It is known as *Condensate*.

In some instances when a gas reservoir is uplifted to shallower depths by earth movements, condensate may separate in the reservoir itself. In this case, it is known as *retrograde condensate*. The material sometimes lies at the eutectic point and does not quite know if it is a gas or a liquid.

Gas-condensate fields occur in the deeper parts of a basin, where oil has been cracked to gas, leaving a large liquid fraction dissolved in the gas.

Liquid hydrocarbons, such as butane and pentane, may also be extracted from gas by processing.

The treatment of Natural Gas Liquids in the statistics is particularly confused. In practice it is usually convenient to treat *Condensate* with Crude Oil, but to distinguish the *Natural Gas Liquids* produced in dedicated plants. They may draw their feedstock from several different fields, making it difficult to attribute the liquid to a particular field.

NON-CONVENTIONAL OIL

As already mentioned, the term *Conventional Oil*, which is in common usage, is poorly defined, being used in widely different senses by different authors. The term *Regular Conventional* (or simply *Regular Oil*) is used here to avoid the confusion. As so defined *Non-Conventional Oil* comprises the following categories:

– oil from coal (as produced by the Fischer-Troppsch process)
– oil from "oil shale" (actually immature source rock)
– oil from bitumen (defined by viscosity as occurring in tar-sands)
– Extra-Heavy oil (defined by density, as occurring in tar-sands)
– Heavy oil (defined with an arbitrary density of 10 – 17.5° API)
– deepwater oil (>500m water depth)
– polar oil

Together they comprise a substantial resource, and have two essential characteristics: they generally become viable only in a high price environment; and they have different depletion patterns. The production of the heavy oils rises only slowly to a long, low plateau before eventually declining, whereas the production of deepwater oil are constrained by the availability of floating production equipment and severe economic and technical constraints.

Coal and Shale Oil

Oil was produced from coals in Germany during the Second World War under the so-called Fischer Troppsch process. South Africa too turned to it during the trade embargo, and it still contributes some thirty percent of the country's needs.

Oil can be distilled from certain organic-rich claystones, popularly termed "shale". The organic material consists of kerogen that has not been converted to oil. Accordingly, the liquid so produced is, strictly speaking, not a natural oil at all, and should perhaps better be classed as part of the coal domain, but it is mentioned here because it is often considered in connection with oil.

The most advanced exploitation of shale oil was in the Piceance Basin of Colorado. It is reported that deposits yielding more than 10 gallons per ton are potentially commercially exploitable.[9] The process carries a high environmental cost in terms of the disposal of waste, some toxic, and the large amount of water consumed. The waste material has a very fine particle size and occupies more space than it did before it was processed. It is unstable when heaped up in tips, adding to the environmental costs and difficulties.

Although considerable investments were made in US shale oil projects in the aftermath of the Oil Shocks of the 1970s, none proved eventually commercial and have been abandoned. There are similar deposits in many other countries, including Australia, Brasil, Russia, Zaire, and China.

Australia, which has limited oil resources, was latest country to try to exploit shale oil near Gladstone on the coast of Queensland.[10] The project had been encouraged by government tax relief, and production was expected to start two years from commencement, rising to 14,800 b/d in the eighth year, but eventually the venture failed like all the others.

The resource is very large and may yet indeed prove a valuable source of oil in the future, when all other possibilities have been exhausted.

Oil from bitumen, tar sands and heavy oil deposits

When oil migrates to shallow depths on the margins of basins, it is weathered and attacked by bacteria, which remove the light ends, leaving behind sticky viscous materials known variously as bitumen, asphalt and tar. These substances grade into heavy oils, and there have been difficulties in knowing how to classify them precisely. Their characteristics vary, depending on the composition of the oil from which they were derived and the subsequent alteration processes. Extra-Heavy Oil and Bitumen are defined respectively on density ($<10°$ API) and viscosity (10 mPa-s). Asphalt is a type of bitumen with a gravity around zero degrees API,[11] occurring generally in what are termed tar-sands.

The two largest deposits occur in the Athabasca region of Canada and along the Orinoco River in eastern Venezuela, which are each estimated to have over one trillion barrels in place.

The Canadian deposit is at the more advanced stage of exploitation, being mined in mammoth open pits, as well as being exploited by wells with in-situ steam stimulation. A new development has been the use of horizontal wells for steam injection as well as production. The mining operation involves stripping off as much as 75 m of overburden, the present economic limit; separating the bitumen with steam, hot water and caustic soda, and then diluting it with naphtha. After centrifuging, liquid bitumen at 80⁰C is produced, which is then upgraded in a coking process and subjected to other treatments, eventually yielding a light gravity, low sulphur, synthetic oil. A huge work force is engaged in the operation. The process is economically viable, although the many efforts to reduce costs have not met with unqualified success. So far, the plants have been fuelled by stranded gas, which is now being depleted, and the Albertan government has expressed concern that the massive amounts of water used in the process are depleting the aquifers. A promising new technology involving catalysts is being pioneered. It is important also to remember that the deposit is not homogenous with many subtle but important variations in reservoir and oil characteristics. So far, only the more favourable locations have been exploited.

In Venezuela, the deposit, which has an average gravity of 9.5° API, is somewhat deeper and therefore hotter which reduces its viscosity. The oil is extracted with steam stimulation and chemical dilutants from reservoirs at depths of 150 to 1200 m. It is estimated that about 270 Gb could be recoverable.[12] Typically, patterns of five wells are drilled on a regular grid. Steam is injected through the peripheral wells, driving the oil to the central well, which can produce initially at up to 600 b/d, for a period of a few years until the catchment is drained. New methods of extracting it directly with the help of horizontal wells and submersible pumps are being applied, expanding the catchment area to increase flow rates to 1400 b/d, even without steam injection. In the 1980s, Venezuela commenced marketing a product made from bitumen, known as Orimulsion, which has been used as a commercial boiler fuel for electricity generation. It consists of an emulsion of 70% bitumen, processed to a particle size of 20 microns, mixed with water and 2000 ppm surfactant. It has a relatively high sulphur content, which can, however, be largely removed by conventional scrubbers in power stations. Production was expected to rise to about 600,000 b/d by 2005. Almost half the investment is in processing facilities. Exports to Europe have dwindled, however, partly because of the emissions, but new markets may be found.

It is obvious that the resources of the tar-sands are enormous. The largest are the Canadian and Venezuelan deposits, but there are many others around the world, including two large ones, known as Aldan and Siliger in Russia.[13]   Minor deposits also occur on the margins of most productive basins, although not readily identifiable in the data. Although they may be economic to produce, at least to a certain scale, they use a large amount of oil in steam generation and are very environmentally unfriendly both for producer and consumer. Undoubtedly, production will rise in the future, but probably to a low ceiling, constrained by the sheer scale of the operation, and only when *Regular Conventional Oil* is much scarcer than now.

Other Heavy Oil deposits
The classification of what constitutes heavy oil is somewhat arbitrary. Canada has a high cutoff at 25° API, because the cold climate affects the flow properties, whereas Venezuela prefers 22° API. Some classification apply a rounded 20° API but here a slightly lower one at 17.5° API is preferred because there are several fields in normal production with oil just below that level   There is a very large number of heavy oil fields, which are found in virtually all producing basins, generally at shallow depth. Heavy oil has a high viscosity, and production rates are low, commonly requiring the pump. The production profile consequently rises slowly to a long low plateau before declining gradually. Many heavy oil deposits have been neglected, or produced slowly, in the past, because their economics compared unfavourably. A larger proportion of what will be produced in the future will be heavier oil: one estimate suggests as much as 37% of the *undiscovered* will be heavy.[14] Deepwater finds also tend to hold relatively heavy oil because of low geothermal gradients below the thick oceanic cover.[15]

Overall production of all categories is here estimated to rise gradually from the current level of about 2 Mb/d to a plateau at almost double that amount lasting from around 2020. Naturally, if the World enters a deep depression as a consequence of the decline of *Regular Conventional Oil* the demand for heavy oils will also sink, because they will always be relatively expensive to produce.

Enhanced Recovery
The proceeds of secondary recovery methods, such as waterflood and gas injection, are included in *Regular Conventional Oil*. Additional tertiary methods can be employed, including the injection of steam, polymers, carbon dioxide, nitrogen and miscible gas to sweep the oil through the reservoir. Steam injection is by far the most commonly used procedure. Heavy oil deposits are the primary candidates for enhanced recovery, largely because the initial recovery is low. Furthermore, progress is being made in improving the efficiency of such techniques, increasing recovery

and production rate.

Another novel approach involves actual mining, although little more than a curiosity at present. Several shafts are being drilled through the shallow Corsicana field in Texas. They are connected by underground galleries which collect oil drained from the overlying reservoir.[16] The first oil in Texas was found near Corsicana in 1893, and this mining development a hundred years later is a commentary on the full cycle of depletion. The gushers are over, and the Texans are reduced to almost digging out the last few drops with shovels. It says a great deal.

Infill Drilling

A field with homogenous reservoirs can be efficiently drained by fairly widely spaced wells: the norm in the United States being one well per 40 acres, described as a 40-acre spacing. Fields with inhomogeneous reservoirs can however require a closer well spacing. Successful infill drilling campaigns on 20-, 10- and eventually 5-acre spacings can be conducted on appropriate fields, after the initial 40-acre development has been completed, especially in cases where that was undertaken long ago without the benefit of modern technology. Such activities are common, for example, in parts of the Permian Basin of Texas, where much oil was by-passed in the initial development due to the particular nature of the reservoirs and the primitive stimulation, which involved "shooting" the wells with nitroglycerine. There is nothing magic about 40-acres, although it was a normal spacing in the United States.

The recent introduction of so-called 4D seismic facilitates early infill drilling. Seismic monitors are placed permanently over a field and can track the movement of oil, so that infill wells can be optimally placed. As this approach becomes more routine, infill drilling becomes a standard procedure as part of the initial development. The prceeds are treated together with *Regular Conventional Oil* production.

Fig. 9-7. Shooting a well with nitroglycerine was a dangerous activity. Here the explosive is being poured into the "torpedo" which was lowered into the reservoir to shatter it and thus improve production.

Oil and Gas in Polar Regions

The frontiers have been pushed back over time. The North Slope of Alaska was considered out of reach in earlier years, but

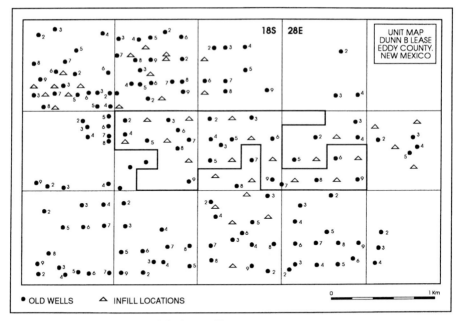

Fig. 9-8. Basin of Texas and New Mexico

has since delivered some major finds. No doubt the hostile environments can in due course be conquered but there remain serious geological constraints.

Antarctica is not very prospective in geological terms, and is in any case closed to exploration by international agreement. The Arctic regions are more promising with some very large sedimentary basins, although the evidence to-date suggests that they are primarily gas-prone. This is due to vertical movements of the crust under the weight of fluctuating ice-caps in the geological past, which tended to depress the source-rocks into the gas generating window and also disrupt the seals. The oil occurrences in Alaska and Western Siberia are therefore rather anomalous, having been partly spared some of the vertical crustal movments by virtue of their particular tectonic settings. Furthermore, Polar Circles are arbitrary cut-offs, which have changed over geological time.

In this assessment, we estimate that polar oil production has two cycles. The first peaked in 1988 at about 2 Mb/d, mainly coming from the Prudhoe Bay in Alaska, and the second, at about double that amount is tentatively expected to follow around 2035 if major discoveries are made in Russia, which is most uncertain.

Deepwater Oil and Gas
As discussed in Chapter 4, the development of the semi-submersible rig in

1962 brought the continental shelves of the world into range for routine drilling. Technological progress has gradually pushed back the frontiers since then, so that oil is now being found and produced in deep and very deep waters. But, as usual, it is not simply a matter of technology for it is necessary to take into account the special geological conditions of the deepwater domain. A cutoff at 500 m water depth is a convenient barrier from both technological and geological standpoints.

While much of the Planet is covered by deepwater, only a few areas have the remarkable combination of geological circumstances to yield oil. Briefly, rifts developed in divergent plate tectonic settings at times of global warming, principally in the Gulf of Mexico and along the margins of the South Atlantic. Oil source-rocks were deposited in the lakes that initially filled the rifts, and were later sealed by layers of salt from the evaporation of brief marine incursions. Later, sands and clays were washed down the continental slopes by turbidity currents, which can be compared with submarine avalanches, They were in due turn affected by deepwater currents that removed the fine-grained material to leave excellent reservoirs, that were ponded behind uplifts on the seafloor. In certain locations, the source-rocks were heated sufficiently by burial to yield oil, which was able to migrate upwards to collect in the reservoirs, often in stratigraphic traps. The tortuous migration path meant that the oil is generally rather heavy, imposing production constraints in the deepwater environment, which tests technology and operational management to the limit. In some cases, the source rocks are on the borderline of maturity, relying on local faults or salt deposits to lift the geothermal gradient sufficiently to permit generation.

Deltas may also extend into deepwater in other parts of the world, which may be also locally productive, although the source rocks in such an environment tend to be lean and gas-prone.

Clearly, it is economically difficult to produce gas in the deepwater domain, although there is some scope for shipboard processing plants by which to convert it into methanol and other liquids.

Despite these serious constraints, the deepwater is an important new domain. Production by field rises rapidly to reach a plateau set by the limits of the floating production facilities, and, as is normal everywhere, the larger fields have tended to be found first. Here, the total endowment is estimated at some 70 Gb, mainly in the Gulf of Mexico, Brasil, Nigeria and Angola, with production being expected to reach a peak of about 7.5 Mb/d by 2015, before declining fairly steeply, as illustrated in Figure 9-9.

SUMMARY

It is not easy to quantify the amounts of Non-Conventional Oil that may occur in the several categories described above. By far the greatest potential is for heavy oil and synthetic crude made from tar sands. Enhanced recovery applied to heavy oil may also make a useful contribution. Macgregor, 1996, provides data on oil-in-place, with: Canada 1.7 Tb; Russia 1.6 Tb; Venezuela 1.5 Tb, compared with Saudi Arabia at 500 Gb at the head of the list.

About 14 Mb/d of *Non-Conventional Oil* are in production to-day (2004). The amount is expected to rise to about 22 Mb/d by around 2015 before returning gradually to present levels by around 2050.

Deepwater Production

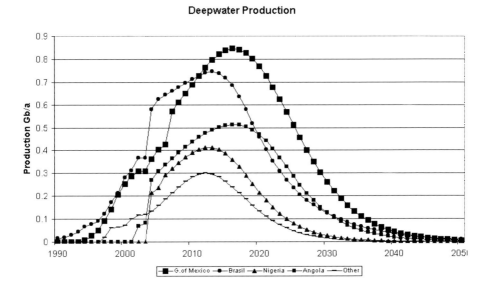

Fig. 9-9. Deepwater production profile

NOTES

1. The term *Regular Conventional Oil* (shortened to *Regular Oil*) is here defined to exclude oil from coal and shale, bitumen, Extra-Heavy Oil, Heavy Oil (10-17.5oAPI), deepwater oil (>500m), polar oil and Natural Gas Liquids produced from plants. The term *Conventional Oil* is poorly defined in the literature, so it is necessary to qualify it.
2. Statoil, 1996, makes this very misleading statement in a report.
3. Adelman, and Odell, in several papers listed in the Bibliography, make the underlying assumption that the resources are near infinite and that production is driven by economics and politics. Odell in the 1970s accused the companies of deliberately understating the North Sea potential as a conspiracy to raise prices. He was correct about the understatement, although probably not for the motive. But that was at a very early stage of exploration and on the basis of much less data and knowledge than are now available.
4. See Laherrère J.H., A.Perrodon and C.J.Campbell, 1996, for a comprehensive analysis of gas based on the authoritative Petroconsultants database.
5. See Moody-Stuart M.., 1996
6. Gas is equivalent to oil as follows

calorific	1 barrel of oil = 6 000 cu ft
value	1 barrel of oil = 10 000 cu ft (at \$1.8/ Mcf & \$18/b)
	10 tcf gas = 1 Gboe (rounded)

7. A good account of coalbed methane is given by Preusse A., 1996.
8. An account of gas is given in McCabe et al. 1993 in USGS Circular 1115.
9. Ion D., 1980, published a useful account of the availability of world energy in 1975, in the aftermath of the First Oil shock, which makes interesting reading as to the attitudes of the time.
10. See Petroleum Review, 1966, for a discussion of the Australian shale oil development.
11. Cornelius C., 1987, discusses the classification of heavy oil, bitumen and asphalt.
12. See Abraham K., 1997, for a useful description.
13. Ulmishek G.F, 1993, speaks of huge remote Siberian deposits; and D.S.Macgregor, 1996, lists them.
14. Masters, 1994.
15. See Imbert P.,1996.
16. See Hart's Petroleum Engineer International, September, 1996 p. 15. for a description of this novel approach.

Chapter 10

ECONOMISTS NEVER GET IT RIGHT

I AM NOT AN economist. So perhaps I am being unfair to write a chapter with such a title. A man in a greenhouse should not throw stones. All the same, I am not alone in expressing scepticism of the economist's skill in predicting or understanding events, as the economic forecasts of most countries confirm. They do not seem any better at dealing with the oil business. I have witnessed many false economic evaluations of prospects, and have seen companies fail in their endeavours by allowing themselves to fall too much under the influence of economists. Certainly, the great international oil companies were built without the intervention of economists. I read economic reports that seem incapable of grasping resource constraints, and I am therefore less than convinced that they are working from the right premises in their work. At the heart of the problem is their unswerving faith in the open market as the supreme arbiter of sense or folly, good or evil, when in reality the natural resources and condition of the Planet are paramount.

Where I used to live in southwest France are to be found many castles built during the hundred years war between the English and the French in the 11th Century. They are amazing structures even by today's engineering standards, but when one pictures the conditions under which they were built, they are truly remarkable. Each stone had to be quarried, and transported to the site by horse. It had to be laboriously faced with a hammer and chisel, which was at least a day's work. Lime had to be slaked and cement made. It took forty years to build a castle, and even if the workers were not paid much, they certainly had to be fed. An unfinished castle was no use to anyone, so it represented an investment with a pay-out of something like fifty years. It was a project that would never fly under modern discount economics, yet the castles were undeniably valuable investments, in many cases still standing, centuries later.

In the past, people certainly operated under longer term economic theory than is practised today, whether building castles or planting orchards for their grandchildren. In fact, the idea of discount economics is quite modern, being credited to Donaldson Brown,[1] the finance director of General Motors in the 1920s. He wanted to measure the relative

Fig. 10-1. A medieval castle could not be built under modern discount economics

performance of different factories, and developed a system of discounting future value, so that an asset ten years hence was deemed to have no current value. He was living in a rapidly expanding economy, when time was money. The World's experience over the past Century has been one of unprecedented growth, made possible by the abundance of cheap oil-based energy. So, an economic theory that gives weight to the time-cost of money was perhaps the best one for the epoch, even though the Roaring Twenties did end in the Great Depression. It is certainly the wrong theory for the epoch that now dawns, when the essential supply declines from natural causes.

If I understand it right, the famous Keynes proposed a theory whereby governments could pump up sagging economies by printing money, which fed investment and in turn created growth. It seemed to work for a while, and might have been a good response to the Great Depression, but it will surely fail if there is an inadequate supply of cheap energy to deliver the expansion the new liquidity conveys. In the same vein, some economists[2] even suggest that high oil prices stimulate the economy by delivering new liquidity, most of the price in fact being profiteering from shortage unrelated to the actual costs of production, being thereby a variant of Keynes' printed money.

Fig. 10-2. Donaldson Brown, who invented Internal Rate of Return

Brian Fleay in his brilliant book[3] describes this short-termism as Neoclassical Economics, being the science of human relationships in the market place, which ignores the role of energy and natural resources.

At the heart of economic thinking is the well known idea that supply and demand determine everything. If wheat prices rise, farmers plant more in the next sowing, and the system readjusts. Inventories are treated as burdens to the system, to be held as low as possible. In essence, economic theory is built around human agency, which does indeed cover most aspects of life. The cost of coal is deemed to be nothing more than the cost of the miners and the capital investment: the resource itself being there for free. If the reserves were infinitely large, perhaps there is no need to consider them other than as a gift from Nature, but there are some danger signals in this proposition. Economic theory extols "just-in-time" management. It may indeed represent current efficiency, but adds to the risks. I remember a strike by truck drivers in France when the refinery gates were blocked by demonstrators. Within no more than a few days, the supermarket shelves were emptying.

Hotelling[4] in 1930 wrote a classic paper on the economics of depleting a resource. He concluded that there was no real charge attributable to the resource itself, suggesting that it should be essentially priced at the discount rate. He recognized that it was a finite resource, but thought that it would be subject to a natural substitution if it began to fall into short supply. Thus, firewood was naturally superseded by coal, coal by oil, and oil by gas in a well ordered progression under understood economic principles. It was not really a finite resource, but simply finite at a certain price.

This thinking still infests the economic community. They write,[5] as I have already quoted:

> *"Total hydrocarbon resources with the potential to produce fuels are so large that they can be considered infinite for the purpose of this analysis"*

In one sense, they are right, because if you include infinitesimally small accumulations, down to a molecule lurking somewhere out of reach, I suppose that the resource is effectively infinite. But they don't really mean that. To their way of thinking, if you want more oil, all you have to do is drill more wells; or increase the price; or invent some technology; or improve the terms or tax. Thus, Peter Odell,[6] a leading proponent of this line of thought, claims that the North Sea will not have been fully explored until every formation down to granitic basement has been exhaustively tested. He further argues for the abiotic origin of oil, which he suggests is perpetually flowing from the deep interior of the Earth.

With this false notion of an infinite resource, the economists spare themselves having to think about the concept of *Ultimate* recovery. That denial itself excludes the very idea of depletion: you cannot deplete an infinite amount. It explains how they like to extrapolate the growing production of the past into the future. They show graphs with production rising to a certain future date at the end of their forecast period, oblivious of the implication that it must fall like a stone on the day after, if it is to respect the resource constraints. Further, they can glibly refer to the Reserve to Production Ratio, saying that the reserves support present production rates for a given number of years, without recognizing the absurdity of the implication that production would then have to drop to zero at the end of the indicated period. It really is very surprising that they can sustain such specious arguments, when the simple observation of an individual well or oilfield provides ample evidence of decline and depletion. In the United States, the tax man even grants a depletion allowance in recognition that it is a dwindling asset, being classed as capital not income. Yet, reputable economists and reputable organizations still persist in this idea of a semi-infinite resource.[7] Worse than that, they influence governments into pursuing very short term energy policies.[8] The high prices of oil that marked 2004 were widely attributed to political tensions in the Middle East and burgeoning demand in China, ignoring the underlying resource constraints and their impact on supply.

While on the subject, it also seems to me that economic measurements are far too restrictive because they ignore the substantial and critical economic input of non-traded activities, such as housework, raising children, home carpentry or reaching self-sufficiency in the vegetable garden. The readings from the false measurements may influence policy: for example, encouraging women to go out to work in the active economy when they were happier and more useful in the passive one. If a hurricane devastates an area, the ensuing reconstruction is deemed a positive economic activity, raising Gross Domestic Product. It is a strange way to measure things. Whether or not we can do it in practice, all the accounts have to balance. Many people live in the most gruesome architectural surroundings because the system does not attribute value to intangible things like happiness or beauty. All this may have to change in a post-consumeristic world brought about by the decline of oil – perhaps even for the better.

RISK

Much of the practical work of economists in the upstream sector of the oil industry is concerned with the management of risk. It is thought that there are economic trends and tools that can improve the judgment of oil men in making their decisions, but it is well to realize that the flawed underlying perceptions of the economist in relation to resource constraints permeate also their short-term calculations.

The industry likes to depict itself as having to face exceptionally high risks, listing:

Exploration risk – they may be looking in the wrong place;
Geological risk – the geological interpretation may be wrong;
Contract risk – the lawyers did a bad job;
Government risk – the terms may be changed;
Political risk – war, sequestration;
Development risk – the engineers get it wrong;
Natural risk – the hundred year wave;
Terrorist risk – somebody blows it up;
Tax risk – the rules change, even retroactively;
Environmental risk – they spill some oil and have to clean it up;
Corporate risk – their stock suffers, or they are subject to a takeover bid;
Commercial risk – prices fall or costs rise
Labour risk – the workers strike.

Personally, I think that all these risks are greatly exaggerated, both absolutely and especially in relative terms. Most businesses work with lower margins and higher risks: an arbitrary change in government policy cutting subsidies can bankrupt the farmer after years of work; the arrival of a supermarket puts long established and successful small traders in the town out of business; the lifting of trade barriers may destroy efficient local enterprise.

What distinguishes the oil industry are the huge sums involved and the very high marginal tax rates. The latter amply cushion most of the risk to which they are exposed, because most charges are taken as operating costs to be set against taxable income.

CASH FLOW

The primary economic challenge in exploration is to model actual or anticipated cash flow. Figure 10-3 shows a typical study either of an actual development project or of a hypothetical one, undertaken to see if a possible discovery would be profitable.

The parameters are quite simple:

– Profit is revenue less investment, operating expense and tax.
– Revenue is production multiplied by oil price.
– Investment is the cost of the facilities, including the drilling; and operating costs are the running costs of labour, insurance, tariffs on pipelines, and contracted services.
– Tax is tax, including royalty.

Year	Prod. Mb	Oil Price	Gross Revenue	OPEX	Op. Income	CAPEX	TAX	Net Cash flow	Cum. Cash flow
0	0	20	0	0	0	65	0	-65.0	-65.0
1	4.6	21	96.6	40	56.6	0	0	56.6	-8.4
2	4.2	19.5	81.9	40	41.9	0	10.4	31.5	23.1
3	3.9	18	70.2	40	30.2	0	9.4	20.8	43.9
4	3.5	16	56.0	40	16.0	3	4.0	9.0	52.9
5	3.1	16	49.6	35	14.6	0	4.5	10.1	63.0
6	2.6	15	39.0	33	6.0	0	1.9	4.1	67.1
7	1.9	16	30.4	31	-0.6	0	-0.2	-0.4	66.7
8	0	17	0	0	0	5	-1.6	-3.4	63.3
TOTAL	23.5		423.7	259	164.7	73	28.4	63.3	

Fig. 10-3 An economic evaluation of a field

The calculation may be made in terms of *money of the day* in which estimated actual future costs are taken, or *real money,* in which the effects of inflation are discounted. The next step is to calculate what is called *discounted cash flow* to determine the present value of future earnings. Thus, $1,000,000 in five years time at a 10% discount rate is worth today

$$\$1M / (1+ 0.1)^5 = \$620\ 921$$

The sum of each future year's discounted cash flow over the life of the field gives the *Present Value* (PV), from which may be calculated the *Rate of Return*. There is also the concept of *Payout*: how long to wait until the investment is recouped, and the project moves into profit.

This describes the simplest outline of the procedure. There is great scope to make it ever more complex, by addressing multiple scenarios and risking each element using statistical probability theory, and so forth.[9] Companies normally have what is called a hurdle rate of return, namely the minimum return that they can accept under their investment policy. This is on the face of it all good stuff. The geologist provides his estimate of reserves; the engineers feed in more about the numbers of wells and production rates; the construction people estimate what it will cost to build the thing; and a committee of economists is dragged out to pronounce on future oil prices. The calculator whirrs, or the Monte Carlo Simulator spins comparing every daft combination, and out comes the answer: the project flies or it does not.

If it does not, well, the geologists can invent some more reserves, or the construction man can have second thoughts about the costs. So, if those involved want it to fly, they can usually massage it into shape. They are often under pressure to make it work, whatever their personal judgment, because they may be bidding in a competitive situation where there is much more at stake than the specific project. They may be drilling for critical new information, on which no immediate economic value can be placed. Failure to participate may create a bad impression with the host government, which would have wider significance. The stock-market too encourages companies to explore, naturally being ignorant of the real geological risks, as beautifully illustrated by Shell's recent experience. This company found that it had used up its traditional stock of under-reported reserves from past discovery, which forced it to mark down its reserves, creating a financial furore, which cost the Chairman his job. The reason why it had exhausted its stock of unreported reserves was the disappointing results of exploration despite an expertise, second to none. Instead of facing the reality that there is precious little left to find, the new Chairman has bowed to stockmarket pressure to increase expenditure in exploration – a strategy which will come to haunt him when it fails to deliver for natural causes.

If the project is made to fly, the proposal is now blessed with a notional number showing it to be sufficiently profitable, and it passes up the managerial hierarchy, each level having less and less knowledge of the actual situation.

I remember a golden case when, having discussed a project in Sumatra for some time, the chairman of the committee, looking at his agenda, turned to me saying "that is all very interesting, but when are we coming to the Indonesian proposal". Evidently, he was not well travelled. In reality, all that really sinks in at that stage is what the magic rate of return number is, and what is left in the budget.

The management desires a notional level playing field and excludes local tax situations so that they can pretend to fairly compare the rate of return from investing in a refinery extension in Texas versus an exploration well in Norway. Thus, they fail to notice that 85% of the risk of the well in Norway is borne by the Norwegian taxpayer, who is willing to accept that the cost of putting it in the wrong place is deductible from taxable income.

To be fair, the economic analysis does force those involved to think about all aspects of the project, which they might not have done otherwise. Even so, it is a dangerous tool. It tends to influence companies to search for oil where the best deal lies rather than where there is the best chance of finding oil. It has sustained so-called frontier exploration for many years longer than justified by the geology. New areas, on which knowledge is limited, commonly have good concession terms, and pass the system much more easily than sound prospects where the conditions are onerous.

The one factor that really affects the economics, however they are conducted, is oil price. On that, the economists have little to contribute, because it has been largely politically contrived[10] and because they do not accept the finite nature of the resource. They are not therefore in good position to assess the distribution of oil in the ground and accordingly cannot take into account the growing control of the resource by a few critical producers, which must surely influence the price more than ordinary economic factors. Virtually none of them anticipated the surge in price of 2004 because for them there can be no capacity limits.

Companies tend to have committees to assess future oil prices, mainly comprised of economists. They read the Wall Street Journal, and consult Salomon Brothers, thinking in terms of supply and demand trends. Consequently, they normally come up with one or more bland scenarios, whereby oil price is above or below inflation by so many points. There is talk of the gentle ramp. Their record in forecasting has been abysmal.

But if all this seems rather negative and dismissive of the economist, in fairness I do admit that it is difficult to see how else centrally controlled global companies could run their affairs. They clearly have near limitless opportunities to explore for oil, which is not, however, by any means the same thing as finding it. They can invest in many different things upstream or downstream, buying reserves or other companies. So, they do need some yardstick by which to choose, and perhaps the economic analysis, in a very general way, did provide a comparison, even if far from reality itself. Their bland oil price assumptions were also understandable as it is difficult to plan for a crisis, even if crises are a normal fact of life. The system more or less helped the management avoid serious mistakes, even at the expense of not getting much right either. Above all, it shielded them from responsibility by allowing them to justify their actions. That is often their motivation as they approach pensionable age. The discontinuity that now breaks will call for radical changes in the way companies operate and the economic systems that follow.

It appears indeed that this system has begun to self-adjust as the companies have downsized themselves. The new business unit or the *Asset Team,* as it is called, may be sufficiently close to the actual business to have fewer real options. They no longer need to be guided by hypothetical economic analysis, as local circumstances, many either beyond their control or offering an obvious preferred choice, dictate their actions.

I have the suspicion that I have been too harsh and sceptical, and have possibly overstated the argument. Perhaps the economists understand perfectly well, but are not free to express their views.[11]

You would think that the depletion of oil-based energy, which has driven economic experience for most of this century, would be fundamental to economic theory. But that does not seem to be the case. It may be that the premise behind discounting has been eroded, and that value, if not enhanced value, does somehow attach to future real assets, especially critical reserves of oil. Oil supplies forty percent of all traded energy and will continue to do so for some time to come. That, one would have thought, should transcend the short-term profit and loss account, at least until some viable substitute for oil is in place, if there is one.

I think that the economists need to go back to the drawing board and figure out some new principles. In fact, some have. For example, Odum[12] and others propose that all value should be perceived to come from Nature in a thermodynamically open system embedded in the environment. For them, the human economy is like a living organism. Its metabolic processes require a continual one-way flow of complex organized energy and matter from nature to replenish and maintain the economic structures inherited from the past; to build new ones; and to provide the goods and services that humans need or desire. Degraded energy and matter are returned to the environment. Above all, the new principles should grasp the finite nature of oil, which has become the main provider of energy in the modern world. These are hopeful directions,[13] but meanwhile, we should accept the pronouncements of neo-classical economists on oil supply and price with grave misgivings.

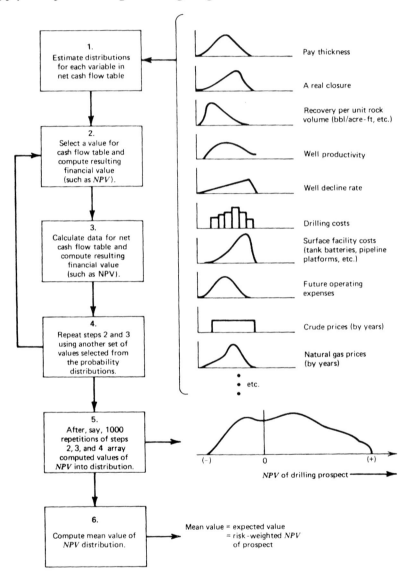

Fig. 10-4. Economic analysis procedure (Ikoko)

RICHARD HARDMAN – AS I SEE IT

Q Our paths first crossed in Colombia in the early 1960s when we were both working for BP. You had arrived from Libya. What had been your earlier career? And what were your impressions as a young petroleum geologist, fresh from Oxford?

Fig. 10-5. Richard Hardman, explorer, executive and former President of the Geoligical Society

A *Yes, I had been working in Libya doing wellsite work and subsurface studies, namely understanding the geology as revealed by borehole data. Before that, I had spent some time in Kuwait, working on one of the world's largest oilfields.*
BP in those days was much influenced by its Iranian experience. Its geological interpretations always had a certain Iranian imprint, with the shallow geology somewhat detached from the deeper zones, as was the case in Iran where the salts of the Lower Fars Formation provided a glide plane. As a young geologist, I naturally knew no different, but with hindsight it may be that BP was then somewhat less technically oriented than were the French or American companies. I am not saying that this was necessarily a disadvantage, because gifted and dedicated amateurs can often perform miracles.

Q When you came to Colombia, you embarked on a project of field mapping in the Eastern Andes, which was to contribute greatly to the understanding of the Llanos, which has recently yielded some giant fields, including Cusiana and Cañon Limon. It was a new experience. What did you make of it?

A *As I look back, I realise how privileged I was to have had this experience. Very few modern petroleum geologists have worked in the field. It was a special experience that taught self-reliance, self-discipline, and indeed management. I was on my own, and each day had to decide what to do: which outcrops to investigate; how long to spend on them; and how to deal with the logistics and labour force. It was physically demanding, climbing the mountains and forcing your way through rough country to search out the critical sections. But there was more time: one grew to know the country and get an intuitive feeling for the rocks and fossils: to see patterns and relationships. As the survey progressed, my knowledge of the geology and structure evolved: I began to know what I needed to know and where to look for it. I also learned to match effort against reward: I could not hope to cover everything, so I had to try to prize out the key information by being selective.*

Q So Colombia was a formative and critical chapter in your career. But you left BP and went to work for Amoco in London when the North Sea was opening. Why did you do that?

A *There was a mixture of motives. We had been living like gypsies, moving from country to country on short-term assignments. My family needed more stability, especially with schooling becoming more important as the children grew older. Also, I did not exactly have a positive response from BP about my career prospects: the local manager was not encouraging. I thought it would be fun to see how the Americans did it, and the North Sea was just opening. It seemed a good idea to be in at the start.*

Q Were your expectations met?

A *I suddenly found myself earning double, so that wasn't bad. But I encountered for the first time the hierarchial nature of a major American company, which was very different from the club-like atmosphere of BP.*

Q Can you explain more about this? I certainly understand what you say about the Club-like environment of BP: they still have reunions for old Libya hands, and retired BP people like to reminisce about the fate of their erstwhile colleagues.

A *Amoco had what I call a long-chain structure. The explorer at the front-end made his proposals through a long chain of command. In principle, each higher level should have added expertise and experience. But the reality was otherwise: these people were effectively filters, who had more to gain in the corporate politics by taking the low-risk easy option than by grasping the real opportunities. As a result they missed the main chance. It was a frustrating bureaucracy. But at the same time, I liked the pragmatic approach of the Americans I worked with: they would say "if that's what the Boss wants, that's what we better give him". Technically it was an invaluable experience, as I had a chance to be part of the pioneering exploration of the North Sea. It was an exciting epoch.*

Q Then in 1976, you were transferred to Norway as Exploration Manager, and played a key role in the discovery of the Valhall Field.

A *It was indeed a challenge. I recruited a small team of young Norwegian geologists, and we began to solve the intractable problems of Chalk geology. Valhall looked like a dough-nut on the seismic records. What was the hole in the middle? Some people thought it was a salt intrusion, but our studies eventually led us to conclude that the escape of gas had affected the seismic velocities in the overlying strata, so what looked like a hole was in fact the higher part of the structure. We were vindicated when Valhall came in as a major field. I became immersed in the subtleties of chalk geology, which remains a particular interest to this day. I was recently able to turn this knowledge to good effect, discovering a chalk field in Denmark which had been rejected by another company that had forgotten its chalk lessons.*

Q I am interested in the idea of there being certain insights that companies have, lose and sometimes recapture. One searches for new ideas, only to find that people have been there before. Only the other day, I found a depletion model by Shell in a book[14] published in the 1970s which is almost identical to that I now propose: little is new.

A *Yes, there are basic geological insights that tend to have lives of their own. They may form in the minds of explorers as no more than tentative ideas which are at first rejected, but they should not be allowed to die. It is one of the reasons why it is of such cardinal importance to document studies in written reports. Later generations can often learn from early work, and use it as a foundation for new hypotheses as new information comes in. The world has now been very thoroughly explored: some work has been done almost everywhere. In the rush of modern business, the ideas of the past are often ignored, but there may gold-mines in the technical files. BP's Colombian discoveries are a case in point. We proposed the Llanos prospects in the 1960s, but they were then deemed economically hopeless, being located on the wrong side of the Andes for export. Indeed the Company pulled out of Colombia altogether. But then later, they were searching for new frontiers and turned up our old reports, which led them to try again. This time they succeeded. It has taught me one thing. Geologists must use relentless logic to support their ideas: but they must not give way easily to all the pressures that stand in the way of testing them. The economist can always find compelling arguments for not exploring or for not making the highest bid for the best acreage. There is a polarity in all of this: ventures either succeed well, or they fail absolutely. It has to do with nature: the average, which tends to underlie economic analysis, does not bring rewards. It takes a kind of intuitive judgment to spot the winners, and to have the courage to go after them: the mealy-mouthed average does not deliver.*

Q From Amoco you moved to Amerada Hess in London, becoming the Director of Exploration. Now you were able to implement your ideas about how things should be run. What did you do?

A *Amerada was blessed with a dynamic leader in the form of Leon Hess, its founder. He knew his own mind and did not need to surround himself by self-serving committees. We did not exactly design a management structure, but circumstances and shared attitudes created it. The decision process was inverted in what I call a short-chain structure. The senior management determined the budget and the broader policy, and then instructed the local office to go out and do it. In effect, it delegated authority down the line, providing enormous challenges and a sense of satisfaction to those on the ground. We became a dynamic group and one of the most successful North Sea explorers. We were small compared with the major companies, who in many cases acted as our operators. It meant that much of our effort had to be dedicated to persuading the operator to follow our wishes: in many cases to save them from their own ineptitude. We often found ourselves expressing views held by the operator's own technical staff, but which had been frustrated in the long-chain command structure. It was a sort of diplomatic inductive process: to somehow get the ideas tested, despite all the back-sliding.*

Q I do understand. I remember when I became your successor in Norway, I often found myself powerless to implement what I wanted to do. I sometimes found it a great help to discuss proposals first with partners, including Amerada, so that they could help push it through the management filters. Long-chain companies often prefer to be voted into action, as it shields the chain from responsibility. You have had a very successful career, and I suppose that the crowning glory is to have been elected President of the Geological Society.

A *Well, I don't know that it is a crowning glory, but it is certainly a privilege to tread in the footsteps of some of Britain's greatest geologists. I would like to leave a mark on the Society by bringing it more into the public arena, and giving it the standing it deserves.*

Q What sort of public issues do you have in mind?

A *At first, you would think of geology as a rather academic subject, but it isn't. It has a much wider and philosophical role. After all, we live on this planet, and geology provides us with a means to know it better. The whole issue of resources and their depletion is a critical one for Mankind, and geology is at the heart of the debate. There is the growing environmental awareness and concern on which geology has much to say. Lastly, there is the issue of professional standards, qualification and training, which has to be modernized to meet today's needs. The publishing division of the Society sets an example of a highly efficient operation and what can be achieved from what was once a scientific journal. Now we have to think about manning levels in the profession: the provision of appropriate training and facilities. One thing I would like to do is to encourage field work to be again the core of geological training not only for professional geologists but for those who take it as a subsidiary subject before moving on to other walks of life. The hammer can often tell you more than the electron microscope, and it costs less. I cannot imagine a better background than a sound grounding in field work as we discussed earlier.*

Q Good Luck! I am sure you will succeed in changing things for the better, bringing together your industrial background and the political insights that come from it, with your basic love of geology. No doubt some of the old guard would like the short chain to become a noose, but if you succeed in this direction your achievements will have the recognition they deserve.

NOTES

1. Brown D., 1977, describes his experiences as an industrialist in America's great age of growth.
2. McKillop.
3. Brian Fleay's book "The Decline of the age of oil" gives a brilliant expose of the economics of growth and the consequences of the coming oil shortage for Australia in particular.
4. Hotelling. H., 1931, wrote the seminal work on the economics of resources.
5. Statoil, 1996.
6. Odell P.R., 1996.
7. Mitchell, 1996, Statoil, 1996. Lord Browne the CEO of BP and his Chief Economist, apparently oblivious of the nature of their own business, mislead both investors and governments with these specious arguments.
8. European Commission Green Paper.
9. There are several articles on this subject in Doré A.G. (Editor) 1996. Quantification and prediction of hydrocarbon resources. Publ. Elsevier ISBN 0-444-82496-0.
10. By "political" I mean the whole role of the Middle East, and OPEC, including the sequestration of the companies' rights, to which may be added the creation of State oil companies with their own agenda. The special US/Saudi relationship is an example of the political factors that fundamentally affected the price of oil.
11. See an excellent paper by Haldorsen (1996). He demonstrates his understanding of the short span of the age of oil and humorously shows a traffic stop-sign overlying the notion that oil prices are set to rise radically. He comments:" Today very few companies maintain the 'hockey-stick' [radical increase in price] approach to oil price forecasting. The industry has become very cautious: too cautious, some will say, particularly so far as long term oil price is concerned. Some scientists note that the world's hydrocarbons may run out in 50 years as the world's population doubles and fission/fusion still cannot power cars". He is absolutely on the right track, but then demonstrates how to ignore his own words of wisdom.

12. Odum H and E. Odum, 1981, Energy basis for man and nature; McGraw Hill, New York, explained more fully by Fleay, 1995.

13. Having written this chapter I now come upon a reference to a new economic theory by Prof. Laughton which comments: "Because of oil companies' use of the highly flawed DCF/IRR approach, the world is littered with under-utilized petroleum production facilities ..." . Evidently I am not alone in questioning neo-classical economics. (see Adam (1997).

Chapter 11

OIL TRADING AND OIL PRICE

As I HAVE SAID several times already, the oil industry has been plagued throughout its history by an environment of cycles, or in other words "boom or bust". It reflects the very nature of oil production and the fundamental pattern of depleting a reservoir. The cycle starts, figuratively speaking, when the pressure is high, and the resulting flow floods the market. The flood kills further efforts, but leads eventually to a natural decline, shortage and high prices, which prompt a repetition of the cycle. There is a polarity about oil: it is either there in profitable abundance, or not there at all. The cost of the marginal barrel[1] has been generally very low. That, anyway, is how it was, so long as there were plenty of new finds to be made to fit the economic cycles. Different patterns are now emerging as capacity limits are breached, and as new discovery shrinks.

But superimposed on the patterns in each individual province is the very uneven global distribution of discoveries and reserves. As each new major territory was opened up, it held sway for a few years: first, the United States and Russia; then Mexico and Venezuela; North and West Africa; and finally the North Sea. The Soviet Union too had its turn, although not quite as much part of the global market as were the others. Overshadowing them all was of course the Middle East with its huge endowment that gave it a lasting disproportionate influence.

From Rockefeller's day onwards, there has been a need to somehow regulate production to stabilize prices, and provide an environment under which the required long term investments could be made. Over the past fifty years, it has boiled down to finding a way to control Middle East production because it was the largest, and even had the potential to be larger yet.

TEXAS RAILROAD COMMISSION

It is now hardly more than of historical interest to look back at the operations of the Texas Railroad Commission that pro-rationed Texas production. It was established in 1891 to supervise the railroads. Since much oil was moved by rail, it had the ability to also regulate oil supply. Rockefeller, after all, had used rebates on the railroads and his ownership of tank-cars as a means of monopolizing the market. When the oil price collapsed in the Great Depression, the Governor of neighbouring Oklahoma, none other than "Alfalfa Bill" Murray, declared an emergency and sent in the State Militia to shut down the oilfields until the price improved. His action was followed by the Governor of Texas, who closed the East Texas fields. When the medicine worked, and prices did begin to recover in the following year, the Railroad Commission was given stronger powers to curb production by allowing operators to produce for only so many days a month. There was, however, widespread cheating and smuggling.

In 1933, President Roosevelt appointed Harold Ickes to be Secretary of the Interior, with a mandate to bring order to the oil industry as his priority. He resolved to reduce the country's production by three hundred thousand barrels a day, and to secure a workable system of allocation between the producing regions. Within four years, he had substantially succeeded in making a national strategic policy work, even in a place as capitalist as the United States.

Fig.11-1. The first train carrying oil from Oklahoma

Pro-rationing continued until about 1971, by which time US production had peaked, and the country needed all it could produce, meaning that there was no more surplus capacity to control. It had little to do with investment, technology, supply or demand, and everything to do with the depletion of a finite resource. It is a cautionary tale now being enacted on the World stage.

THE POST-WAR WORLD

Since the United States was the dominant producer in the inter-war years, the activities of the Texas Railroad Commission had a global impact, controlling the price of the marginal barrel. But after the Second World War, things changed with the colossal growth of overseas production from Venezuela, North Africa and the Middle East, the latter rapidly emerging as the dominant force.

Now, the mantle of regulator fell upon the seven major international oil companies, known as the "Seven Sisters". Five were American; one was British; and one was Anglo-Dutch. These huge integrated enterprises controlled everything from the wellhead to the forecourt, and despite their competitive relationships were sufficiently monolithic to exert a practical unified control.

As peace returned, they had a pricing system based on the marginal barrel in the Gulf of Mexico, from which differing freight costs were discounted. Although the marginal barrel was in the Gulf of Mexico, it was not the cheapest barrel, as production costs in the Middle East were much lower.

Discounts were allowed against the nominal Gulf of Mexico price. The major oil companies nevertheless accepted this marker, and commonly exchanged oil with each other around the world to meet their individual marketing needs, paying a cash adjustment to cover only the freight charge. Other companies were at a disadvantage because they had to pay full list price plus freight.

The companies were at the same time under pressure from the producing governments which wanted the maximum revenue they could get. The national affiliates of the major companies in the producing countries paid a royalty, normally equivalent to one-eighth of their production, as well as tax on profits. Since most oil was traded between the affiliates of the same company, the parent was able to rig the pricing structure to its best global advantage, so it was difficult to determine exactly what profit was attributable to the production or sales in any particular place. It led to the concept of a "posted price", at which the companies declared themselves willing to sell oil to any comer on an arm's length basis. It was used as the basis for tax to the producing government, but the posted price was a tax reference that did not represent the real price of oil, which was known only to the companies as they traded it amongst their affiliates around the world. It is well to remember this when considering the historical oil price trends: actual prices were probably mainly less than reported.

As product prices fell with the growth of production in the years after the Second World War, the companies sought to reduce the posted price for crude to better bring it into line with the product market. Naturally enough, the producing governments objected as it reduced their revenues.

OPEC

Venezuela's oil minister, Perez Alfonzo, was particularly strong in his view that the companies were denying his country a fair share. He was resolved to try to form a "global Texas Railroad Commission" to pro-ration production, and he set about enlisting support from other producing countries. Success came on September 14th, 1960, when his efforts were rewarded in the formation of a new organization by which Saudi Arabia, Venezuela, Kuwait, Iraq and Iran agreed to cooperate to support the price of oil. It was known as the Organisation of Petroleum Exporting Countries, or OPEC. But even in its early stages, it proved rather ineffectual, because of differences of policy between its members, leading it to set up a secretariat in Vienna in 1965 to try to strengthen its executive function.

Middle East crude was being lifted on a posted price that purported to yield a fifty per cent take to the government: in reality it was often closer to eighty percent, since the posted price was above the market price. But it was not as simple as that because of the inevitable distortions of tax. Under double taxation agreements, the companies were able to charge the tax paid to the producer as an expense against taxable income in their home country.[2] In effect, the consuming taxpayer was paying the tax levied by the producer government. Furthermore, in some cases, the parent company was able to charge an unrealistically high price for the oil it sold to its affiliate in the consuming country, which therefore operated at a notional break-even or loss-making level, paying little, if any, tax. The scope for manipulating inter-affiliate profits and losses, and freight charges, was huge, providing the parent with much of its profits. While the companies did not make inordinate apparent profits for their shareholders, they were able to build up huge feather-bedded establishments, treating all costs as a deductible expense against tax somewhere. The German government finally moved to stop the abuses of the apparently loss-making marketing affiliates in its country, and began to levy tax on profit based on its own assessment of a deemed fair cost of imports, irrespective of what the affiliate paid its parent.

With the establishment of OPEC, the companies lost the ability to set "posted prices", which were now imposed by the producing governments. But nothing much else changed, as most oil was still traded by the affiliates of the same major companies as before. This arrangement lasted until the early 1970s when the national affiliates of the major companies were expropriated in the main production countries, or their rights sequestered. The expropriation itself was, naturally enough, motivated by a desire of the host governments to secure a higher revenue from their oil, although there were also nationalistic influences. They realized that most of the oil in their territories had been found. They were not overly concerned to lose the foreign investments for further exploration and development as they rightly concluded that *the past was worth more than the future*.

Trade on the surface continued much as before. First, the companies and, later, the governments, looked to long-term supply contracts as a form of security. But in fact much did change beneath the surface, and especially in the minds of the oilmen: they no longer felt in control of their main sources of production. It led to two reactions: to pump like hell; and to do whatever was possible to find alternative sources that they could control, even if such was much more expensive oil.

The search for alternative supply during the 1960s and 1970s was remarkably successful, as there was still much left to find, and the eternal pattern of "boom and bust" was repeated. Flush production from the new giant fields, found early in the new provinces such as Alaska and the North Sea, began to flood the world. Furthermore, the Soviet Union had brought in major new oil provinces during the 1950s, and was now beginning to export its flush production.

I remember sailing in Dublin Bay in the early 1970s. We would often tack under the stern of a rusty tanker at anchor, flying the hammer-and-sickle, which had been discharging oil for the Ringsend Power station. To witness this ship from behind the mysterious Iron Curtain was somehow impressionable at the time. It was helping to flood the world with cheap oil.

The Yom-Kippur War, and the decision on October 17th, 1973, by several Arab countries to apply an oil export embargo, was another turning point in the evolution of the market. The organization announced that its members would begin cutting production by five percent a month, and absolutely embargo shipments to the United States and the Netherlands, which had declared particular support for Israel.

For four brief years, OPEC was in command, and the World hung on Sheik Yamani's every word as he jetted in his burnouse from Riyadh to the Intercontinental Hotel in Geneva, making oblique comments and veiled threats. His power did not last, however, for OPEC's revenue surplus of 1974 had been transformed into a deficit by 1978. By then, the founder members had been joined by Algeria, Libya, Nigeria, Gabon, Indonesia, and Ecuador as well as other smaller Gulf producers. Ecuador, Gabon and Indonesia probably joined as a matter of prestige, since their relatively small exports were not very critical. Ecuador and Gabon have since left, and Indonesia can't be far behind having become a net importer. It has nothing to gain from belonging to an exporters' club.

It was a short reign because the demand for oil began to fall as a result of improved efficiency in the industrial countries, as exemplified by the jet engine and other factors. Demand for OPEC oil fell further in the face of competition from the new flush production elsewhere. Many of these finds were offshore, where the high investments prompted a rapid build up of production to secure a quick pay out.

By March 1982, OPEC, having discovered that setting arbitrary posted prices did not work, had no recourse but to cut production. Complex negotiations were held to resolve how to apportion the quota amongst the members: two key factors being population and oil reserves. Both were subject to exaggeration in efforts to secure higher quotas.[3] As discussed earlier, it transpires that some of these countries started to report *Original* not *Remaining Reserves*. In other words, they reported the total found, without deducting production. It explains why the numbers have barely changed since, and it also probably made sense to have a relatively fixed number to avoid difficult re-negotiation of quotas.

But in the meantime, new companies with particularly Libyan production to sell, were entering the European market. Unlike the established "seven sisters", they had no ready-made refining and marketing organization to absorb their oil. They started to trade product on the Rotterdam Spot market. This in turn put pressure on the local affiliates of the "sisters" who found themselves being undercut. Before long, they too began to behave like independents, reducing their ties with their parents and using their wits to buy and sell on the spot market.

During one baleful epoch of my life in the late 1980s, I was obliged to attend the Board meetings of a marketing company in Oslo. I well remember arriving to glum faces as the autumn cold began to grip. One said "last year, it was October 15th"; another chimed in "but the leaves turned early this year". What they were debating was when to start lacing their diesel with paraffin. They were no longer taking premier product from their own refinery, but making spot purchases of Polish oil, much of which was contaminated with water and solids. If it sat long enough in the tanks, the junk would settle out, but if not, there was a sort of freezing emulsion that clogged everything. At next month's meeting, the faces were even glummer. They had got it wrong, and an entire bus line in north Norway had been immobilized as the diesel froze. I was always worried when I saw one of their tank trucks approaching an airliner I was travelling on.

Before long, the spot market for product was extended to crude oil.[4] For the first time, prices became transparent, although the major OPEC producers tried to resist the spot market, which they felt undermined their sovereignty. Not only did they not sell on the spot market themselves, but they prohibited resale of their oil bought under contract. In general the spot market came to control the price, whether the OPEC countries traded on it or not. Even term contracts were commonly linked to spot prices.

In 1985, Sheik Yamani tried a new formula: the net-back, in which the producer was paid in relation to the price at which the refiner sold his product. It was a boon to the refiner who was assured of a satisfactory margin, no doubt further enhanced by imaginative accounting. Refinery output increased, and crude prices and tax revenues to the producers fell further. The system did not survive for long.

As North Sea production grew rapidly in the 1980s, prices again sagged: ultimately to a low of about $6 a barrel in 1986. The Middle East share was down to sixteen percent. The market was not, however, the only factor, for one of the causes of this collapse was related to President Reagan and Mrs Thatcher's resolve to bring down the Soviet Empire by economic warfare. They persuaded King Fahd to open the tap,[5] as already discussed in Chapter 4. Many of the US independents and several contractors faced bankruptcy as a consequence. Stripper wells in the United States were shut in.

It was another watershed: security of supply, which had caused much concern in the 1970s, ceased to be an issue. What mattered now were prices and price volatility. Price volatility is what makes money for the stockbroker, and in 1986 the market saw the increasing role of financial traders, familiar with the well known principles of paper transactions, futures and derivatives. They were called the Wall Street refiners, who traded in paper more than the physical delivery of oil. Oil became traded as any other commodity, like pork-belly futures, and Exchanges developed in New York and London to do so. The brokers needed a benchmark price against which to trade the various oils around the world. West Texas Intermediate[6] became the benchmark for North America, and Brent from the North Sea for most other transactions. In addition, Dubai crude had a role for Eastern Hemisphere trade.

A convoluted market evolved in which physical trade was supplemented by forward markets and futures. There was a wide variety of crudes and products with complex inter-relationships. The forward market provided a form of low-cost hedging, whereby the companies could minimize the risks of volatility. Brent Crude from the North Sea was the main marker, with other crudes and products being linked to it via differentials based on quality and freight charges. It was itself, however, a relatively small market, controlled by a few privileged members, and its days are now seriously numbered as Brent physical production declines steeply. Modern communications made it a highly efficient market, so that virtually any transaction could be hedged by another. Hedging to minimise risk was also accompanied by rank speculation having the opposite effect.

For the first time, prices became truly transparent, as they moved up and down, with the benchmark reflecting a huge pyramid of hedged future transactions. In large measure, the hedging became a mechanical procedure run by sophisticated computer programmes aimed at minimizing the exposure by buyers and sellers to perceived or computed market risk.

The industry will always have some physical stocks: much oil is tied up in transport in pipelines and tankers on the high seas which is a form of storage; and refiners cannot exactly match the changing seasonal demands for different products so they have to maintain some product in storage. When stocks are low, prices tend to be in what is termed *backwardation,* meaning that the price of physical delivery is above the price of future delivery, because a greater value attaches to oil available for immediate needs. The opposite situation, known as *contango,* arises when stocks are high.[7] This relationship naturally assumes no fundamental supply shortfall, being no more than a short-term response to stocks and market.

The large traders are laying off price risk without really speculating about what the future price will be. It is now being reversed as a shortfall in actual supply is perceived. A combination of high physical stocks and *contango* is the market's warning signal, meaning that judgment for the future rather than mechanical hedging is beginning to affect prices. The reporting of price also has become more efficient, with Argus and Platt's providing credible services, used in all sorts of transactions, and as measures of performance.

Prices continued to languish during the latter part of the 1980s until the Gulf War successfully supported them by the removal of about two million barrels a day, an outcome that poorly enforced OPEC quotas had failed to achieve. The Gulf War itself was a great boost to the new trading system. It was a precise crisis, to which a foreseeable positive outcome soon became apparent as Desert Storm got under way. At the height of the crisis, prices reached about $40 a barrel, but, with an end in sight, it was tailor-made for hedging on the futures market. The market gained credibility that it could enforce the performance of the contracts for which it was responsible.

There can be no doubt that it is an efficient and transparent market representing a great improvement over the murky world of inter-affiliate transactions of the past. But I am not sure that efficiency is the same as wisdom. I think that mechanical hedging effectively is an extension of today's environment, and leaves little room for prudent physical storage against the unexpected. Could the whole pack of cards implode? One can lend to the future, but one cannot borrow from the future if the tank is empty. The Futures Market says little about the actual future price of oil but it gives an indication of the relationship of present stocks and needs, and above all the perceptions thereof.

During the winter of 2004–2005, physical stocks were again relatively low as production capacity limits have been breached. Prices soared above $50 a barrel, retreated and then rebounded to over $55 in the past few days. Evidently there is nothing particularly magic about the actual level as traders and speculators test the limits of the daily range in their quest for profit. Already the mind-set comes to see $40–$50 as a norm, although it would have been almost unthinkable a few months ago when prices were half that high. Some attribute it to a temporary aberration related perhaps to events in the Middle East and burgeoning demand from China, but the reality is that it reflects the onset of the physical decline in supply. The underlying trend in prices can only be upwards, but there will almost certainly be also temporary collapses as demand declines in response to economic recession, itself caused by the high prices. It seems reasonable to envisage a series of increasingly severe vicious circles in the years ahead. At the moment, there is plenty of slack in the system in our affluent and very wasteful society, but it will become increasingly difficult to tighten our belts in the years ahead as essential needs, such as fuelling the tractor, can no longer be met.

Charting the short-term future is an unenviable task.

Crude oil prices since 1861			Crude oil prices since 1861		
US dollars per barrel			**US dollars per barrel**		
Year	**Money of the day**	**$ 2004**	**Year**	**Money of the day**	**$ 2004**
1861	0.49	10.08	1898	0.91	20.17
1862	1.05	19.42	1899	1.29	28.59
1863	3.15	47.26	1900	1.19	26.38
1864	8.06	95.22	1901	0.96	21.28
1865	6.59	79.53	1902	0.80	17.05
1866	3.74	47.16	1903	0.94	19.31
1867	2.41	31.82	1904	0.86	17.66
1868	3.63	50.35	1905	0.62	12.74
1869	3.64	50.49	1906	0.73	14.99
1870	3.86	56.35	1907	0.72	14.25
1871	4.34	66.91	1908	0.72	14.79
1872	3.64	56.12	1909	0.70	14.38
1873	1.83	28.21	1910	0.61	12.08
1874	1.17	19.10	1911	0.61	12.08
1875	1.35	22.70	1912	0.74	14.15
1876	2.56	44.38	1913	0.95	17.74
1877	2.42	41.96	1914	0.81	14.93
1878	1.19	22.78	1915	0.64	11.68
1879	0.86	17.04	1916	1.10	18.66
1880	0.95	18.16	1917	1.56	22.53
1881	0.86	16.44	1918	1.98	24.36
1882	0.78	14.92	1919	2.01	21.53
1883	1.00	19.80	1920	3.07	28.39
1884	0.84	17.26	1921	1.73	17.92
1885	0.88	18.07	1922	1.61	17.80
1886	0.71	14.58	1923	1.34	14.55
1887	0.67	13.76	1924	1.43	15.49
1888	0.88	18.07	1925	1.68	17.76
1889	0.94	19.31	1926	1.88	19.68
1890	0.87	17.86	1927	1.30	13.87
1891	0.67	13.76	1928	1.17	12.65
1892	0.56	11.50	1929	1.27	13.73
1893	0.64	13.14	1930	1.19	13.20
1894	0.84	17.92	1931	0.65	7.91
1895	1.36	30.15	1932	0.87	11.80
1896	1.18	26.16	1933	0.67	9.57
1897	0.79	17.51	1934	1.00	13.82

Fig. 11-2. Crude Oil prices

Crude oil prices since 1861			Crude oil prices since 1861		
US dollars per barrel			**US dollars per barrel**		
Year	**Money of the day**	**$ 2004**	**Year**	**Money of the day**	**$ 2004**
1935	0.97	13.08	1976	12.38	40.17
1936	1.09	14.54	1977	13.30	40.53
1937	1.18	15.19	1978	13.60	38.51
1938	1.13	14.84	1979	30.03	76.40
1939	1.02	13.58	1980	35.69	79.99
1940	1.02	13.45	1981	34.28	69.59
1941	1.14	14.32	1982	31.76	60.72
1942	1.19	13.49	1983	28.77	53.30
1943	1.20	12.83	1984	28.78	49.78
1944	1.21	12.73	1985	27.53	47.18
1945	1.05	10.79	1986	14.32	24.14
1946	1.12	10.61	1987	18.33	29.83
1947	1.90	15.73	1988	14.92	23.27
1948	1.99	15.28	1989	18.23	27.02
1949	1.78	13.80	1990	23.73	33.54
1950	1.71	13.13	1991	20.00	27.11
1951	1.71	12.17	1992	19.32	25.41
1952	1.71	11.90	1993	16.97	21.74
1953	1.93	13.33	1994	15.82	19.84
1954	1.93	13.27	1995	17.02	20.75
1955	1.93	13.32	1996	20.67	24.43
1956	1.93	13.12	1997	19.09	22.14
1957	1.90	12.47	1998	12.72	14.80
1958	2.08	13.30	1999	17.97	20.16
1959	2.08	13.20	2000	28.50	30.96
1960	1.90	11.86	2001	24.44	25.75
1961	1.80	11.12	2002	25.02	25.77
1962	1.80	10.99	2003	28.83	28.83
1963	1.80	10.86			
1964	1.80	10.72	1861–1977 US Average		
1965	1.80	10.55			
1966	1.80	10.23	1945–1983 Arabian Light		
1967	1.80	9.96			
1968	1.80	9.56	1984–2003 Brent dated		
1969	1.80	9.07			
1970	1.80	8.56			
1971	2.24	10.23			
1972	2.48	10.96			
1973	3.29	13.68			
1974	11.58	43.38			
1975	11.53	39.59			

Fig. 11-2. Crude Oil prices (continued)

BP: AFTER TWO LIVES, THE OLD LADY CONSUMATED MARRIAGE

The title of this section in *The Coming Oil Crisis* suggested that the old lady contemplated marriage. Indeed she did, and we can now congratulate her for having successfully consummated it.

British Petroleum, or BP, began its life as the Anglo-Persian Oil Company which was formed in 1909 out of a syndicate created by William Knox D'Arcy (see Fig. 3-18), who had signed a concession for 480,000 square miles of the Zagros Foothills in Iran on May 28th, 1901. After initial disappointments, he struck oil at Masjid-i-Sulaiman in 1908. Under the influence of Winston Churchill, the British government took a 51% interest in the Company just prior to the First World War, both to secure oil for the British Navy and to establish a stronger British presence in the Middle East.

For fifty years, Iran was to be central to BP's enterprise. Discovery followed discovery, and the Company used this sure supply to build up markets around the world, especially in Europe and the British Empire. Much has been written and conjectured about its political role in Iran. The Iranians are almost united in seeing it as a dark force behind the scenes conspiring to deny them their proper rights and aspirations, being widely thought of as a front for the British Government. The truth is probably that it was doing no more than trying to hold on to its rights in a difficult and uncertain country. As the provider of the bulk of Iran's revenue, it was a hot seat to occupy.

The Company strengthened its position in the Middle East by taking up a share of the Iraq Petroleum Company when that was formed in 1928, followed five years later by securing a fifty percent interest in a concession to Kuwait. In both of these steps, it was much assisted by the British Government.

Whereas the major American companies grew up in a highly competitive environment and achieved their size by acquisition and merger in a country where there was a huge number of independent wildcatters that needed support to bring in their discoveries, BP operated under long-term concessions where there was no competition. The Iranian and Kuwait operations were staffed by BP technical personnel, and the Iraq Petroleum Company had its own staff, although using BP personnel on secondment. These were very large scale, single-minded operations, which did not in fact call for a huge exploration establishment. The geology of Iran was itself on a grand scale, with large clearly identifiable anticlines offering prospects, obvious to the most casual observer. As a consequence, BP's exploration staff was made up of quite a small group of people, who knew each other intimately. It was more like a club than a corporate hierarchy.

At the head of it was first G.M. Lees and later Norman Falcon, who died some ten years ago at the age of 92. From Cambridge, he had joined Anglo-Persian as a field geologist in 1927, and together with J.V. Harrison had made the pioneering geological map of Iran, which remains the bedrock of modern studies. He was a highly respected man who was elected a Fellow of the Royal Society in recognition of his scientific achievements. He was Chief Geologist from 1957 to his retirement in 1965. In those days,

Fig. 11-3. Norman Falcon, legendary Chief Geologist of BP

the position of Chief Geologist was a highly respected one with ready access to the Chairman. Today, if the position exists at all, it is probably concerned with training and vacation schedules: such is progress. Being spared the pressures of wheeling and dealing, so central to exploration in the American environment, BP's pioneering geologists were able to dedicate themselves to the science and practice of exploration. They had the rather 19th Century style of the dedicated and gifted amateur, but they were very effective.

Although BP relied during this epoch wholly on its Middle East production, it did maintain a modest interest in other exploration possibilities. J.V. Harrison went on roving commissions to Latin America, investigating Argentina, Colombia and Mexico. The Company tried to establish itself in Colombia in the 1920s, but withdrew without drilling. It mounted another abortive endeavour in Papua-New Guinea. It had little incentive to find more oil as its own production greatly exceeded its market, making it a substantial seller of crude oil. The other prospects it saw around the world must have compared very unfavourably with what it already had. It sometimes had local marketing incentives to make a gesture to exploration, not to mention local income against which to expense it.

This chapter came to an abrupt end in 1951 when the Company was ousted from in Iran, the foundation stone of its enterprise. Its directors were cast in an imperial mould, and probably assumed that at the end of the day they could rely on the British Government to protect their rights. Surely, Britain, which had only recently emerged victorious from a world war would not allow a major asset to slip from her grasp. She did, and probably not to the dismay of her ally then seeking a greater stake in Middle East oil.[8]

Now began the second successful life of BP. It embarked on a vigorous campaign of exploration throughout the world to replace what it had lost in Iran. Its supply situation was not critical as it still had Iraq and Kuwait, although they were to be lost too within twenty years. Norman Falcon led his team to find new fields, and he was remarkably successful.

One of the first new efforts was in Canada through the acquisition of Triad, a small company with a large land-holding. BP's team was led by Dan Ion, who incidentally had strong interest in world energy resources and their depletion,[9] and A.N.Thomas. It was a new experience for them to work in the highly entrepreneurial environment of North America, making deals and trading rights, which probably sharpened their commercial wits. It also brought them to investigate the Arctic, which thirteen years later was to lead to the remarkable Prudhoe Bay discovery in front of the Brooks Range of Alaska. With reserves of about 13 billion barrels it is by far the largest field in North America.

The story of its discovery makes interesting reading. BP, again with its Iranian background, at first concentrated on the foothills of the Brooks

Fig 11-4 Peter Kent, Harry Warman and Alwyn Thomas, credited with the discovery of Prudhoe Bay

Range, and having teamed up with Sinclair, as an American partner, drilled a number of dry holes in this province in the mid 1960s. It then turned towards the less disturbed platform to the north finding a huge structure at Colville, but the results of drilling were discouraging. The structure was an early uplift and lacked a reservoir development, a feature nicely called a "bald high". Sinclair decided to pull out, but a combine of Arco and Esso turned attention to the region, approaching BP to see if it would sell, which it nearly did. Interest was focused on a structure at Prudhoe Bay which was smaller than Colville but with better prospects. Arco and Esso made a successful high bid for the crest in a Lease Sale that was held in 1967, whereas BP submitted lower bids for the entire structure, ending up with the flank. Arco realized that it was a critical test of the Alaska province and persuaded all the companies to contribute to the cost of the well in proportion to their holdings. The well was drilled on the crest and tapped the gas cap of a large structure, but most of the oil lay on the flank in BP's territory. BP's development of the field was another remarkable pioneering endeavour, reminiscent of the Company's early days in Iran. Credit for the discovery goes largely to A.N. Thomas, Harry Warman and Peter Kent. It was a remarkable coup for an outsider to find the largest field in North America, the home ground of the major US companies.

Another venture was launched in Libya in the 1950s, where one of the obstacles were the minefields left over from the War. Whereas many companies concentrated on the Sirte Basin near the coast, BP headed into the interior, where it brought in the giant Sarir Field in 1961. With reserves of 4.5 billion barrels, it is the second largest in Africa.

Fig. 11-5. Sir Peter Kent, whose early work led to Britain's small onshore fields

The Company also joined forces with Shell to explore respectively East and West Africa. The latter soon yielded major discoveries in Nigeria. Other exploration ventures were conducted in Papua-New Guinea, Trinidad, Abu Dhabi, Australia, Algeria, Senegal, Colombia, Sicily, Switzerland and the United Kingdom itself, of which some were successful. The Company continued however to be a victim of the imperial past of its mother country, being driven out of Libya and Nigeria for political reasons unconnected with the Company itself.

North Sea exploration commenced in the 1960s to follow up the trend that had given the giant Groningen gas field in Holland in 1957. BP naturally took up a key role, making the first offshore discovery in British waters with the West Sole gasfield. That was in turn followed by other major discoveries including the giant Forties field. The Company was also active in Norway, but rather as a late starter having been put off by the outrageous terms of the Fourth Round of 1979. It farmed into the licence that yielded the Ula and Gyda Fields, and it was a partner with Shell in the Draugen Field off mid-Norway. The most recent and somewhat unexpected discoveries West of the Shetlands cap the list of a very successful exploration campaign.

In charge of exploration during much of this period was Sir Peter Kent, and the achievements were made by what was still quite a small closely-knit team, although eventually a more American style emerged with faceless committees, economists and bureaucracy.

Mrs Thatcher's policy of privatization in the 1970s allowed BP to buy out the former state company, BNOC, further strengthening its position. The Government also disposed of its fifty-one percent shareholding. There was a fear that this might have been acquired by Kuwait in an ironic twist of fate, but that attempt was thwarted, and the shares were widely placed, many being taken up by American investors. The investment bank Goldman Sachs was left holding a rather large block of shares, which turned out to be an excellent if unintended investment.

One colourful incident was the unheard of dismissal in the late 1980s of the Company's Chairman, Bob Horton, a determined man with a dramatic voice who risked borrowing to pay the dividend in the conviction that oil prices were about to rise. They didn't, and after a generous handshake, he went on, surprisingly, to run British Rail.

Having flirted with Statoil in a global "alliance" for exploration cooperation, an interesting third life for BP was in the offing with a possible marriage with the American giant, Mobil, which was never a strong explorer, but had a huge global market. An article,[10] entitled "oil buccaneer", described how Mobil was buying high risk upstream oil and gas interests in remote places as what sounds like the last throw of the dice in an attempt to replace its declining production in Indonesia which provided half of its upstream profits. It was a telling insight. But in the event, Exxon snatched Mobil at the altar steps, and BP turned to Amoco, a singularly unsuccessful explorer, described in Chapter 5. Indeed, she indulged in a degree of polygamy, taking Arco too.

She was not alone: Chevron merged with Texaco; Exxon with Mobil; Total with Elf and Fina, leaving only Shell as a surviving stalwart spinster, trying to find oil in the barren ground. The once mighty Seven Sisters have been reduced to four. As the opportunities dwindle, and their throughput declines, there is room for fewer and fewer players.

BP has become substantially a financial institution, sharing a Chairman with Goldman Sachs, on whose board its Chief Executive also sits. It once led the industry with its early warnings about depletion, but now prefers facile imagery in the best of financial traditions. Indeed, it goes so far as suppressing scientific papers by its technical staff, some of whom have been sufficiently public spirited to reveal the true position.

The change of posture could have serious strategic consequences for Britain in the years ahead, when the country might very much want to have a national company of this size to meet the challenges of the times. Think too of poor Ireland where the valiant efforts of its national company Aran were sold down the river to Statoil. The global market may soon have much to answer for.

NOTES

1. The price of oil is heavily influenced by the marginal barrel, that is to say the often relatively small amount by which the supply is above or below the demand.
2. Nasmyth J., 1996, gives an excellent review of the period 1945-1985.
3. Barkeshli E, 1996, an Iranian official, confirms that reserves were exaggerated in the "quota wars".
4. See excellent articles by J.Rieber for a fuller discussion of the evolution on markets.
5. Schweizer P, 1994, and R.V.Allen, 1996, describe the C.I.A conspiracy with Saudi Arabia.
6. West Texas Intermediate is a posted price which did not match realised prices, as listed in Figure 11-2.
7. See Long D., 1996, Verleger P, 1996 for useful discussions of stocks and the futures market.
8. See, Kinzer S. for am account of how the CIA engineered a coup bringing down the Iranian government, which paved the way for the so–called Consortium, by which American companies were able to access what had been BP's exclusive concession.
9. He published a valuable book in 1978 on the subject.
10. Economist, 30th November 1996.

Chapter 12

THE EXPERIENCE OF NORWAY
Was oil a blessing or a curse?

THE MILLIONS OF viewers around the world, who watched trolls emerge from clouds of freezing mist at the ceremonies marking the 1994 winter Olympic games at Lillehammer in Norway, must have thought that this was a fairy-tale country. Indeed, anyone who has visited, or still more lived there, will agree that here is an earthly paradise. That is exactly what I thought after living in Norway for ten years. It was a marvellous place in almost every regard. Its natural beauty is truly outstanding: from its mountains to the fir-clad islands and channels of its rocky, protected coast. Its towns too with their neat, white clapboard houses and tidy gardens, with the immaculate Volvo parked outside, were pictures of quiet unostentatious dignity. But most of all, it was the fine people with their striking pure features, reserved manner, stalwart and simple characters. They are, or at least were, almost universally honest. My secretary, Anita, once dropped a gold watch on a train and advertised its loss. A few days later the finder telephoned to tell her at which police station it was to be found. I am, or rather was, an enthusiastic admirer of Norway. But things have changed, thanks to oil-based affluence, and not really for the better.

Every country faces a dilemma over its policy towards developing natural resources, especially oil: *who* will benefit and *when*? This generation or the next? So far, most interest has attached to the issue of *who?*, but at the end of the day it is *when?* that matters more. Apart from *who?* and *when?* is also the small matter of *by how much?* Let us therefore zero in on Norway to see how it dealt with its oil industry,

Fig. 12-1. The beautiful coastline of Norway

Fig. 12-2. The fine featured Norwegians became victims of their success

whose arrival came as a great surprise to everyone. They started right but eventually were misled by short-term money when they began exporting their birthright at a rate only surpassed by Saudi Arabia. How were they misled? It is instructive to investigate this important oil country. No criticism is implied: most other countries are much worse when it comes to protecting their resources.

HISTORY

Norway is a mountainous country on the western seaboard of Scandinavia, populated by some four million people. Life was difficult on this rocky coastline that extends for 2000 kms into the Arctic Circle. From southern Norway to North Cape is farther than to Rome. The natural conditions have left their mark on the people.

It is an ancient land, although the modern State came into existence only in 1905 after more than two centuries of domination by Sweden and Denmark. It is a sort of Ireland of the North. Independence sprang from a cultural and political flowering in the 19th Century, which mobilized a sense of fierce nationalism and common purpose, still manifested by an abundance of flagpoles in the country.

The sea and the mountains were its resources: the sea for fisheries and shipping; the mountains for hydroelectric power. Norsk Hydro, the largest private company, was formed with French capital in the early years of the last Century to produce artificial fertilizer, fixing nitrogen from the air by hydro-electrical power. That itself was a major economic discontinuity. The shipping business had humble beginnings during the 19th Century. Much of the fleet was run

Fig. 12-3. The Norwegian flag flies everywhere
Photo by Egil Eriksson

by owner-captains, who traded on their own initiative, sharing the proceeds with their crews. It was one of the foundations for the spirit of cooperation which is such a marked national trait.

From these beginnings grew a number of shipping dynasties: the Olsens, Bergesens, Smedvigs, Wilhelmsens, Uglands, to name the better known. They competed with entrepreneurial drive abroad, carving out for themselves profitable niches in world trade, yet at home they lived modestly, being very conscious that they were the breadwinners of their communities. The capitalist of the high seas changed his coat on entering home waters.

Fisheries, too, provided opportunities and hazards. Fortunes were made in good years, but unpredictable changes in currents and water temperature sometimes meant that the herrings did not to arrive in the fishing grounds on the appointed date. Fishing for cod in an open boat off North Norway in mid-winter was no fun, yet they did it through a blend of independence and inter-dependence that is such a national characteristic.

Norway was plunged into the trauma of War by an unprovoked German invasion in 1940, but not before a veteran curator of an ancient fort on the Oslo Fjord managed to cause the sinking of a battleship of the grand fleet by firing an old cannon, ironically made by Krupp. Much of North Norway was fired by the retreating Germans at the end of the War under their "scorched earth" policy. The disgrace of Mr Quisling's puppet government was redeemed by an active resistance movement.

The experience of War left an indelible mark on the country's subsequent history. Post-war reconstruction had to be closely controlled by the government. Control once grasped was not easily surrendered, and Norway remains one of the most intensely regulated countries on Earth. While regulated, there remains however an intense individualism: to reconcile these opposing threads is one of the conundrums of Norway.

Norway has had a long history of left-leaning Government because co-operation was more important than competition in facing the natural challenges of living in a land of barren rock, where the sun barely rises above the horizon in mid-winter. It was not however doctrinaire socialism built on envy and class-warfare as in Britain, but a much more positive version. Taxes became very high, but most people were happy to pay them. In fact, the actual burden was not as high as it seemed because of the facility to deduct interest on loans. It was an extension of control: everyone was in debt to the banks which also controlled their lives. There is a very serious office, called the *Ligningskontor*, which supervises tax collection under the logo of an unblinking eye.

They also had a thing about alcohol. It was a state monopoly to which people previously went with sheepish unmarked bags. There was however no shortage of drunks, one of whom, Oystein, a sailor back from the sea, lived next to us and had a party, lasting for several years, paid for by social security. It was not perhaps the best form of national cooperation, but he was at heart a nice chap all the same.

The nation's wealth grew in the twenty years following the Second World War. Poverty was eradicated, and high standards of living were attained. The average Norwegian found himself living in housing that would be the envy of the elite in most countries, and many had country cottages or pleasure boats as well. The banks probably owned them, but the people used them, and led happy simple lives, heading into the mountains with leather backpacks at weekends. They did not have much ready cash for a pint of beer, but their surroundings and lives were superb. It was a cooperative society in which everyone was supposed to have a useful role, which once recognized would be supported through thick and thin by community and government action.

How did oil come to this model land? It came from Holland, where, in 1936, Shell exhibited an operating drilling rig at an industrial fair in The

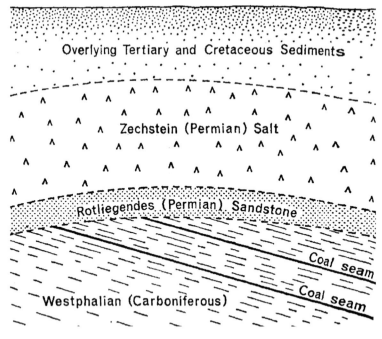

Fig. 12-4. Geology of the giant gasfield in Groningen, where gas from deep coal is sealed beneath salt

Hague. To everyone's intense surprise, it struck shows of oil, which stimulated exploration that resulted in several small finds. Another chance discovery was made in 1957, when a weekend communications failure led to an unintentional deepening of a well at Groningen. It fell into a giant gasfield in Permian desert sandstones, which no one had previously thought of as a possible objective for oil or gas. It did not take long to extend this productive trend offshore and into British waters, where more large gasfields were found.

By the mid 1960s, the industry was turning its eyes northwards, wondering if the same trend, which relied on the natural coking of deeply buried coal, could extend into Norwegian waters. This interest was greeted with incredulity in the country, where no one could imagine oil or gas beneath the stormy waters of the North Sea. But in 1963, Norway proclaimed sovereignty to its continental shelf with the agreement of the neighbouring countries,[1] and on April 13th, 1965, offered licences to explore and produce petroleum to the industry. The business was managed by the Ministry of Industry and Handicraft, whose very name illustrates how far Norway then perceived itself to be from the world of oil.

It was a successful offering in which seventy-eight blocks[2] were let under twenty-two licences. There were six major applicants: Shell; Esso; the Phillips-Fina-Agip Group; the Amoco-Noco Group; Texaco and Chevron (then Gulf), and the French companies, Elf and Total, in partnership with Norsk Hydro.

The first discovery was made by the Phillips Group in 1968 with a modest find, called the Cod Field, near the UK median line. Later in the same year, the Amoco Group drilled Well 2/11-1, which, although aimed at a deeper objective, unexpectedly found oil shows in the Cretaceous Chalk. This is the same rock as that of which the famous White Cliffs of Dover are built, and had not hitherto been considered an oil objective. A few months later, the Phillips Group brought in the giant Ekofisk discovery on a Chalk prospect in the neighbouring Block 2/4.

The remarkable combination of geological circumstances responsible for Ekofisk makes it almost one of the wonders of the world.[3] The Upper Jurassic source-rocks are at peak generation below it, and the oil, having nowhere else to go, migrated upwards along self-made fractures. The Chalk, which normally lacks adequate permeability and porosity to hold oil, just here had been deposited as submarine slumps, analogous to avalanches, in which the porosity had been exceptionally preserved. Lastly, early faulting and deep-seated salt movements gave closed structural traps to hold it. This remarkable

Fig. 12-5. The miracle of Ekofisk
Photo by Egil Eriksson

Fig. 12-6. Myles Bowen, who opened up the rifted Jurassic prospects of the North Sea

discovery, along with its satellites found later, hold about three billion barrels of recoverable oil; one-eighth of Norway's total discovery.

The Government had no particular policy towards oil in 1965, other than to find it. The terms were normal, with a one-eighth royalty and standard corporation tax on profits. Companies with adequate technical and financial qualifications were welcome, working either singly, as in the case of Shell and Esso, or as groups.

Another round of licensing was held in 1969, but did not deliver any particularly interesting results.

Meanwhile in Britain, across the median line, important finds were being made by the industry which had negotiated comparable rights from the government on similar terms. My friend, Myles Bowen, whom I had first met in Borneo in 1956, led his company, Shell, to test a structure close to the Norwegian median line, finding another giant field in Jurassic deltaic sandstones,[4] which was named the Brent Field.

Britain was then a socialist country bent on bringing just about everything under public ownership. Once the oil had been found, it was, in those distant days, soon regarded as a national birthright being exploited by undesirable capitalists and foreigners. A state oil company, the British National Oil Company, was established. In a disgraceful chapter of bad faith, the rights previously freely granted to the companies were clawed back under duress to give to this new State entity.

Already socialist Norway saw what was happening across the border, and decided in 1973 to set up its own State company, Statoil, with the idea of protecting the national interest in the Norwegian extensions of this new trend which had already been found in Britain. To its great credit, however, honest Norway did not renege on its previous contracts.

A large highly prospective structure offsetting the Brent Field had been identified on the seismic data. Although the fledgling State company was now in place, the government felt that it was not qualified to take on a project of this scale, and accordingly granted rights to a large group of companies, run by Mobil, with Statoil having a fifty percent interest. It was a great achievement for Mobil to be appointed Operator[5] of what was to become the North Sea's largest field, Statfjord, but it probably got its rights by agreeing to transfer its operatorship to Statoil at a later date, little thinking that the option would be exercised: it was.

This development coincided with the First Oil Shock when oil prices increased five-fold. Norway realized that it had become an oil nation in no uncertain manner. It moved to improve its administration of the business by setting up the Norwegian Petroleum Directorate in Stavanger and the Department of Energy in Oslo. *Directorate* was the operative word: it decided to do just that – direct and control. In many other countries, government supervision of the oil business is often

Fig. 12-7. Arve Johnson, behind a smiling face was a forceful man with global ambitions
Photo courtesy of Statoil

entrusted to retired oilmen, with a mandate to follow developments in a substantially passive sense. But in Norway, where virtually every aspect of life, apart from the offshore shipping business, was managed by the government, it was no great step to decide to run the oil business too. The Norwegian bureaucrat is used to being in charge.

Farouk al-Kasim, an Iraqi, who had trained in petroleum engineering at Imperial College in London and had married a Norwegian, was recruited to take charge. It must have been assumed that anyone hailing from the Middle East would know about oil. He was extremely effective. Working with him was Egil Bergsager, an imaginative and optimistic geologist from Haugesund. These two, under Frederik Hagemann, the Director, cooperated closely with the Ministry in Oslo, which was more concerned with the legal and political aspects. Together, they set about planning Norway's oil policy, aided by a staff of mainly young men and women, straight from university, who found the growing control of this enormous and expensive business a heady start to their careers.

At first, the momentum of the past continued with the third round of licensing, when various tracts were infilled without any spectacular results. The newly established Statoil began to recruit its staff, and to intervene forcefully under the abrasive leadership of Arve Johnson, who was the protégé of the Labour Party.

In 1979, came the Second Oil Shock when oil prices rose to almost $50/b. Norway reeled under the promise of new riches of unprecedented proportions. Meanwhile, new seismic surveys had been shot over the northern North Sea, and it became obvious that Norway had some superlative prospects with every chance of yielding giant fields.

THE GLORIOUS FOURTH

Against this backdrop, Norway moved to organize the Fourth Round[6] of licensing under new terms. And new terms they were.

The first move was to license a large prospect, close to the Statfjord Field, applying a national solution: 85% went to Statoil, and the balance going to the two other national companies, Norsk Hydro, of fertilizer fame, with nine percent, and Saga, with six percent. Saga, representing ninety-six shipowners and industrial companies, was the flagship of private enterprise in Norway, which while in eclipse was far from dead. The Directorate, however, was not delegating any of its powers even to nationals, and imposed as a condition of the licence that the group had to drill eighteen appraisal wells to test different structural compartments of what was a much faulted structure. Each well probably cost around $15 million: so the commitment cost over $200 million. It was worth it, for the giant Gullfaks Field was the outcome. It has reserves of almost two billion barrels, which, at the then oil price of around $30 a barrel, meant that a gross revenue of around fifty billion dollars would be earned before expenses. There was plenty at stake.

The government was however concerned about the impact of the huge floods of money that promised to flow ashore. It rightly recognized the great dangers that this could bring to traditional industry and the finely tuned Norwegian social structure. It very wisely decided to move cautiously, and hold the pace of exploration and development down. It further decided to strike the toughest bargain possible with the foreign companies. Norwegians soon became known as "blue-eyed Arabs". It was a time of perceived oil shortage and high prices: the companies had little option but to accept. The hard-nosed Texaco and Chevron did however pull out, and BP was a less than enthusiastic applicant.

The terms were truly outrageous, and could only have been countenanced in such unusual circumstances:

1. Statoil was granted a 50% interest in all licences, but its costs were borne by the other partners;
2. Statoil voted its full unpaid interest in the management of the venture, which with the support of the other Norwegian partners conveyed total control;
3. Statoil had the right to increase its interest to 70%–80% in success;
4. Heavy drilling commitments were imposed (in some cases with as many as five, six and even more *wildcats* being required);
5. Checkerboard licensing was organized whereby offset blocks were

retained for subsequent privileged allocation.

6. The government formed the groups, often deliberately putting together companies with differing technical views in moves calculated to maximize the management difficulties. (Normally, companies work together building up mutual understandings and relationships, and develop their knowledge for their mutual advantage, but this was denied them under a policy of divide-and-conquer).

Whereas the companies had invested huge amounts over years to assemble the data, knowledge and judgment that pointed to the prime blocks, Statoil was able to enter for free, being born with a sugar-coated silver spoon in its mouth. And like many spoilt children became rather aggressive and uncompromising in its privilege.

But all was not quite how it seemed in those days of national socialism: the costs borne by the foreign companies in carrying Statoil were in fact deductible as expenses against their own taxable income. This in effect meant that the government was investing huge amounts of national money, in the form of lost revenue, in a new state enterprise without having to seek parliamentary approval for the annual budget. Whereas the Minister of Health had to crawl on his hands and knees to extract budgetary support for a new hospital, Statoil had *carte blanche* to spend as much as it wished, free of control or constraint, in a hidden manner that virtually no one in the country appreciated. And spend it did. It was a form of daylight robbery: not against the foreign companies, but against the long suffering Norwegian taxpayer.

In command of this robbery were the mainly young administrators in the Directorate, the Ministry and Statoil itself. They were relishing the huge power they now wielded to insist that this reservoir be tested or that new draconian safety code observed.

Ironically, the enormous benefit that the foreign companies enjoyed in being able to deduct all the expenses of this extraordinary game was not always recognized, because exploration budgets were commonly administered apart from fiscal budgets to ensure that projects around the world were compared on equal technical merit. Norway soon earned for itself the reputation of being one of the most difficult countries in the world to work in, although in reality it was one of the best. They were strange times; but about twenty international companies hung in there, not wanting to be left out of what was perceived to be one of the world's last great oil provinces, as indeed it was.

Of course, Norway was Norway. The robbery was not perceived as such, and the robbers had the high-minded intent of protecting the national interest. The well-known themes of foreign exploitation and abuse by multi-national companies were flowing through the veins of the young

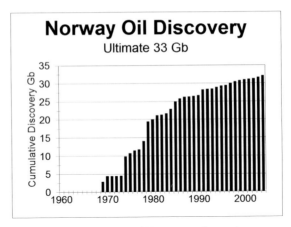

Fig. 12-8. Discovery of Norwegian oil over time

socialists. But, as I will explain, although the policy found a lot of oil, it did not protect the national interest in the long term.

In the short term, however, the Round was a resounding success as one giant discovery followed another. In this brief span of time most of Norway's oil and gas was found as illustrated in Figure 12-8.

The ball had started rolling, and it proved difficult to stop. At first, Norway and Statoil needed access to the industry's knowledge and expertise, and great efforts were made in the licence terms to provide for "technology transfer". It was a remarkably successful policy, and within a few years the dedicated and hard-working Norwegians had learnt it all. Actually, much of the business is not particularly high-tech: all sorts of humble people manage to drill oil wells, but money helps, especially if it isn't yours.

Along with this, a number of Norwegian shipping companies went into oil drilling and even manufacturing rigs. They include Fred. Olsen,

Fig 12-9 A Norwegian Seismic Vessel

Wilhelmsen, Odfjell and Smedvig.[7] The national oil companies gave them enormous support by awarding contracts, often on better than commercial terms. GECO, a seismic contractor, was established by several enterprising Norwegians, largely from the expertise of the American GSI.[8] It too prospered with contracts for the huge surveys that were shot over the extensive Norwegian shelf.

The Fourth Round in 1979 was Norway's golden hour. With hindsight, it may be asked if Norway might then have been wiser to say "thank you and good-bye" to the foreign companies, who had brought their knowledge and expertise, and had trained up the three national companies to high levels of efficiency. The foreigners could have produced the fields to which they had contractual rights: honest Norway would never renege on them. But the exploration of the remainder of the shelf could have been left to the three national companies.

They could have gone about it slowly and seriously, and they could have aimed to deplete their resources gradually, so that they might last as long as possible. Furthermore, they could have used the oil revenues cautiously, such that the economy would not become too dependent on oil. Remarkably enough, it would not have cost the nation any more, because, as already explained, the foreign contribution to exploration was largely deductible from the taxes that were levied on past production. Levied they were and at a high rate providing a marginal rate of 85%. Not only were the foreign companies able to deduct the expenses incurred in Norway, but they were able to claim the cost of the whole hierarchy of time-writers in regional offices, home offices and research establishments who reckoned they were in some way engaged in work linked to Norway, however peripherally. I well remember having to sign off these monthly inter-affiliate charges, although most of my time was spent trying to limit the damage to our endeavours arising from interference by the regional office in London. The taxman is an accountant. He is satisfied if the money was spent under the rules, but is not qualified to ask why it was spent.

Had a national policy been adopted, the three national companies, and especially Statoil, would have been responsible and answerable for their exploration budgets. But the government did not see it this way, and instead set forth to licence the rest of the shelf in successive rounds. Statoil continued for many years to have a mandatory 50% interest in everything, but the more excessive terms were gradually relaxed. Statoil's share of abortive exploration drilling probably cost the unconscious taxpayer at least ten billion dollars in lost revenue. Statoil was not to blame since it was forced to take a 50% share in all licences, whatever its own assessment of the prospects, and the government imposed the drilling obligations. A further flaw with this arrangement was that the foreign companies had a contractual right to exploit whatever they found, even though their investments were in fact substantially paid for by the Norwegian taxpayer. It meant that Norway had lost control of the pace of development.

EYES TURN NORTH

The Fifth Round in 1980 found Norway's two other productive basins: Haltenbanken off mid-Norway and Hammerfest off North Norway. The outrageous Fourth Round terms were still in place, and surprisingly, the foreign companies accepted them, not wanting to be excluded. There is a sort of mindless momentum in the operation of major oil companies: if the Norway case had been successfully argued for the Fourth Round, the case for the Fifth was plain sailing: all the maps having been duly massaged to offer the same grand prospects. Besides oil prices were still high, generating large tax obligations, against which the exploration expense could be charged. In fact, the risks of the northern shelf were infinitely higher than applied to the North Sea; and the world circumstances of perceived shortage after the Second Oil Shock in 1979 were changing. Evidently, some of the foreign companies had come to understand the Norwegian system, and to realize just how much they were being subsidized, especially when taking into account the risk factor.

Haltenbanken turned out to be a fair success, delivering about ten billion barrels of oil, making it one of the world's last major *Petroleum Systems* to be found. The Hammerfest Basin never looked too promising, with the zone of oil generation being narrow and somewhat removed from the main structures. It proved to be gas-bearing, but gas in this remote location would not become commercial for a very long time to come.

CLIVE NEEDHAM

Fig. 12-10 Clive Needham

Q Clive, your personal experience has mirrored what the industry is going through now, and will increasingly do in the future: careers as such are over as everyone becomes a contractor or a consultant. But, first could you say something about how your career evolved before you made the change.

A *I do agree that it appears that lifetime careers with one company are over, but this is true of many* industries in the 90's. It is not exclusively a facet of oil field life. We have all become contractors now – even within oil companies: some have gone to a purchase and supply basis with internal customers.

The change is for the worse, in my view, because the romance has gone out of the oil business, particularly on the exploration side. I admit that a younger geologist may disagree and say that I am displaying alarming signs of an exponential slide into old bufferism. I'll leave that for the moment and recap briefly my career, which may explain the view that my early days in the business had their romance.

I graduated, none too auspiciously, in 1974 from Portsmouth Polytechnic. I only mention this because in those days the major oil companies did not recruit direct from such institutions, and so it was necessary as a first step to start out as a contractor. I worked for Geoservices, a French mud logging company (still in existence, unlike many of the oil companies that were around at the time), known those who either worked for it or employed it as "Frog Log". Mud logging in itself is not a glamorous activity. In essence, it is the means by which a well is monitored; its progress charted; and what it is finding – oil, gas or often nothing – determined.

For a new graduate, the job offered variety, travel and exposure to a wide range of interesting individuals. It provided a very good background and insight to the business: it is a good idea to do it for a while and then move on.

So, where is the romance? Seeing the Rockies from the top of the Regent 25 rig in Alberta, nearly 50 miles away across the top of the never-ending pine woods. Being asked in the mud logging unit by the company representative: " Do you guys have any beer in here?". "Of course not", we reply, horrified innocent lads that we were. "Well, you'd damn well better have some the next time I come to the unit". A complete travesty of all known Company regulations, but we were fifty miles from the nearest road.

Mud logging is not a career, it is experience. I joined Amoco in fairly typical manner for those times (1978). They had two rigs drilling and had lost most of their wellsite geologists through transfer and natural wastage. They were short staffed. I was working on their rig at the time and knew their wellsite procedure. So they took me on and I worked at twice the rate of pay, with half the time offshore that I did as a Frog Logger.

Working as a Wellsite Geologist for Amoco was great fun. I loved it because I had a lot of autonomy on the rig. In those days, you worked in co-operation with the Company rig superintendent, but you only reported to London (by Telex, at 6.00 am every day). It was genuinely good fun. What spoiled the job was being transferred out of the field into the central London office and finding oneself reporting to pasty faced young lads who had never seen an oil rig but cut a dash in a suit at executive committee meetings.

So it was that I came to answer an advertisement by Getty Oil in the Sunday newspapers. I was interviewed at the then Getty offices in the grounds of Sutton Place, the Elizabethan home of the late John Paul Getty. This made a big impression, driving in through the wrought iron gates and up the winding drive, crossing the humpback bridge over the Wey river (complete with traffic lights) to the house itself. It was here that J.P. lived with his 'companions' and even a lion – and ran a multi-billion dollar business.

Getty was a great company to work for. Small in the UK, but with enormous income, courtesy of the Piper, Claymore and Heather fields and $30/b oil. Our job was simple – go out and find more of the stuff anywhere in Europe and Africa.

I worked for a splendid Chief Geologist from the US, Barry Faulkner, who was as rounded a character as he was in profile. Early one morning, he suspended in the office of a geophysicist colleague a tyre on a rope with a bunch of bananas attached to it – to indicate what he thought of him. I got the blame for this incident, but survived.

Well, we did look everywhere. After an early spell on the North Sea, I was put onto new ventures in the Mediterranean and had to travel to Spain, Malta, Sicily, Tunisia and other such spots in the pursuit of new acreage to explore. It certainly was great fun and a privilege; and it provided a huge range of experiences.

The business was different then. Note that I am not saying it was better in the sense of how effective it was, just that it was much less of an industrial process than it has become. Many people will say that is exactly why it had to change – the high and escalating price of oil hid the many inefficiencies and the poor finding record of many major companies, once the giant fields had been discovered in the North Sea. The price crash of the mid-eighties plus the hunt for oil on Wall Street rapidly led to a shake-out that has produced what we have today – a shrunken number of companies who have had to adapt to survival with oil prices below their real value of twenty years ago.

My time with Getty ended as Chief Geologist for a new venture in China, a grand title considering I was also the only geologist in the local office for a while. Getty had taken a view that they wanted to be in the first round of western companies to enter China and hence have the best opportunity to pick up the most prospective acreage. Unfortunately, geology doesn't work like that, and what had been statistically correct in the North Sea did not prove to be the case in the South China Sea (as for example BP learnt to its cost: four licences and seventeen straight dry holes as I recall).

The competition for acreage was such that Getty was forced into a joint venture with Japex for the licence they got (15/33) and a joint-venture company was formed with staff from both parents. The Pearl River Oil Operating Company (see what I mean about romance) operated from the nineteenth floor of the White Swan Hotel in Guangzhou and I had a Japanese boss. I rapidly learnt the cultural differences between East and West.

The critical factor was the interest, the excitement perhaps, of being *involved in such a new venture in a place like China. That it came to an early end for me was indicative of the changes that were overtaking the industry at this time. Oil was cheaper and easier to find on Wall Street. You just hired a bunch of suits in a merchant bank and off exploring you went – just down the road. Texaco decided to capitalise on the well documented difficulties of Getty Oil at board and family level and appeared as a late entrant to a proposed merger with Pennzoil. There is no need to dwell on this further, except to note that Texaco's total market capitalisation a couple of years later was less than the sum it paid for Getty Oil. I felt that the time had come to go.*

Q We then found ourselves together building up a new exploration office in Norway. I think we succeeded beyond all expectations. We turned around a company that had failed to secure new licences for many years, and put it at the head of the pack, even securing an operatorship, which is no mean feat in Norway. You were responsible for the technical work which was a key element in the growing success. What do you think were the most impressive technical developments of those days in Norway?

A *We did indeed find ourselves together, working from above a garage in Hillevagveien, such was the commitment that Petrofina had put into re-investing its spectacular returns from Ekofisk. Petrofina S.A. was nick-named "the Anonymous Society" by the Norwegian authorities having had such a low profile in the country. We set about the task of improving the technical reputation of the Company and establishing credentials to justify the award of the sort of acreage that we aspired to. We did this by starting to compete on the basis of being 'the best non-operator in Norway' (should I acknowledge the authorship of this golden strategy – you often used to observe that I hadn't quite mastered the art of fawning!). Some thought this a negative strategy to adopt, an acceptance of always being second-best. However, it proved itself correct because the competition among the non-Norwegian companies for operatorship was so intense and it was better to deliberately aim for the No.2 spot in the most attractive blocks.*

The major tool we employed was regional geochemistry. There were a number of reasons for this. Firstly, much of the new ventures activity was aimed at large tracts of open acreage in the Barents Sea and off Mid Norway. At that stage of exploration in the mid-eighties, it was essential to be able to screen large areas in order to reject those of low potential before focusing the more expensive evaluation effort on the best targets.

It is a truism that oil can be found in areas of with wildly varying structural styles and reservoirs, but it is never found if there is no source-rock. A source-rock is any rock formation that contains sufficient organic carbon to be able to generate and expel hydrocarbons under the combined effects of temperature and time. If oil is not generated, there can be no discoveries.

As source-rocks need to be volumetrically significant to generate a sufficient hydrocarbon 'flux' for migration and entrapment to occur, it makes sense to search first in unknown areas for evidence for the presence and effectiveness of source-rocks. In this way, large areas can be coarsely screened before looking for the correct combination of coincident source-rocks, reservoirs, structures and seals.

At that time, major technical advances were being made in two areas of research that combined to provide the oil industry with a very powerful tool for acreage evaluation. The first was the growing understanding and everyday use of geochemical modelling to look at the maturation path undergone by a source-rock in the course of its geological history. The second was the new understanding about the way in which the crust of the earth behaves when subject to the primary tectonic stresses that have shaped the oceans, continents and, critically, the sedimentary basins.

To take the second first, the understanding of extensional tectonics, literally the stretching of the continental crust in response to the separation of continents and the opening of oceans (e.g. the North Atlantic) had led to a general recognition of the relationship between the degree and rate of extension and the formation of the major sedimentary basins in which the bulk of the worlds oil is found. What was also recognised was the relationship between basin subsidence and the heat flow through the earth's crust from the thermonuclear heat of the interior to the surface.

A seminal paper was published in 1980 by Douglas Waples, acknowledging the earlier work of a Russian scientist, Lopatin, in relating the maturity of coals to both the temperature to which the coal bearing strata had been subject and (crucially) the duration of exposure. Waples extrapolated this thinking to hydrocarbon source-rocks (which coals are too) and stated that there is a good empirical fit between the behaviour of source-rocks under thermal stress over geological time and the second order reaction equation of the chemist Arrhenius. This simple exponential relationship has allowed a whole generation of geologists to undertake burial profile analysis for various points in a sedimentary basin, and pronounce with reasonable accuracy on the likelihood of hydrocarbon generation (and even assess the likelihood of gas versus oil) whilst knowing little more than the outline stratigraphy of the basin and a generalised figure for the temperature gradient (often taken to be 33–35°C per kilometre of burial).

The regional exploration geologist is looking for a "well cooked" (note, not over-cooked) source-rock with the additional appetisers of reservoir and structure to contain the oil.

We (and all our competitors) therefore spent much time and effort on the understanding of the burial and thermal history of the relatively unexplored mid-Norway Atlantic margin and the very poorly understood Barents Sea.

It was also necessary to play the political game of 'supporting' Norwegian R & D and even industrial efforts. They were really little more than disguised subsidies by the State, as the offshore industry by this stage comprised only serious taxpayers who could recycle exploration related R&D to their profit and loss accounts. Thus many a wild idea was funded at the expense of the Norwegian exchequer. I seem to recall we had a few of our own, such as a heart foundation for corpulent oilmen. All was done in the name of providing an individual character to our applications for the prime acreage. Still, it was all good fun sifting the ideas that came forward and trying to judge their impact on the NPD and Ministry.

We succeeded at this game too. We got into our first choice of acreage in the Barents Sea, the ill-fated PL 136, Blocks 7219/9 and 7220/7. I remember this because it was the biggest disappointment of my technical career. A beautiful, giant (one billion barrel plus) prospect made up of a clear rotated fault-block with Jurassic reservoirs and the all important Upper Jurassic source-rock adjacent to it and modelled to be in the oil generating "window". I remember being in your office when we got the LWD resistivity curve faxed in from the rig over the top of the reservoir. The resistivity of the reservoir was lower than the overlying shales: no oil. Even worse, it turned out that there had been oil, as residual oil was present, and that the structure had been filled and then it had leaked at some point in Late Tertiary time owing the late stage uplift of the western Barents margin, about which little was then known. The structure had previously been up to 1,500m deeper than its present depth of burial. When it had been uplifted, the relative reservoir pressure ruptured the seal and the oil migrated out again. It was rather like finding your well baited trap had caught your prey, he'd eaten the bait and left a rude note as he got out again.

We came to regard the endeavour as if it was ours and didn't curtsy enough to our masters. They resented our independence and success, and eventually pitched us out. At first, I was relieved to be free of the stress but then became unreasonably resentful at the injustice of it all. It left a scar, but I turned to consulting which in fact is a much better life. How did you react.

In many ways, Norway was the high point of our careers. The problem is, to really succeed you have to be committed and you have to be have individual ideas. This can often result in a clash of ideology with a conservative and nervous management. Our management in distant Brussels awoke to the costs of the Barents Sea exploration campaign (about $25 million per well dry hole cost) and the fact we had failed to find anything other than residual oil. Then the finger started to be pointed and resentments recalled at what was seen as a cavalier approach to exploration and, particularly, to them. There is no doubt there is truth in that, but a cavalier approach, a bit of flair, was required to make us competitive with the likes of Conoco, BP, Mobil and Shell. Dull conformity was not going to work. Our failure was that of the Barents Sea itself – we implicitly said it was going to work and we failed to deliver the 'expected' success.

The first casualty of this was you. You were initially moved sideways to the role of 'Consultant' and then, wisely, decided to resign with pride intact. I was actually promoted initially, to be Exploration Manager. I was then moved to a new venture office in Ho Chi Minh City, Vietnam, where it was my turn to fall. The failing this time was telling the truth. Giving an accurate assessment of the exploration potential of the acreage that had been acquired in the Gulf of Thailand. Telling management the average reserve size of the identified prospects was one fifth of what they had been 'promised' proved a great disappointment. Though if the Exploration Manager cannot give an honestly optimistic view, I think they should have just put someone in the job who would say yes.

Q You are also back in the business, this time as a contractor. You face the same issues but from a different perspective. How would you sum up the main differences.

A *I initially did not want to have anything further to do with this silly game – but it is addictive and I soon realised it was the only game in town that I had any capacity for. I started consulting for Western Atlas, a major oilfield service company, and then joined them as a employee. I believe the very fact that Western recruited me, a non-specialist in their areas of expertise (dominantly wireline and seismic data acquisition and processing) reflects the structural change in the business. Ten years ago, they would never have wanted me and I would not have wanted to work for them. They provided data for oil companies to work on and nothing more. This was a stable, symbiotic relationship that everyone understood and was happy with.*

Two factors altered the arrangement: the restructuring of the oil companies and the greater complexity of the data now produced by service companies like Western. I can give two examples: 3D seismic data; and the workstations that integrate the seismic data with the well logs that depict the acoustic or resistivity properties of the formations in the wellbore.

Oil companies had decreased their technical staffs at a time when the volume of data they were receiving was increasing exponentially. There was only one outcome: they had to come to their suppliers for not only data, but also an interpretation of the data. Western Atlas increasingly find themselves asked for a technical assessment of the geology of the structures over which they have acquired data. Service companies are nothing if not aggressive and keen to expand their business, with the result that Western started to recruit people with oil company backgrounds who could speak the new (to the service company) language of interpretation.

It works quite well. An interesting facet is to see how the service companies cope with the new world of the oil companies for whom they now work. I will name no names, but some (who will thrive) have restructured themselves because they believe in it, are committed to and genuinely want to have a successful partnership with their service suppliers. They are willing to recognise that they do not have a monopoly on expertise and will accept the technical support of their supplier. Others, from the gadarene end of the oil company spectrum have cut staff, mouthed the words fed to them by their management consultants and have then carried on as before. They are failing to gain the benefits of downsizing and partnership with their service companies; and their staff appear increasingly demotivated as they have less and less people to do the work. Their success rates will fall and they will decline.

The main threat to the service companies is this failure of the oil companies to recognise the true meaning of their own restructuring. They want innovation, support and R&D to be supplied by the service sector. They make a great song and dance of concern for quality, safety, the environment and service over cost when seeking tenders for work. Yet they invariably seek to trim costs below the level of profit that will allow innovation to be funded.

As a species, oil companies do not yet fully comprehend the real requirements of partnering-alliancing for the future. They want the benefits, whilst preserving the master-servant culture of the past. There remains some way to go before mutual benefits will routinely flow from the new arrangements. Otherwise, the oil companies should staff up and do the job themselves – if the stock market will let them take on those costs.

All the evidence I can find suggests that the North Sea production will peak in a year or two and then decline rapidly. Every effort will be made to extend plateau production, but it simply means that the end when it comes will come quicker. There may soon be more work to do in removing exhausted platforms than in installing new ones. Does that make sense from your vantage point?

I did some work in Norway during the mid-1990s and it was a surprise to observe how far production decline had to go down the exponential decay curve before prompting a reaction. The old story of the dinosaur with a brain so small that it took seven minutes to react to a kick up the backside came to mind.

If stable (falling in real terms) oil prices are to remain, then most oil companies can face nothing but a prolonged decline, with an early exit from exploration for the least successful. In some ways, this started some years ago – remember Burmah. It could be argued that this is occurring in microcosm in Aberdeen even now. I should explain: when exploration commences in a new area and there is appreciable success, leading to production, the oil companies establish themselves and draw in a wide variety of service suppliers from the large – Western Atlas, Halliburton etc – down to a range of small specialists. As the area reaches maturity, the oil companies lose the efficiencies of scale in production: rising costs and declining productivity take their toll. They start to look elsewhere for their future because they find that declining production takes the same or more management and technical time as peak production.

They can invest money elsewhere for a higher rate of return than they can hope to get from the declining years of a large field. Their interest wanes and other specialist companies – scavengers if you like, start to feed off the decaying carcasses of the older fields. This has happened in Aberdeen with the advent of Oryx and Talisman, to name but two. Who could have imagined Chevron moving out of Ninian House – but they have done so as they smoothly moved out of the Ninian Field and handed the responsibility for the field decline and abandonment to others.

The large players have already mentally gone from the mature area of the North Sea – how many have real executive decision making powers in Aberdeen? I mean by that, raising capital and employing it in a risk sense. I think it is a round zero. It is incidentally the one area where I can understand the fears of the Scottish National Party: Aberdeen is just a service industry town now. If the SNP should come to power, they will not be dealing with the powers in the industry when they talk to people in Aberdeen. It is well down the slope towards withdrawal by the large oil companies. Sure, they will be there for many years to come, but the real decisions are being taken on other areas, elsewhere.

This gives tremendous potential for the service companies to take on the management role of the declining resources. And why not, when they often actually do the work and fully understand the means by which it is to be carried out. After all, most of us watch television and control the channels but we know (or care) little for how the visual image is created. We are happy to pay others to take care of that, so long as the programme is satisfactory.

The future in the North Sea may well lie with contractors, but the real decisions will still be made by the sources of capital. Where the financial implications of the decisions are in the realms of loose change, they can easily be left to others, with a spending authority cap.

The oil companies' problem is that they control such a small part of the world's remaining reserves that they will be increasingly marginalised as their reserve base declines. The evidence of this is around us now – Shell's cash mountain (have they really nothing to spend it on?), the flat market performance of some companies through a bull market (Petrofina). The only long term hope is in relationships with the Middle East, if that is possible, and a rise in the price of oil, justifying exploration and development in difficult areas – deep water, hostile environments etc.

The question of the future level of oil prices is, I know, a subject dear to your heart. I will stop here and let others with better crystal balls consider that.

Postscript: After his experience in the contracting business, Clive Needham returned to the main stream taking up an executive position with Agip, the Italian oil company, moving from Aberdeen to London, when it closed its Aberdeen Office in the face of declining North Sea activity.

STATOIL: A STATE WITHIN A STATE

Time passed, work continued, and Statoil grew ever larger. It built what in architectural terms must be one of the ugliest buildings in northern Europe as its corporate headquarters, and began recruiting staff.

Fig. 12-11. Statoil's character may be reflected in its architecture: large and square

It now employs something like eleven thousand employees and enjoys an income of over ten billion dollars, almost all coming from the giant fields given to it by the taxpayer in 1979. This puts it in the top dozen oil companies.[9]

Arve Johnson, the aggressive Chairman, was thrown out when what were described as serious budgetary over-runs on refurbishing the Mongstad Refinery were discovered. He was replaced by a young economist, named Harald Norvik, who seemed to want to run the Company as an efficient enterprise like any other, which of course it is not. It seems to have been rather a hot seat as subsequent Chairmen were also deposed for various reasons, including a bribery scandal in Iran, which incidentally shows how international operations, where bribery and corruption are endemic, are outside the natural domain of a Norwegian national company.

In 1987, came the penultimate throw of the dice, when the Government decided to open up the huge Barents Sea, far within the Arctic Circle off northern Norway. It had a disputed border with the Soviet Union, which may have been one of the reasons for establishing a presence, especially as some discoveries had been made on the Russian side. It was obvious from the first surveys that this was an entirely new province having virtually nothing in common with the North Sea. In particular, the well-known Upper Jurassic source-rock which made the North Sea work, appeared too shallow to generate oil.[10] It was necessary to hope for some older source to come to the rescue. The geological masseurs soon went to work and came up with hypotheses that a Triassic organic clay, seen outcropping in Svalbard, might do the trick, or that very deep-seated Devonian or Silurian rocks, which generate in Russia, might somehow not have been as burnt out as their great depth of burial suggested. In structural terms, there were some huge structures that deserved testing, even if the chance of success was remote.

It was already remarkable that the Almighty should have granted even God-fearing Norway one trend as prolific as the North Sea: it was too much to expect a second.

The environmentalists and fishermen, who always oppose the opening of new areas, were once more against it, but the Government found a solution by classifying the exploration as a form of *research*, which is a national passion. Several licences were let in this and subsequent rounds, and the data were shared so that everyone could know the outcome. That was not good. It appears that the whole region was uplifted by about 2000m on the melting of an ice-cap in the Oligocene, about 25 million years ago. The resulting pressure differential caused the seals above the reservoirs to leak: in the same way as fish brought from great depth almost explode at the surface. There were numerous sniffs of gas, but not much sign of oil. In short, it was a bust, although a large gasfield, named Snohvit, was found. Again, the wisdom of bringing the foreign companies into it at all is to be questioned. The three national companies could equally have done the "research" at a slower pace. If they had held the negative results confidential, they could have perhaps farmed out their rights to uninformed foreign companies when they realized that it was no longer promising. If necessary, they could have offered new terms whereby the foreign companies spent money that was not deductible from Norwegian tax.

Most such money could in fact be claimed as a cost against tax in the foreign companies' home countries, adding still another convoluted distortion to the situation. Under double taxation treaties, the taxes in Norway, notwithstanding all the many deductions, were a charge against home country tax.

Exploration continued in the North Sea, and a number of modest finds were made, without there being any particular surprises.

In 1995, they cast the last throw of the dice by opening the Voering Plateau, a deepwater area, north of Haltenbanken. The most obvious interpretation is that a thick sequence of Cretaceous claystones rests directly on oceanic crust, in which case the essential Upper Jurassic source-rocks will be lacking. If so, it will be another bust. However, the imaginative geologists have managed to propose that some Upper Jurassic might have been preserved on very tenuous seismic evidence, or that there might have been structural inversions, or still even more obscure interpretations. It is certainly worth throwing this last dice to find out, however high the odds, for there can always be surprises and anomalies. The issue is: with whose money?

Figure 12-12 shows Norway's estimated endowment of oil; and Figure 12-13 shows the production profile based on these numbers. Production has peaked and will decline steeply.

The Norwegian shelf has now been culled nineteen times with successive licence rounds, most of which were aimed to test the more promising prospects identified at the time. It stands to reason that there cannot be much left except smaller finds within known provinces. There is just a chance for some new deepwater areas or sectors within the huge Barents Sea that for some reason escaped the regional effects that destroyed the prospects but the odds have lengthened considerably.[11]

Norway	Gb
Produced	19
Reserves	11
Yet-to-Find	3
Yet-to-Produce	15
Ultimate	33

Fig 12-12 Status of Oil Depletion

It is more difficult to assess the country's gas resources. As discussed in earlier chapters, gas source-rocks are more widely distributed in Nature than are oil source-rocks, but gas needed a stronger seal to hold it in the reservoir. The production profile is also very different, being much more influenced by the pipeline infra-structure than is the case with oil. It is difficult to draw the economic thresholds. About 150 Tcf of gas have been discovered so far in Norway, which is equivalent to about 25 billion barrels of oil in calorific terms. In other words, almost half of Norway's known hydrocarbon resource is gas, but of this much lies in just

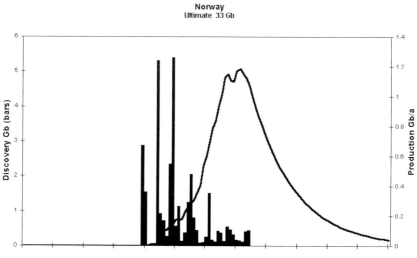

Fig. 12-13. Norway's production profile

one super-giant, known as Troll Field. I could imagine that another 20 Tcf might be found over the next twenty years, and some of the remote fields in the far north might become viable to produce by then. Gas prices are likely to rise in sympathy with oil prices, and Norway's gas is undoubtedly a major asset for the future. It will to some extent take the place of oil as its production declines. Managing the gas business is an ideal role for Statoil.

It is evident that Norway developed its oil too quickly both for its own, and the World's, good. Production rose rapidly to a peak reflecting the earlier peak of discovery under that eternal relationship. Norway, with its small population, could not absorb the oil, and therefore exported it. Today, Norway exports just over a billion barrels a year, 93% of its production, namely much more than Kuwait. The North Sea contributed greatly to the so-called "glut" of 1986, when the share from the Middle East swing producers sank to 16% and prices collapsed, as I discussed in earlier chapters. The United Kingdom was equally to blame but had more of an excuse in that its 60 million inhabitants needed the oil. The OPEC producers tried to stiffen prices by cutting their production, however ineffectually, while Norway and Britain pumped like there was no tomorrow. OPEC is not an altruistic organization, and its idea of increasing prices had no more than its own benefit in mind; but had it succeeded it would have curbed demand and thereby given the World a little precious breathing space to adjust to the coming shortfall. Norway and Britain, in ignorance of the world resource situation, may have thought that they were contributing usefully by undermining the price of oil. But if so, it was at the expense of their future. It is more likely that they did not know what they were doing in terms of resource depletion. I remember hearing Lord Lawson, Mrs Thatcher's Chancellor, speak at a conference in the mid-1990s. He spoke, beneath a self-proclaimed halo, of how he had rejected advice about depletion to build an exciting new profitable industry, creating employment and wealth for the Nation. Yet, a few years later, Britain passed its peak production and faces total exhaustion within twenty years. The Chancellor had evidently not understood the strange irony of depletion : *the better you do the job, the sooner it ends.*

There has been a general move to privatise State companies everywhere outside the Middle East. It is claimed that this will improve efficiency. Perhaps it will. But these concerns are being returned to private hands when there is little more oil to find around the world. Now, plans are afoot to privatise Statoil as a final insult to the unconscious taxpayer by transferring ownership of the enormous and then undervalued assets of the giant fields, found in the early 1980s, to foreign investors and speculators without an ounce of responsibility for the company or its mother country. They were a free gift by the taxpayer of some $100 billion[12] and should remain in national hands. Another mistake was to allow Statoil to move overseas in the pretence of investing the Norwegian tax-payers' money overseas in new exploration when there is so little left to find. It will never make money doing that. It is not that Statoil does a bad job: just the wrong job.

Having licensed large interests in its fields to foreign companies, Norway is contractually bound to allow them to be exploited as fast as their operators wish, barring some national emergency when it has powers to restrict production. Having gone to enormous pains to control the discovery of its oil, it effectively gave up control of the exploitation of this precious resource. It would have been wiser to have done it the other way round: to be lax in finding it and strict in producing it.

Statoil had privileged access to the prime positions, which was a benefit far exceeding the advantages of its favoured contract terms, but instead of delivering the proceeds to the widows and orphans in the fjords, it is bent on a programme of global expansion. The eleven thousand employees are well paid and want to continue that way. Nothing suits them better than global travel and all sorts of exciting experiences on the shores of the Caspian. It is impossible to go to a conference anywhere in the world and not find a Statoil man pronouncing on one subject or another. The management too can swell their breasts with the thought of their new global stature forgetting that their company's income owed little if anything to their endeavours, the assets having been a free gift from the taxpayer.

Norway has of course enjoyed the prodigious wealth that peak oil production brings. In the 1980s, Stavanger was a quiet place on a Saturday afternoon. It would be unkind to describe it now as a new Babylon, even though beautiful, fresh-faced, teenage girls dismount from the pillion seats of motorcycles at midnight to head into the countless dives that now surround the harbour: some, later, to become single-parents, well husbanded by the State. The shops are full to capacity, and the streets groan under the weight of affluent consumers. In Oslo, a breed of clipped-speaking Yuppies enjoys the affluent life. Crime, only recently almost unknown, is prospering too: the country retreats are burgled and the yachts are stolen. It is becoming less of a fairy-land.[13] But even this excess could not spend it all, and the Government has built up an oil fund of 120 billion dollars, the largest single investment in the world, which it has placed in the hands of the likes of Goldman Sachs to manage. The idea is that it should provide a form of pension for the future when oil income declines, but a more likely outcome is that it will be lost in the stockmarket crash that likely accompanies the onset of the Second Half of the Age of Oil.

All of this is an understandable consequence of a heady flow of easy oil money, but it is a party that is nearly over, as oil production peaks, and heads sharply into decline. There is not much time to prepare for a new cold northern wind.

NEW POLICY OPTIONS

One can think of a few easy policy decisions to save what is left:

1. Change the licensing system, such that the national companies are given exclusive access to any remaining prime tracts;
2. Disallow the deduction by foreign companies of exploration and related expenses from taxable income.
3. Open highly speculative areas to foreign companies to explore with their own money (itself mainly deductible in their home countries), accepting that they will be free to produce what they find. To attract them, it would be necessary to greatly ameliorate the terms. It might be enough to rely on royalty alone against which nothing can be deducted; scrapping tax on income altogether. It is a notional income because little is likely to be found. There is a big difference between looking for oil and finding it.
4. Stiffen, wherever possible and mainly in the area of deductible allowances, the taxes paid by foreign companies on existing production, offering at the same time to have Statoil buy these interests from willing sellers, no longer liking the new deal. The international companies are forced to have a short-term view of the future as their shares are traded on the World's markets, and might sell out for a quick buck. Several believe their own underestimates of future oil price rises.
5. Prohibit Statoil's overseas adventures, and trim it down to perform an efficient operator function on existing fields and those to be acquired in Norway.
6. Strictly control Statoil's budget and demand the highest possible dividend.

In fact, the State has already moved some way in this direction, as in latter years it has retained a direct economic interest in the oilfields it licensed, so as to recover some of its control. This economic interest is, however, administered by Statoil, no doubt at a direct and indirect fee and in a way advantageous to Statoil. A return has yet to come from the investment: it is not a good compromise. The State, as a shareholder, should aim Statoil to protect the national interest as part of a longer term policy, and not imagine that it will somehow grow to be another global "Shell", when the Shells of this world are heading for decline along with oil production, as this company's own recent experience of having to downgrade its reserves confirms.

Having recovered control of its oil and gas, Norway should do everything possible to slow the rate of production to make the resource last as long as possible. The Norwegian Petroleum Directorate already makes detailed resource estimates, which although I think over estimate the undiscovered potential, are valuable and no doubt can be revised as new knowledge comes in.[14]   By recognizing that it is a finite resource, Norway should make a major effort to husband the proceeds for the days that are to come.

Norway's gas now comes into focus as Britain's supply heads into steep decline. It will be under increasing pressure to increase exports to Europe. By all means, it can keep the existing pipelines full, but it should resist pressure to build new ones for export.[15]

In addition, Norway is ideally placed both physically and mentally to dedicate its highly privileged oil income to research into all forms of renewable alternative energy. If the scenarios herein depicted are anywhere near correct, there is a growing future for such new forms of energy: from the windmill on the exposed headland; to the tidal and wave power plants using the energy released by the Atlantic rollers crashing against the rocky cliffs or on dedicated offshore structures; and to the lightweight hydrogen-powered or photo-voltaic vehicle. More than anything else, Norway could pioneer social engineering in tele-commuting; tele-cottaging; job-sharing; more simple non-consumeristic pleasures for less drudgery at the office; improved life styles. It is called voluntary simplicity: a subject now being studied at Oxford.[16]   Norway is already more than half way there with its highly regulated society and its only recently abandoned belief that life is for living rather than mindless consumerism. It could in fact set an example.

Norway in recent years has excelled in many areas: its footballers win matches; its negotiators bring accord to the Israelis and Palestinians; its singers win contests; its skiers win slaloms, and its walkers win marathons. It has also done a marvellous job in conserving its fish stocks by adopting the simple but sensible policy of restricting the number of fishermen and keeping their boats small. In the 1980s, cod was almost wiped out by over-fishing but now has recovered.[17]

It is well set on the path to excellence but its greatest challenge is about to come. The World could learn a great deal both from how Norway harbors its oil and gas, and how wisely it uses the proceeds of its windfall. It *was* a windfall in the form of just a few hundred metres of Upper Jurassic Clay deposited 150 million years ago.

A POSITIVE POSTSCRIPT

A new electric car, called the City Bee, has been designed and is beginning production.[18] As proposed above, it is exactly where Norway's research should be aimed. New oil licensing terms have been introduced whereby the companies have the right to withdraw without drilling if the surveys are negative. It is at least a step in the right direction: reducing direct government interference in the conduct of exploration. Statoil too is withdrawing from blocks whose prospects have been downgraded, most of which it would never have entered in the first place but for government policy.

There is much more to do, but these are promising and very important steps. Once the resource constraints are better understood, I am confident that the government will react appropriately. They may not understand yet but I hope that they soon will.

DR. LEIK WOIE THE PREVENTIVE DOCTOR

Fig. 12-14. Dr Leik Woie, a Norwegian cardiologist interested in preventive medicine

QLeik: it may surprise you to be interviewed about oil. There are many oil experts in Norway, so why should I talk to you about your country's oil policy? I will answer that question myself: it is because it is not a technical or even an economic issue, but one of national interest on which your views are as valid as any other thinking citizen. Tell me how you came to be who you are?

A *That is impossible: no one either knows who they are or how they become who they are. But I have had certain experiences in the national life that make me very conscious of government policy and its effects. I have worked, as you know, in the health service.*

QI can understand: as a doctor you face life and death decisions in your daily work; you have judgments and tests to make; but at the end of the day you have to rely on a sort of instinct to know what is best for your patient. Doctors are sometimes accused of being little more than the sales representatives of the drug companies, but with your interest in preventive medicine, you are clearly not guilty of that.

A *It is true, the phenomenal progress of modern drugs means that there is always something to prescribe, often at great expense. Long ago, I was shocked to treat young patients suffering from heart disease. I came to realise that in many cases their condition was almost self-induced by them having lived in the wrong way without knowing it. I began to see that preventive medicine would not only bring enormous benefits to many people, but would be very cost effective for the government. I tried to influence the politicians to move in this direction, but with only limited success. It is easier for ministries to pay drug bills on prescription than take the more difficult decisions that arise from preventive medicine. It is easier for them to say "you have to be sick before you are treated" whereas I say "you have to be treated to avoid becoming sick"*

Q I have often compared explorers with doctors. We use seismic surveys as you use X-rays; we do geochemistry in place of blood tests, and we use the drill instead of the catheter to find out what is buried under the seabed. We can make our scientific evaluations, but after years of experience, we develop an intuitive nose by which to get the diagnosis right. Does that make sense?

A *I suppose it does, although I had not thought of it that way. To continue the analogy I can see that you begin to see relationships and patterns in the size of discoveries and the timing, which almost fit statistical theory. In medicine, we do exactly the same: we need to know the statistical chances but we need to know when to override the statistics and apply our judgment. Sometimes we save the patient in this way, but sometimes we fail, like you drill a dry hole. We may fail, but often that is no surprise. We knew that it is an almost hopeless case, and we are simply doing the best we can without much hope of success. We also know of near miraculous cures, which come about for reasons we don't understand. I suppose you also come upon the surprise good discovery.*

Q Yes, there often are surprises, sometimes even good ones. But our patients are getting old: the oil basins on which we work are mature and over-mature: the hopes of giving them the elixir of life or the recipe for eternal youth have faded. We are dealing with an oil life-span, in much the same way as you deal with a human life-span. What about government policy?

A *I certainly see the parallels. The demands on the health service must inexorably rise as the population ages. We need a realistic policy. Think of government tobacco policy. Tobacco should be banned on health grounds, yet those who smoke say that the tax they pay more than covers their health costs. Nothing is straightforward.*

QAs an ordinary citizen what do you think about Norway's oil experience?

A*First, I think it was a miracle that we should have been so blessed by having such a valuable resource. I think that the management of such a huge and important resource and benefit is a very serious thing. I think the earlier policy of a slow development to make it last as long as possible was the right one.*

As a physician who cares about preventive medicine, I would like to be sure that my country does not suffer from a hangover and indigestion from the oil-feast we are enjoying. I am concerned that my grandchildren should think well of me in later years. I would prefer them to think of me as someone who stood out against the mindless consumption of their birthright rather than as someone who condoned the outrageous plunder of the resource.

It has come as a surprise to me that well informed oil people can still say that the resource is so large that it can be treated as infinite, when all the evidence indicates the contrary. The fact that Norway's larger fields were found first confirms the proposition even here. In medicine we use basic and fundamental research to answer questions. The same approach should be applied to assessing the world's oil resources: I am not impressed by the make-believe whether put out by politicians, economists, journalists, engineers or doctors

I am shocked when I hear that the people in charge fail to understand elementary arithmetic: production obviously eats into the reserves, and simple statistics show that the rate of discovery is falling. It is like drilling into a Swiss cheese, even without the advantage of seismic surveys, you will always hit the largest holes first.

QWhat do you think about the role of Statoil? This huge enterprise was given free access to prime rights which belonged to the nation. Had the same rights been given to foreign companies, the Norwegian people would have received directly their rewards at a very high rate of tax. When the job was done, the foreign company would have left. But Statoil gets ever larger and is now taking the Norwegian taxpayers' investment into foreign areas where it no longer has the privilege it enjoys at home. Do you think the Norwegian taxpayers knows and agrees with what is being done?

A*I think that most Norwegian taxpayers are satisfied with the work of Statoil, but I am afraid that he does not understand his indirect investment in Statoil. Drilling for oil in an unknown province is a risky business, and* *according to your research there is not much left to find. I also read in the newspapers that the findings so far in 1996 have been very disappointing. I am therefore sceptical that Statoil or other companies will succeed in finding big reserves in the Norwegian sector or overseas. The Barents Sea drilling has been an expensive experience.*

As far as I understand, Statoil's technical achievements are excellent. Like all Norwegians I am a shareholder in Statoil and I would like to see my company dedicate itself to Norway's long term interest by concentrating on the Norwegian shelf. Can they extract more oil from the wells? Can they make marginal fields more profitable? If it is possible, they should also buy the interests now held by foreign companies in the existing fields, which are surely more profitable than anything likely to be found in the future. I do not think there will be another Statfjord.

I also believe very strongly that we in Norway must realize that the oil income will not last forever. It may all be gone in one generation. We should be doing our best to use this asset to build a strong future in an age when energy is expensive throughout the world. There are all sorts of interesting things we could do, and we are exceptionally privileged both in our social structure and natural resources. We have a tradition in electrical power, which will come again into its own as the prime energy carrier. We should be using our present advantages to lead in developing things like the electric car. Above all, we need to prepare sensibly for the future, rather than live only for today when oil pays our bills.

I have looked into the Norwegian situation in some depth, partly because I know it well. Every country faces, and has faced, a dilemma in controlling its oil industry. It is different from most businesses because it is concerned with depleting a finite resource, which once gone is gone forever. It is also different because it has been difficult to tell in advance whether the resource exists or in what quantity.

Countries have to set terms to attract companies to explore, and they are at least morally obliged to respect the terms if the companies should be rewarded with success. But whereas contract terms should be inviolable, all governments retain the right to change the tax regime, even retroactively.

SUMMARY

Norway began with normal terms and tax that was an equitable basis for both parties at the beginning. Then when it became evident that a trend of major finds on the United Kingdom shelf was almost certain to extend into Norwegian waters, it imposed draconian terms that the industry accepted only because their imposition coincided with a time of world oil crisis. High taxes were imposed but expenses were allowed as a deduction. It led to distortions that are always associated with high taxes. Never were more Rolls Royces to be seen in the streets of London than when the socialists had an 80% marginal tax rate in the 1970s.

The creation of Statoil was a concealed tax-driven construction: Parliament would never have agreed to pour away such colossal sums. As I have already emphasised several times, it was achieved indirectly by having the foreign companies carry its costs, but to deduct same against Norwegian tax. It sounds like a socialist conspiracy, but perhaps it just happened that way.

Britain had the same earlier experience. It opened the shelf to the companies, some working with state enterprises, like the Gas Council and the National Coal Board, who paid their way. The socialists then created BNOC and clawed back rights previously freely negotiated with the companies to give to it. Mrs Thatcher swept all that away. The oil interests of the Gas Council were floated off to become Enterprise, to be later acquired by Shell, and BNOC was bought by BP. Everyone was thinking about how to share the oil profits: no one was thinking of depleting a finite resource or the next generation.

Petroleum Revenue Tax in Britain continued to effectively subsidize exploration by a free wheeling entrepreneurial industry made up both of major international companies and as many new small companies, floated on the stock market. Some of them bought producing properties simply to acquire the benefit of paying PRT and enjoying the deductions that came with it. But eventually in 1994, the government closed that facility, when it realized that exploration was no longer yielding new revenue producing projects sufficient to justify the allowance. The tax deduction was in effect a form of national investment. While some momentum continues, the removal of PRT heralded the end of the exploration business in Britain, and rightly so, since it has found pretty much all there is to find.

Many other countries develop their oil by Production Sharing Contracts. There are several variants but the essential formula provides that the State Company holds the rights and the foreign company takes the exploration risk, but can recover expenses from a certain percentage of the production. The split of the oil after recovery may be as high as 85:15. It comes to much the same thing at the end of the day, but is politically more attractive to the host country, as the foreign company is seen more as a contractor than as an owner of national resources. The facility to recover expenses through a share of production provides the same hidden subsidy for exploration and other indirect costs, as does tax, once production is established.

When the King of Spain let rights to his *conquistadores* to exploit the Spanish Empire, the tenancy lasted for a specific period of time. The

Spanish currency was then divided in units of eight: "Pieces of Eight" as the pirates used to say. One of the eight was for the King: a 12½% royalty. It was found to be a workable tax, which has been inherited by many oil contracts. In some ways, it is the best arrangement as it is not subject to distortion.

The whole issue of licensing terms will no doubt have to be reviewed throughout the World as supply shortages appear. The immediate response to shortage will be to stimulate further exploration, probably with all sorts of tax driven inducements. There may be another brief golden age of exploration. But as they discover that it does not deliver, what will they do then? Will they pump good money after bad? Or will they belatedly come to realize that it is a finite resource, and invest instead in finding ways to use less.

NOTES

1. Norway did well out of the settlement of the median line on geographic criteria. Had the boundary been drawn on the topography of the seabed, the deep Norwegian Trench would have excluded Norway from the prospective part of the North Sea. The Germans came out of the arrangement particularly badly.
2. A block is a designated area of the seabed bounded by 20' of longitude and 15' of latitude (about 525 km2), and a licence conveys the right to explore for and produce petroleum under certain conditions.
3. See Kvendseth S.S., 1988, for an excellent account of the story of Ekofisk.
4. The Jurassic Period occurred between 195 to 140 million year ago. The North Atlantic had not then opened, but a rift system, analogous to the present Red Sea, was beginning to form. Rivers flowed northwards into the Arctic Ocean, as a sort of precursor of the Rhine, and deposited huge deltas. The climate was warmer than today.
5. Companies tend to explore in packs, holding "undivided" rights to a concession. One of the companies is appointed "Operator" to manage the operation; the others being termed non-operators. Each votes his participating interest on important decisions. The arrangement gives rise to endless conflict. The Operator tends to think of his partners as worthless free-riders; whereas the non-operators tend to think of their Operator as a delinquent, overcharging on overhead, not giving their particular venture the priority it deserves, and being guilty of general ineptitude (see comments of Hardman, Chapter 11).
6. A Round of Licensing is the term used to describe the periodic offering by the government of licences to explore and produce petroleum.
7. See Nerheim G. and B. Utne, 1992, for a valuable description of the history of Norwegian shipping and its move into drilling.
8. It ran into difficulties in the mid 1980s, when oil prices collapsed, and was bought by the excellent French contractor, Schlumberger.
9. Oil & Gas Journal, 4.9.95.
10. Although it *appeared* too shallow, it may in fact have been too deep, prior to the 2000m of regional uplift that occurred in the Oligocene, and was not recognized until after drilling. It would explain the preponderance of gas. The remaining hopes are to find some part of the area that was not affected by this uplift for whatever local reason. Like a chicken, exploration often twitches after death.
11. In February 1997, the Norwegian Petroleum Directorate announced an implausible increase in reserves to 40.9 Gb for oil and condensate and 38 Gb for gas, and an undiscovered potential of 22 Gboe of oil and gas. The exact explanations need to be evaluated.
12. Assuming that about half of Norway's 17 billion barrels of oil belongs to Statoil, and was worth about $15/b after necessary, direct and genuine production cost.

13. The Economist has reported that the world's highest per capita purchase of canned music is in Norway, and not all of it is Mozart.
14. See Brekke H and J-E Kalheim, 1996. I believe the reason for the overestimation is the practice of risking every prospect independently. Every prospect will be perceived to have some chance, however high the risk, and the risked reserves of the many notional prospects on this huge shelf add up. In practice, the appraiser will rarely attribute a zero probability to a prospect although, in reality, prospects either succeed or fail; and huge areas may be entirely barren for the reasons discussed earlier.
15. Unfortunately, in early 2005 it failed to resist recent British pressure for a new line, which will now be built to help meet the consequence of Britain over producing its own resources.
16. See *The Economist*, 19 October 1996, p.39.
17. See *The European Magazine*, 5–11 December 1996.

Chapter 13

RENEWABLE ENERGY AND RESTRAINT

IN EARLIER CHAPTERS, I have explained how oil production is set to decline and become much more expensive. For many people, this sounds like bad news: a sort of doomsday message. There can, indeed, be no doubt that the transition to a new low-energy world will be difficult, perhaps very difficult. But it is not the end of the line. Almost by definition, Man can go on and find solutions. As often happens, there may even be a silver lining, such that many aspects of life may improve. Visiting a choked modern city, or an airport where tense-faced passengers are subject to undignified frisking by security guards, suggests that limits have in any case been reached.

The economists like to say that if you want more oil, drill more wells. It is quite possible that the coming crisis will indeed spawn another exploration boom, as people in ignorance of the underlying resource constraints continue to live in the past, hoping that what worked before will work again. It won't. Instead, some new thinking has to come into play.

There are two obvious solutions: use less energy and find alternative renewable sources. In fact, there is a more important third solution of better using what remains. I think the least promising approach is to try to find alternative energy sources to maintain the *status quo*. I do not believe that they can ever be a substitute for the fuels we have come to use in such a profligate way today. That is not to say that they are irrelevant or unimportant: quite the contrary. They will contribute in useful and indeed essential ways, even if falling short of being a substitute for the easy oil-based energy we have known. I will summarise what they are. Clearly each approach has its own best application. It seems to me that alternative renewable energy and the idea of using less are complimentary processes that have to be considered together.

I am thinking mainly of *Regular Conventional Oil*, so in this context the term alternative could include gas, *Non-Conventional Oil*, nuclear power, coal as well as the more restricted forms of alternative energy such as solar power. I am no expert in this subject but will try to imagine some scenarios and comment upon them from this vantage point.

FOOD

Without food, Man dies, so that has to be the priority. The first issue here is: how many mouths are there to feed. Human population has been growing since prehistoric times, but at a very slow rate. It was 300 million at the time of Christ and had managed with the help of primitive yet sustainable agriculture to rise to about one billion by 1850 when the first oil wells were drilled. One of the reasons for the low rate of increase was that people could not escape the consequences of local floods, natural catastrophes and famines, which often triggered terrible global transport. The population then expanded six-fold to present times. Oil furthermore contributed greatly to ever more energy-intensive agriculture. It is difficult to avoid the conclusion that the population will also peak soon after oil does so, although the mechanism by which the subsequent decline will be achieved does not bear thinking about. Some forecasts that ignore the oil factor project that population may continue to expand throughout this Century.[1] That seems most unlikely.[2]

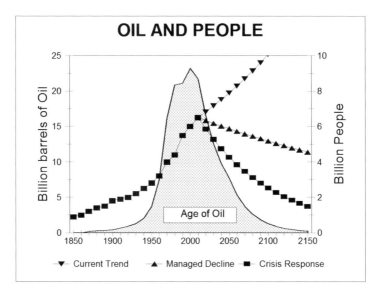

Fig. 13-1. Population forecasts

Reproductive rates in much of the world have however been falling, some faster than others. Worldwide, the level is about three. In Western Europe, it has already declined from 2.8 in the early 1950s to 1.6, which is less than the 2.1 needed to maintain parity. But life-span has lengthened in parallel, so that the falling birth-rate has been compensated for by increased longevity. Longevity is, however, likely to stabilize, and the falling birth-rate will accordingly flow through to a declining population.

Life expectancy in developed countries is now 76 for males and 82 for females compared with respectively 52 and 54 in undeveloped countries. A most remarkable decline was in Russia where male life expectancy fell from 65 in 1987 to 57 in 1994, on the fall of the Soviets, due apparently to alcoholism and environment factors, before recovering to more normal levels today.

Generally, reproductive rates in the Third World are much higher. However, in some countries, such as Colombia, they are already falling as people gradually come to prefer to have fewer children as social conditions change. The urbanization of society probably helps the move in this direction: children no longer being cheap labour to run the farm. In other countries, as in most of Africa, conditions are much worse, and declining birth-rate may be due more to pestilence, war and starvation. Longevity here too is likely to decline, so that a falling birth-rate will have a quicker impact.

Population disparities, however, raise the issue of migration. Already, large numbers of North Africans are crossing the Mediterranean in small boats to arrive as illegal immigrants on Spain's beaches. In 1996, a church in Paris was full of illegal African immigrants on hunger strike rather than accept repatriation. Other hopeful migrants are assembling in Poland at Europe's backdoor. The United States too faces waves of immigrants from Latin America, especially Mexico.

Most of the industrial countries, except Japan, built enormous long-term problems for themselves by admitting large numbers of *gast-arbeider*: Mexicans and Puerto Ricans in America; West Indians and Pakistanis in Britain; Algerians in France; and Turks in Germany. They proved themselves to be hard-working and useful members of society in an expanding economy fed by cheap oil, but they cease to be needed as they age and multiply in a stagnant and declining economy. Since they have no real roots where they live, they tend to degenerate into urban ghettos, being forced to live on crime and social security. They started out as fine people, but not all of them ended that way for no fault of their own.

But, by and large, migration is likely to be curtailed and controlled, especially as the global market declines as a consequence of the oil crisis, and as the world becomes a more protectionist place when nations break up into smaller regional units. The violent Basques and Ulstermen may already reflect a deeper, otherwise so far mainly passive, trend in this direction. It is in fact enshrined in the Maastricht Treaty, which says that no decision shall be taken at any level higher than it has to be. The Mayor of Milhac, the village where we lived in France, already had authority to grant us permission to live in his commune, and thereby his country, for five years.

So, I think that populations in Western Europe, and the developed world generally, will fall naturally to sustainable levels unless subject to massive immigration. Also, the land can evidently support greater numbers of low consumers, having more than enough agricultural capacity to feed themselves. In terms of energy, precious oil supplies will have to be dedicated to agriculture, probably by means of tax allowance on what will then be a high normal cost of this fuel. In addition, successful efforts could be made to reduce the energy content of food by a reduction of packaging and the production of frozen foods generally. The present system of high energy production with intensive use of fertilizers coupled with set-aside, by which subsidies are paid to take land out of production, would have to be reversed. Lower yields, consuming less energy from more land, would be far preferable. The import of fresh foods by air would cease, and there would be more seasonal variety. More soup would be served.

Mention of migration is a delicate issue leading to accusations of racism, which has become formally an illegal activity. Personally, I have no hesitation to admit to being a racist, having an enormous respect for all the many different races amongst whom I have spent my working life. But I think they are all different, such being greatly to their credit. The problems arise only when migrants, or what in reality are economic slaves, are imported to reduce labour costs and then expected to become identical with the indigenous people. In an expanding world there might be room for everyone, but in a contracting one it makes sense for each tribe to look after itself as well as it can. Already, the indigenous Dutch find themselves almost outnumbered in their overcrowded home, being now forced to emigrate in increasing numbers to places like Australia, having been squeezed out by newcomers. There is nothing new about this, as in earlier centuries progressive migrations out of Central Asia filled Europe with wave upon wave of immigrant. The Celts are some of the most ancient people, but survive only on the western seaboard to which they were pushed by later arrivals.

TRANSPORT

Transport relies heavily on oil: indeed in the United States more than sixty percent of the oil used in the country is for transport.[3] It is here that the pending shock will be most severely felt.

There is however huge scope for savings in innumerable ways. To visit a large city, such as London or Dublin, is to observe a scandalous waste of energy, with streets full of near stationary traffic burning up precious fuel at a prodigious rate in the most inefficient and polluting manner imaginable. That obviously has to end. If town centres were closed to private motoring, there would be room for efficient modern tramways and tracks for bicycles, which could become the normal mode of travel over the short distances involved. Indeed, as I recently discovered on a trip to London, the distances on foot are really quite short. In Oxford, they have

Fig. 13-2. Burning up precious resources in a traffic jam

already introduced the rickshaw.

Private motoring is today heavily subsidized by the free marginal use of roads and parking, as well as by the widespread provision of company cars, whose cost is taken as an expense against corporate taxable income.

I remember when I was employed in London in 1979, I was given a large company car of which I had very little need. If I filled it up at a designated garage, I could have an unlimited supply of free gasoline, and I had a credit card for fuel on journeys farther away, which was however treated as a taxable benefit. It was an absurd arrangement, involving much unnecessary administration, all treated as a deductible operating expense. I learn that employees with company cars drive from London to meetings in Scotland in preference to going by train so as to increase their company mileage that increases their tax free personal allowance. It is madness.

Cars should be cheap to own, but the fuel should be expensive to encourage minimal usage and the development of more efficient vehicles. I drive a small Renault Clio, which is more than adequate for all normal needs, yet the roads are choked with enormous cars which must be more of a status symbol than an essential means of transport. A change of attitude could make enormous fuel savings at no hardship. High fuel costs would impose difficulties in relatively under-populated rural areas, where people depend on cars, but that could be solved by appropriate allowances for those in need. The idea of the electric car is probably a good one, although electricity generation and transmission is itself not a very efficient process. Despite this drawback, it can in fact be more efficient than the internal combustion engine, which is in practice often not running at an optimal speed. The solar electric car is already a technical possibility, but it needs mass production to become economically competitive with existing vehicles. I suspect that there is much scope for improving engine design to reduce consumption. I have read about a novel design by a Mr Negre in France.[4] It has three separate cylinders for respectively compressing the air-fuel mixture, exploding it and evacuating it, which are done in a single cylinder in the traditional four-stroke engine. It results in an efficiency improvement of 50%. Furthermore, it can run partly or exclusively on compressed air: fifteen litres being enough to drive it for one hour. The compressed air is partly generated by braking. Going back to the concept of energy concentration discussed in Chapter 1, one could imagine many ways of achieving low-level air compression: for example, by small windmills on the garage roof running day and night. Perhaps useful energy could be concentrated in this way for use in novel new engines that may be developed. There is plenty of scope for inventiveness. The Toyota Prius is another good solution. It has a gasoline driven turbine running at a constant most efficient rate, and an electric transmission. The act of braking generates electricity for later propulsion.

The railway network could be rediscovered and revitalized in developed countries. It should be used wherever possible to replace long-range trucking. The economics of building roads at public expense which subsidizes motoring needs to be re-evaluated in relation to railways. The economics should include a full evaluation of all the costs, including irreplaceable energy costs.

The airline business will go into near extinction as fuel costs soar. Very few people actually need to travel by air. Modern communications make most business travel unnecessary. It is also presently subsidized as a deductible operating expense. Aviation fuel is tax free for some bizarre reason. Tourism too will suffer.

Not much can be done to save marine bunkers for shipping, except by the reduction of world trade. Perhaps the sailing ship to modern design will return.

The Third World will of course be the hardest hit, as it depends heavily on road transport, much in a dilapidated condition and very inefficient.

Much is written about the so-called hydrogen economy as if it were the fuel of the future,[5] being benign in environmental terms. The problem is the manufacture of hydrogen by electrolysis consumes more energy than it delivers.

ELECTRICITY

Electricity is a most convenient source of energy for lighting and operating power tools but it does suffer from huge losses in generation and transmission.

Much can be done to save electricity and to restrict its use to the purposes for which it is particularly suitable. There is great scope for savings. The widespread installation of modern lamp bulbs which use a fraction of the electricity consumed by traditional incandescent bulbs could make an enormous impact. Inverted tariffs could encourage more careful usage.

Fig. 13-3 The new bulb saves electricity

I have worked in offices where the lighting was left on 24 hours a day because it prolonged the life of the bulbs, which cost more to replace than the fuel they used. People should walk up stairs rather than take the lift: it might also help their health.

Various alternative energy initiatives could contribute greatly: improved insulation; solar water and space heating; and even wider use of geothermal heat are obvious examples. Air conditioning and space heating are commonly excessive. I once worked for a company that offered its staff a pullover in preference to turning up the heat, although in this case the move was motivated by parsimony rather than energy saving.

Obviously electricity generation from oil and gas needs to be phased out as soon as possible, to be replaced by that produced from by wave, tide, wind, hydro and solar sources. Biomass, including particularly wood-pellets, made either from dedicated planting or simply the normal thinning of commercial forests offers great possibility. Coal too will likely make a come-back, including its derivative, coal-bed methane, and no doubt efficient methods could be devised to remove any adverse emissions from the smoke stacks. The nuclear option is a contentious issue. It has its adherents in France, who have lived happily with nuclear energy for many years, but others fear accidents and worry about the waste disposal issue. It remains possible that a new research effort could deliver acceptable new nuclear systems.

Again, the greatest difficulties are in the Third World. Half the world's population still has no access to modern energy, and their perceived needs are increasing, quite apart from the growth in population itself. Much of the electricity as is available is generated from oil, and is widely subsidized already. The oil crisis will be a crippling blow in this area, especially in the sprawling urban centres of population which get larger every day. It seems a hopeless situation.

There is more hope in rural circumstances, where small scale alternative energy projects can have wide application. Even such simple steps as improving the efficiency of cooking stoves can have a great impact.[6]

Writing about energy shortages is depressing: it seems so hopeless. But it may not be as bad as it seems. I am not after all speaking about the end of oil production for a very long time, nor is it going to get much more expensive to actually produce, even if it costs much more to buy as a result of profiteering from shortage. So, there is some breathing space, and the anticipated radical increases in price will concentrate the mind wonderfully.

Savings in energy usage have to be made, but since we are now so profligate they can probably be achieved relatively painlessly. It is a case of aiming in a new direction, rather than having to take any extreme measures. There is still some time to make the adjustment although not too much.

The most effective weapon is the tax system, which at present is inefficient and commonly gives unintended benefits. I have already discussed the hidden subsidies on exploration that are provided by allowing the expense to be deducted against high marginal tax rates. It is the same in many other parts of the economy. Business expenses, however incurred, are a form of subsidy that distorts the reality of the costs. One way to solve that would be to remove corporate tax altogether so that there would be nothing to deduct expenses from. Barker[7] of the prestigious Cambridge Econometrics has already shown that increasing fuel taxes while at the same time reducing national insurance contributions (a tax on employment) is a viable option for the United Kingdom. At the very least, the general principle that the real costs of doing things are properly metered and allocated should be observed more effectively. The environmentalists have already been pressing for changes in this direction, such that the cost of pollution should be charged to those who cause it,[8] including the consumer. It should be the same with energy use. Those who use it should pay for it in full. As Timson points[9] out, market forces aimed at reducing the cost of energy do not encourage energy efficiency.

I also think that the tax system should be used to impose shareholder loyalty by penalizing short-term trading. It goes to the heart of the matter. Ownership implies responsibility and a view of the future. World trade driven by a shiftless investment community, with thoughts of nothing but a quick buck, is a disastrous formula, encouraging the management excesses and abuses, which have become universal, and general feckless behaviour. So far, cheap oil has paid for it all, but not for much longer.

It seems to me that the most sensible approach to achieve a sustainable energy supply is to adopt the principles of the so-called Rimini Protocol, described in the next chapter.

RON SWENSON: A CALIFORNIAN EXPERT ON RENEWABLE ENERGY

Fig 12-4 Ron Swenson

QRon: we have been in touch for some time. Obviously, the scope for alternative, or renewable, energy as it should better be called, depends heavily on the price and availability of conventional fuels. But first, how did you become interested in the subject?

A*I had an experience in High School which showed me that the big defence contractors were operating a great boondoggle, so when I went to university, I began searching for solutions that contributed to "living-ry" not "weapon-ry". Then, from 1965 to 1968, I taught Cybernetic Systems at San Jose State University. For two months, Bucky Fuller was a visiting professor in one of the courses I taught, and he brought me to see the view that humanity can learn to do more with less, and thus discover that there is enough for everyone.*

QFrom what you say, renewable energy is already a viable option which in several areas successfully competes with under-priced conventional fuels.

A*Hydro-electric, a renewable energy source, was at the forefront of electricity generation a hundred years ago, and still provides a large percentage. Wind energy is already an economic success, typically costing less than oil, coal and nuclear energy - that is if environmental costs are included in the accounting. Solar thermal (by the Luz Company) has had a good economic performance. These are important signposts.*

QIt seems to me that the United States faces a very difficult energy future. Already it is importing more than 50% of its oil on a trend that can only rise. Prices are going up, and there is a lot of evidence that the world will soon face another oil price shock. How will America cope?

A *The meek will inherit the Earth. An oil crisis will not affect people who have never had it, as for example, the descendants of the Incas in the highlands of Peru who still have an indigenous lifestyle. But the United States citizen, who is very spoiled and very naive about the forces which affect our future, is ill-prepared. The likely reaction will be to blame a Middle East dictator, who is perceived to be diabolical, such as Saddam Hussein or the Ayatollah Khomeini. Then people will turn on each other. In '73 and '79 people were shot at the lines in gas stations. Then the environment will be ravaged, and only then will people wake up. The cities will be the worst hit.*

Q Solutions are there, but what is lacking is the understanding and awareness. Above all, we face the problem of lead-time. An oil shock arrives overnight, but it takes ten years to build a nuclear power station, and no one wants one in his backyard. It seems to me that the emphasis today must be in terms of education and spreading information.

A *I agree. I'm putting out the word as best I know how. Surprisingly, it's just as hard to wake up people in the solar industry - they've been under dog for so long, they don't know how to assert themselves. I do exhibits, talks; and we have an extensive website on renewables [www.ecotopia.com]. We are developing alternatives in high profile projects. Because young people like cars, solar car racing is an especially good way to educate and motivate. I've been involved in such a project in Mexico for the past four years. We competed in SunRayce'95 in the USA and participated in the World Solar Challenge 1996 in Australia.*

Q Tell us more about what the practical applications are.

A *Practical renewable energy is, above all else, a function of mass production. Wind turbines are already practical in areas of consistent high winds (Patagonia, the great plains of the USA, New Zealand etc.). Solar thermal electric has proved economical at large sites in the southern California desert. Photovoltaics is now often the best choice in remote sites, and will be economic in large fields when there is serious production. The new technology of photovoltaics in thin film or concentration gives substantially better energy performance and economics than the older style of flat-plate PV panels.*

Q It seems to me that electricity, however convenient it is, is a very wasteful use of energy due to the huge losses in generation and transmission.

A   *On the contrary, electricity is the future, especially for urban transport. It's clean, and delivers more overall efficiency, because a large power plant has a much higher efficiency than an automobile engine, which in traffic is often running at much less than its optimal performance. Long distance transmission is not required for photovoltaics, so it makes a lot of sense to generate power at the load.*

Q   What about the environmental impact of renewable energy ?

A   *For photovoltaics and solar thermal concentrators in rural or desert areas, there is the impact of extensive land use, although depending on the structural design of the frames, land may remain relatively free underneath the solar panels. Obviously, if installed in forest areas trees would have to be felled.*
    Wind generators have had some impact on raptors (eagles, hawks), which are sometimes hit by the turning turbine blades. They are also sometimes considered unsightly, and noisy if placed near residential areas.
    In the case of solar, there is less need for long transmission lines, which are both inefficient and unsightly, because the receptors can be installed right at the load.

Q   In America the automobile is regarded as almost a national birthright. The country is built around it. Gasoline prices are already absurdly low, about one-third as much as in Europe. Some people think it unfair that America should be burning up a disproportionate share of the world's resources. It is precisely in the transport sector that the crunch will be felt most severely, yet it seems the most difficult problem for which to find a solution in terms of alternative energy. Is that so?

A   *In urban areas with short distances, solar panels on electric cars, or charging them from rooftop solar panels will be quite workable. For longer distances, solar powered railroads with a third rail or catenary lines are feasible. Buses and trucks with motive power based on renewable principles will be able to handle long distances with biomass fuels or hydrogen as an intermediary.*

Q   The continental climate of the US means that many places face extremes of heat and cold. That seems a particularly good area for the application of renewable energy?

A   *The climate for solar energy is good around the world. Seattle has hydropower, a form of solar energy; the Sahara Desert has direct sunlight for photovoltaics; Patagonia has wind; the Philippines has the potential for sustained and managed biomass production. The United States has it all too. There are bioregional differences, and so there will be some transmission from one region to another. I can imagine a shift in the industrial base from the coal-rich eastern states to the sun-drenched Southwest. Perhaps it will be a case of Phoenix vs. Pittsburgh vying for the industrialists' investment !*

Q I suppose what is needed is a dual approach: first to use less conventional fuel so it lasts longer and second to bring in as much renewable as possible. America is a great believer in the capitalist system and market forces, but perhaps the depletion of a resource calls for government intervention as the problem cannot be handled by market forces alone. The market lives in the short term, whereas what is needed are longer term solutions. Have you had positive political responses?

A *The political response to date has been very primitive. Environmental issues have so far attracted much more attention than the depletion of the resource. The ozone threat has so far had the best response. The US has not paid enough attention to carbon dioxide emissions, and its failure to lead has meant that places such as China or India which rely on coal for energy generation have not been adequately pressed to control emissions that jeopardize everyone's future.*

NOTES

1. See World Resources Institute, 1996, reporting UN estimates (p. 174).
2. See Stanton W.
3. See U.S.Department of Energy, 1989.
4. The Economist, 16th October 1996 .
5. See Rifkin.
6. Bohnet M., 1996, explains German experience on applying alternative energy in developing countries.
7. Barker T., 1995, from the prestigious Cambridge Econometrics unit gives a compelling argument.
8. Hawken P., 1993, surprisingly an executive of an American mail order company, who you would think was wedded to consumerism, makes some thoughtful and persuasive proposals.
9. Timson, R., 1996, explains that the increasing application of market forces to British energy supply has done little to encourage more efficient usage.

Chapter 14

THE WORLD BEGINS TO WAKE UP

WHEN *THE COMING OIL CRISIS* was written nearly ten years ago, very few people were even vaguely aware of the limits imposed by Nature on the discovery and subsequent production of oil and gas. Such a perception was very evident to an earlier generation of petroleum geologists but it seems to have been lost to the new generation, being naturally far from the minds of anyone outside the oil business.

Part of the explanation relates to the new mindset and working environment of the oil companies. In earlier years, the higher management commonly had an exploration background, or could at least call on objective advice. Norman Falcon, the distinguished Chief Geologist of BP, had a respected place on the Board (see Chapter 11). Those days are long over as financial pressures call for the appointment of money managers and image makers to senior positions. Under the new order, if the Exploration Manager started hinting at the natural limits, as some did, he would be accused of pessimism and a failure to deliver the posture of the dynamic oil-finder expected of him. The exploration departments were effectively relegated to the position of internal contractors, doing what is asked of them.

I well remember an occasion when I had been called back to consult for Amoco in one of its periodic attempts to re-build a position in Norway. I arrived in Houston to meet the team, and help them prepare the applications for concessions to the Government. The team was undoubtedly capable in technical terms but there was a strange lack of direction or sense of judgment. I sat in the meetings to hear the geologist concerned with each area expound his interpretation. At the end of one such presentation, I commented that that particular area certainly did not have what it took. The man looked crest-fallen and apologised that he had evidently not worked hard enough to develop the prospect. I reassured him that he had done a magnificent job in describing a place lacking the necessary geology. His reaction was revealing because it showed that he saw his job, not as using his judgement, but as applying his skill to employ geological mental gymnastics to make a purse of a sow's ear : if the obvious Upper Jurassic source was not deep enough to generate oil, he would invoke long-range migration, or structural inversion such that what was now too shallow had previously been deeper, The scope for convoluted hypotheses was limitless. Judgment as such was not part of the job.

I had seen things very differently. When, as was often the case, I found myself having to propose exploration projects in new areas that did not look promising, the best I could do was hope that common sense judgment would prove wrong. We commonly lacked sufficient information to be absolutely sure, and the only way to secure that was to drill holes. To get the money to do so from the managerial financiers, we had to pretend that there was a good hope of making money. They themselves risked little, because they could take the cost of failure as a charge against tax, so that the unconscious taxpayer funded the dry hole. The problem was that they had many alternative opportunities around the world, against which any particular venture had to compete.

Thus, the higher management, lacking professional qualifications to judge real exploration potential, has been delivered an endless list of similar-sounding prospects for acceptance or rejection based on hypothetical economic and political evaluations that miss the point. It all involved much theatre in the hierarchies of corporate power pyramids and posturing, ending up as little more than exercises in internal or external public relations. It was not so much a case of the blind leading the blind, but rather the blind leading those who had eyes to see but were asked to look the other way. Yet, even in this system, most large valid prospects normally did make it to the top of the pile and deliver easily predictable profitable results, while the cost of the lengthening list of dry holes was happily written off against taxable income.

It is understandable that the economists working in such an environment were misled into thinking that there was no shortage of exploration opportunity. They in turn conveyed this impression to the investment community, who naturally not only believed what they were told, but had a vested interest in doing so because any talk of decline or limits was anathema to their business.

The financial reporting procedures added to the confusion. Companies were not required to report what they found, but their current "reserves" which led to the much used concept of "reserve replacement". For the financiers, it made no difference if reserves were added by discovery, by acquisition or by revising upwards what had been under-reported. They therefore denied themselves knowledge of the actual discovery trend. It was not conspiracy or trickery but rather a matter of mind-set because the underlying notion of natural limits was simply not there. The accounts were designed simply to describe the current status as if there were infinite opportunities in exploration like indeed there are for most other businesses. If you want more potatoes, and the price is high enough, the simple solution is to grow more, and the system readjusts to deliver a normal economic return. Consistent with this way of thinking was the widely used parameter of *Reserve to Production Ratio*. It simply states that the Reserves could support current production for a given number of years with the tacit assumption that more Reserves could always be added as the need arose. It absolutely ignores the issue of depletion, which makes a nonsense of the calculation. It is absurd to imagine that production can be held static for a given number of years and then stop dead, which is implicit in the ratio once the notion of a finite limit is introduced.

In short then, the World approached the end of the last Century in denial about the depletion of the resource on which it had come to depend so heavily. Denial is perhaps too strong a word, as it was not exactly deliberate denial but rather a case of living in the past. *The Coming Oil Crisis*, the predecessor of this book, was not exactly a best seller when published but it did begin to contribute to a new awareness. The voices in the wilderness, and there were several of them,[1] began to be heard. A turning point was an article published in the *Scientific American* in 1998.[2] In addition to the oilmen themselves was what might be called the renewable lobby, promoting solar and wind energy, fuel cells and even nuclear energy. To that point, they had been primarily motivated by environmental issues, including climate change, but readily saw the significance of the depletion of fossil fuels. I began to be invited to speak at conferences, and the word began to spread.

THE CREATION OF ASPO AND ODAC

Professor Wolfgang Blendinger, an ex-Shell geologist, is the professor of petroleum geology at Clausthal University in Germany, and his own experiences in the oil industry gave him an intuitive grasp of depletion. He became interested, inviting me to give a lecture in December 2000 at his university, situated almost literally in the heartland of Germany on the flanks of the Herz Mountains. The lecture was filmed and streamed on the internet reaching a wide audience.

The German department with responsibility for natural resources, namely the *Bundesanstalt für Geowissenschaften und Rohstoffe* (BGR), sent a delegation to the lecture, and over some beers afterwards, I proposed trying to form an organisation to formalise the study of depletion. They suggested a meeting with Professor Wellmer, the Director in Hanover, whom I was able to visit a few days later. He welcomed the idea but suggested that the best approach would be to keep it informal to avoid inevitable bureaucratic delays. We were travelling on to Norway where my old friends in the Oil Directorate, who had initiated the study ten years before, joined with enthusiasm on the same informal basis.

I saw the need to give some identity to this ephemeral grouping and started to write a monthly Newsletter, at first distributed to a handful of

Fig. 14-1. Some key members of the ASPO network (from left to right, top to bottom):

Prof. Aleklett, Prof. Bardi, Dr Bentley,
Prof. Blendinger, Dr Campbell, Dr Gerling,
Mr Illum, Prof. Rosa, Prof. Wellmer,
Dr Zittel

interested people. Little did I imagine that it would grow, as it has done, now having a readership running to thousands. It is noteworthy that in the second issue, dated February 2001, I proposed the outline of a Depletion Protocol, of which more was to follow as described below.

In March, 2001, I received a call from Sarah Astor, the daughter-in-law of David Astor, previously the Editor of the Observer Newspaper. He had perceptively taken the oil shocks of the 1970s as a very serious matter, and was much impressed by a BBC film, *The Last Oil Shock*, which I had helped make. It was arranged that Sarah would meet me in Cork and join me for a bus ride to Dublin where I had to speak at another conference. She said that the Astor Family would like to endow an institute to raise awareness of the issue, which eventually became The Oil Depletion Analysis Centre (ODAC) in London. At first, it was run by Dr Roger Bentley from Reading University, who organised a successful workshop at Imperial College in London and began to analyse the data. Jim Meyer later took over the running of the organisation to concentrate on raising awareness by distributing news items primarily through the website.

Not long afterwards, I received a visit from a professor of nuclear physics at Uppsala University in Sweden, by the name of Kjell Aleklett, who had read the *Scientific American* article and saw the significance of oil depletion in relation to energy policy in Sweden. So, he joined the new organisation, which was named ASPO (The Association for the Study of Peak Oil and Gas). By July 2001, interest had grown widely with new members joining the network, such that virtually all European countries were represented by influential scientists in universities and government departments.

The next turning point came in May 2002 when Professor Aleklett organised The First International Workshop on Oil Depletion in Uppsala, to which about 65 people came from around the world, receiving wide media coverage. It is not necessary to record all the steps that followed. Subsequent annual workshops were held in Paris and Berlin, with the 2005 event following in Lisbon. Professor Aleklett has also organised a website,[3] as did several other national committees, including Ireland (see www.peakoil.ie). Somehow, ASPO has become a voice that is heard, although it is nothing more than a loosely sewn network of interested scientists. Even the Deutsche Bank, Aramco and the US Congress have referred to its position.[4]

In parallel with this endeavour, the late Buzz Ivanhoe organised a Newsletter in the United States through the Colorado School of Mines, which also began to attract serious attention.

CONFERENCES, SUBMISSIONS, BOOKS AND MEDIA ATTENTION

A growing world awareness of oil depletion and the inevitable peak of production began to spread. The ASPO members and their associates found themselves being invited to an increasing number of conferences around the World. There is no point in listing them all as the list is a long one, but it is worth mentioning some highlights.

Jens Junghans and Klaus Illum played key roles in organising a presentation in the Danish Parliament, followed up by a dedicated conference organised by the Danish Society of Engineers. There were the normal spectrum of presentations but they now began to include senior figures from the European Union and Government departments.

In London, Roger Bentley, and others made an official submission to the House of Lords, followed up later when Chris Skrebowski and I gave a presentation to select committees in the House of Commons in July 2004. The net began to widen as presentations were given by ASPO members and associates as far afield as Calgary, Houston, Abu Dhabi, India, Australia, Hawaii and Japan. In Canada, Julian Darley built up the Post-Carbon Institute, with the help of presentations and a website, addressing primarily the responses to peak oil, but taking peak oil itself as a foundation. In California, Kellia Ramarez started carrying the story on a internet news service.

In parallel with this came a growing interest by the media. Dutch Television, Korean Television, three different BBC programmes, Irish Television, French Television and a host of independent film producers started to arrive in Ballydehob. I became used to scrambling over rocks with the Atlantic breakers below, as a fitting backdrop, from which to explain the essence of the oil depletion argument. Amund Prestegard interested Norwegian television in a programme, facing eventual legal conflicts when he declined to change the substance of the message. Maj. Wechselmann from Sweden secured support for a film *Looking for La Luna*[5] retracing my steps from Trinidad to Colombia with lots of colourful oilfield images.

Ironically, it was in large measure the invasion of Iraq that prompted this new interest in oil depletion. Many people perceptively saw that it had an oil agenda, and began to ask just how important Middle East oil was. The BBC went so far as to broadcast a programme entitled *War for Oil*, being produced by David Strahan, whom I had already helped with the earlier programme *The Last Oil Crisis*. But the doubling of oil prices in the latter half of 2004 really began to concentrate the mind, leading to an avalanche of newspaper articles, including no less than the Wall Street Journal which sent a journalist to Ballydehob for an interview.[6] It also stimulated a large number of new books.[7]

THE ESTABLISHMENT BEGINS TO CONFESS

This growing popular awareness began to embarrass the official institutions, which had previously been able to obfuscate and confuse the issue with bland near-meaningless pronouncements, behind which governments could hide. The first to move was the World Energy Council, which in a presentation in Koblenz, not only accepted an imminent peak but spoke of the dire consequences.[8] Then, even the International Energy Agency began to shift its ground. In its 2004 World Energy Outlook, it belatedly accepted the notion of peak but built a scenario whereby it would not arrive until 2030 while in asides, footnotes and oblique comments demonstrating convincingly that it could not be delayed so long. When pressed at a conference in Sweden, they lamely responded to the effect that they simply developed hypothetical scenarios built on "business as usual" assumptions, and did not attempt to forecast what would happen in the real world.[9]

The United Kingdom Department of Trade and Industry published a forecast showing that Britain's oil and gas will be virtually depleted by 2020, although so far the Government itself seems to have failed to notice.

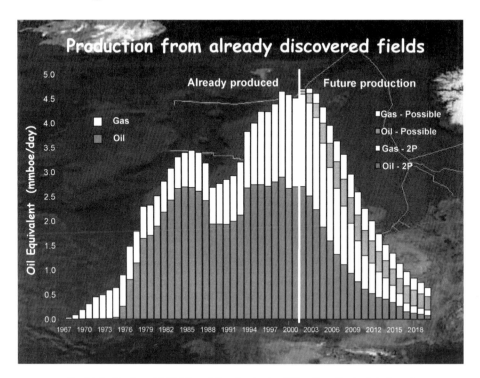

Fig. 14-2. Decline of UK Oil and Gas Production (After Dept.Trade & Industry)

DEPLETION PROTOCOL

Perhaps the most promising development of all relates is the so-called Depletion Protocol. It arose at a conference in London when I was asked to cover not only the problem but offer some ideas for a solution. It did not take long to see that the only real way forward was to cut demand to match world depletion rate. It happened that the then Secretary of OPEC, Mr Lukman, was in the audience, and he came up after the lecture expressing enthusiasm for the idea, which he thought would help reduce the tensions and pressure that OPEC was then under.

Later, I gave a talk at Uppsala University naming it *The Uppsala Protocol* for the occasion, which attracted some media interest in Sweden, but again nothing particular ensued. The next step came on April 5th 2003 when I received an invitation by no less than Mr Gorbachev to attend a conference organised by the Pio Manzu Research Centre in Rimini in Italy, entitled *The Economics of the Noble Path*. It was a remarkable affair at the Grand Hotel, at which philosophers and thinkers addressed the world condition. Armed police patrolled the corridors, and police helicopters hovered overhead. It seemed a good opportunity to propose again the Depletion Protocol, now re-named the *Rimini Protocol* for the occasion. It attracted much interest from the Italian press and television. I later drew up the Protocol in as formal terms as I could contrive as follows:

The Depletion Protocol

WHEREAS the passage of history has recorded an increasing pace of change, such that the demand for energy has grown rapidly in parallel with the world population over the past two hundred years since the Industrial Revolution;

WHEREAS the energy supply required by the population has come mainly from coal and petroleum, having been formed but rarely in the geological past, such resources being inevitably subject to depletion;

WHEREAS oil provides ninety percent of transport fuel, essential to trade, and plays a critical role in agriculture, needed to feed the expanding population;

WHEREAS oil is unevenly distributed on the Planet for well-understood geological reasons, with much being concentrated in five countries, bordering the Persian Gulf;

WHEREAS all the major productive provinces of the World have been identified with the help of advanced technology and growing geological knowledge, it being now evident that discovery reached a peak in the 1960s, despite technological progress, and a diligent search;

WHEREAS the past peak of discovery inevitably leads to a corresponding peak in production during the first decade of the 21st Century, assuming no radical decline in demand;

WHEREAS the onset of the decline of this critical resource affects all aspects of modern life, such having grave political and geopolitical implications;

WHEREAS it is expedient to plan an orderly transition to the new World environment of reduced energy supply, making early provisions to avoid the waste of energy, stimulate the entry of substitute energies, and extend the life of the remaining oil;

WHEREAS it is desirable to meet the challenges so arising in a co-operative and equitable manner, such to address related climate change concerns, economic and financial stability and the threats of conflicts for access to critical resources.

Now it is proposed that

1. A convention of nations shall be called to consider the issue with a view to agreeing an Accord with the following objectives:
   a. to avoid profiteering from shortage, such that oil prices may remain in reasonable relationship with production cost;
   b. to allow poor countries to afford their imports;
   c. to avoid destabilising financial flows arising from excessive oil prices;
   d. to encourage consumers to avoid waste;
   e. to stimulate the development of alternative energies.
2. Such an Accord shall have the following outline provisions:
   a. No country shall produce oil at above its current Depletion Rate, such being defined as annual production as a percentage of the estimated amount left to produce;
   b. Each importing country shall reduce its imports to match the current World Depletion Rate, deducting any indigenous production.
3. Detailed provisions shall cover the definition of the several categories of oil, exemptions and qualifications, and the scientific procedures for the estimation of Depletion Rate.
4. The signatory countries shall cooperate in providing information on their reserves, allowing full technical audit, such that the Depletion Rate may be accurately determined.
5. The signatory countries shall have the right to appeal their assessed Depletion Rate in the event of changed circumstances.

I think it certainly deserves urgent attention by the World governments as offering a mechanism for a managed transition to declining oil and gas supply. Demand would be put into better balance with supply, meaning that World prices would be held low, to be in reasonable relation to actual production cost. This would allow the poor countries to afford their minimal needs. Profiteering by particularly the Middle East producers, which in turn leads to massive and destabilising flows of money, would be avoided. Above all, the consumers would be forced to face the reality of their predicament. Even the Middle East itself would benefit by being forced to prepare by lessening its dependence on oil revenue which is inevitably set to decline in the future as depletion hits that region too.

Interest in the proposal does now seem to be growing. A committee of international politicians considered it at the 2005 ASPO Conference in Lisbon, to be followed by another in Rimini when world leaders are to be invited to address it.

Speaking of protocols, it is interesting to note the changing reaction of what can be called the Climate Change lobby. Its models of damaging emissions are flawed to the extent that they are based on extrapolations of oil demand rather than supply, and at first it seemed as if the protagonists were negative to any notion that the natural depletion of fossil fuels would reduce the impact on the environment. But now they seem to become more positive seeing that the Rimini and Kyoto Protocols actually work in parallel, albeit for different motives.

In short, the World does begin to wake up. How successful it will be in facing the challenges remains to be seen, but at least it becomes increasingly aware of the issue. Some countries may adopt policies to secure oil by military means, which, if successful, would raise the peak and steepen the subsequent decline, making a bad situation worse. Others may begin to find ways to use less, and find alternative ways to live. No one should under-estimate the challenges.

NOTES

1. Notably, L.F.Ivanhoe, Walter Youngquist, Richard Duncan, Jean Laherrère, Alain Perrodon, Brian Fleay, Roger Bentley.
2. Campbell C.J. and J.H.Laherrère.
3. See www.peakoil.net.
4. Deutsche Bank Research December 2 2004 www.dbresearch.de/PROD/DBR.
5. The *La Luna Formation* is the prime oil source-rock in northern South America.
6. See Jeff Ball, *Dire Prophecy*, Wall Street Journal, September 21 2004.
7. See Heinberg.
8. J-M.Bourdaire.
9. Stockholm meeting Dec. 2004.
10. Clare Short.

Chapter 15

SYNTHESIS – WHAT IT ALL AMOUNTS TO

IF YOU HAVE COME this far, you deserve a medal, and so, as a matter of fact, do I. If you have simply browsed, that will have served its purpose too if it has prompted you to think more about the implications of depleting the World's premier energy supply. We are not used to depleting things, being confident that we can always run into the supermarket and replenish our stocks. And, we prefer not to think about the end of the one finite resource we do know about only too well: our own life-span. But in the 21st Century, we will come to experience the virtual depletion of oil, an energy source that has become central to our way of life. We will have to change the way we live as production declines towards eventual exhaustion. I stress that it is the onset of terminal decline that is more relevant that the end of oil itself. It is not too soon to start thinking about what that may entail.

In this last chapter, I will try to sum up the message of this book.

THE FORMATION AND ENTRAPMENT OF OIL AND GAS

Oil is derived for algae that proliferated from time to time in the Earth's long geological history. Gas comes from plant remains, being more widely distributed. On death, the organic material sank to the bed of the sea or lake in which it had lived, or was washed in from the surrounding land. In most cases, it was dissolved or destroyed, and only rarely in stagnant troughs was it preserved and concentrated. The resulting organic-rich layers were buried by other sediments, and with further subsidence became heated by the Earth's heat-flow. After a critical exposure to heat, the organic material was converted to oil and gas by chemical reactions. It is obvious why the circumstances for prolific oil generation occurred so rarely. The conditions for generation normally lie at depths of 2000 to 5000 m. There is not much oil to be found deeper.

Once formed, petroleum, whether consisting of oil or gas, which was under great underground pressure, began to migrate upwards through the rocks in hair-line fractures. In many cases, it just dissipated, but in some instances it encountered a porous and permeable layer, such as a sandstone, along which it could flow, displacing the water that previously filled the pore-spare between the grains of rock. If the conduit led straight to the surface, the petroleum escaped to the atmosphere, leaving behind a sticky tar residue. But in some cases it was folded or cut by faults, providing traps in which oil and gas accumulated. No trap has perfect integrity, and much leaked out over time. Older rocks therefore become less prospective because they have been exposed to leakage for longer.

Much of the petroleum was trapped in the cracks and crevices of the migration paths along which it moved, and much was lost at the surface. So, only about one percent of the amount generated found its way into traps that are large enough to be exploited in oilfields. As mentioned, oil accumulations at shallow depth on the margins of basins were oxidized and attacked by bacteria becoming tar and heavy viscous oil, whereas oil that was overheated on being buried too deeply was cracked into gas.

Gas contains dissolved liquid hydrocarbons that condense at the surface, and can also be extracted by processing, being known respectively as *Condensate* and *Natural Gas Liquids*. Oil also contains dissolved gas, which may separate out in the reservoir to form a gas-cap over an oil accumulation.

EXPLORATION

In earlier years, geologists searched the world for seepages of oil at the surface, and looked for promising structures in the vicinity to trap it. They found most of the prolific basins, and many of the giant fields, in this way.

Later, seismic surveys were developed to explore what lay beneath the surface. The technique involves the release of energy from an explosive charge, or in other ways, at the surface, and recording the echoes reflected back from rock interfaces far underground. By computing the time taken for the echoes to return, it is possible to calculate the depth and configuration of the buried structures.

The offshore was opened after the Second World War, as marine seismic surveys were perfected, such that it became possible to map the continental shelves rapidly and in great detail. Technological progress has greatly improved the resolution of seismic surveys, and the computer work-station has brought enormous computing power to the interpretation. Geologists and geophysicists were able to investigate the oil zones thoroughly, searching for thin and subtle reservoirs, and ever smaller traps.

Exploration boreholes, known as *wildcats*, are drilled to test prospects identified by geological interpretation and to gather information. The technology of drilling has made enormous progress, such that it has become routine to drill 5000 m wells in the stormy waters of the North Sea. The process involves drilling a large diameter hole from the surface, commonly 30 inches in diameter, and then cementing steel casing into it to seal off the formations. The diameter of the hole is progressively reduced as sections are cased off. The drill string, with a bit on the end of it, rotates to make the hole, and a special mud is pumped through the drill string to lubricate the bit and remove the cuttings. The mud is weighted up with the heavy mineral, barytes, to balance the formation pressure: if it is too high, the mud escapes into the formation; if it is too low the formation encroaches on the well and causes the bit to stick.

Geologists examine the cuttings brought to the surface in the drilling mud-stream to identify the rocks the borehole is penetrating. Cores are taken where necessary for closer inspection. Sondes are also lowered down the borehole to record the electrical and radioactive properties of the rocks, making it possible to identify different rock types, measure porosity and determine which zones are oil- or gas-bearing.

The breakthrough offshore came with the development of the semi-submersible rig, in which a platform, holding the drilling derrick, is mounted on two submerged pontoons that lie beneath the wave base, providing a stable structure relatively unaffected by the weather.

Other important developments have been to find ways to drill highly deviated wells to reach far out from a platform. In extreme cases, the wellbore may be up to 90 degrees from the vertical. It can track a thin productive zone, which can be drained rapidly. Also, a single well may have several branches at depth. Various techniques to improve the permeability of the reservoir can be applied, such as injecting acid or fracturing it by injecting fluid under very high pressure.

To produce an oil zone, it is necessary to first seal it off from the overlying and underlying strata, and then let off controlled explosive charges in the well to pierce holes in the casing to allow the oil to flow in. Normally, there is sufficient pressure in the reservoir to expel the oil to the surface, although sometimes it has to be pumped.

In short, advances in technology have made exploration and production highly efficient. The geological processes responsible for oil accumulation are now very well understood. Of particular importance was the geochemical breakthrough of the 1980s that made it possible to identify and map the zones generating oil and gas. It not only showed which trends had potential but it allowed large tracts to be written off as non-prospective, once the critical information had been gathered.

RESERVE ESTIMATION

Before a *wildcat* is drilled, the geologists and engineers have to estimate the likely reserves of the prospect to determine if it has the potential to be commercially viable. They map the volume of the trap, using seismic data and applying their best estimate of the likely reservoir conditions. The first well will reveal whether or not it is oil or gas bearing as well as much more information about the details of the reservoir, but it is usually necessary to drill several appraisal wells to confirm the estimates.

There remains a range of uncertainty in both technical and financial terms. It is normal and reasonable to plan expensive developments on very conservative estimates, which are commonly termed *Proved Reserves* under Stock Exchange rules. They are naturally subject to upward revision over the life of the field: the increase being termed *reserve growth*. It is, however, widely misunderstood, being taken as a dynamic akin to exploration, driven by improved technology, when in fact it is little more than the natural evolution from a low initial estimate.

While, the reserves of a field will be known absolutely only on the day when it is finally abandoned, at which point they equate with the *Cumulative Production*, the estimation of reserves is a straightforward procedure in technical terms. The reporting of reserves is, by contrast, a *political* act, reflecting more the needs of the reporting authority than the actual situation. In the absence of clear and universal definitions, reporting procedures or audit, there is plenty of scope for latitude in reserve reporting. In part, companies treated reserves as a form of inventory which they booked as best met their financial and commercial needs.

Government statistics are often unreliable: the greatest distortion arising in several OPEC countries in the late 1980s when they were competing with each other for production quota, based on reported reserves. It now transpires that they began reporting the total found, not what remains, which explains why the numbers have barely changed for twenty years despite production.

CATEGORIES OF OIL

It is normal classify oil into *conventional* and *non-conventional* categories but unfortunately there is no agreement on the boundary adding to the confusion. Here, we coin the term *Regular Conventional Oil* (or simply *Regular*), defining it to exclude oil from coal and shale; bitumen and extra-heavy oil, heavy oil, deepwater oil, polar oil and liquids from gas plants. Most of the oil produced to-date falls in the *Regular* category, as so defined, which will dominate all production far into the future. It has a characteristic depletion profile with production starting at zero, and rising rapidly to one or more peaks before declining exponentially. The overall peak generally comes close to the midpoint of depletion when half the total has been extracted.

In addition, there are large amounts belonging to the other categories, which will ameliorate the decline after peak but have little impact on peak itself.

Based on a full evaluation of the data, conflicting and inconsistent as it is in many respects, it can be said that the World has now about reached the end of the First Half of the Age of Oil. It lasted 150 years and saw the rapid growth of industry, transport, trade, agriculture and financial capital, made possible by an abundant supply of cheap oil-based energy. The Second Half now dawns and will witness the decline of oil and all that depends upon it.

HOW MUCH HAS BEEN FOUND AND WHEN

The search for oil has been going on for all these years. Almost everything, that there is to know about the geological conditions responsible for it, has been learnt. The World has been very thoroughly explored. Much of it is made up of ancient shields, or oceanic rocks, that are absolutely non-prospective. Almost all potential basins have now been identified, and investigated by seismic means and drilling in fair detail. It is, accordingly, nearly inconceivable that any new provinces remain to be discovered. That said, there are certain new tracts in very hostile environments that are under-evaluated, such as in Antarctica, the Falkland shelf, the Greenland icecap, off Iceland, and in several Arctic provinces. There is no particular reason to think that they are oil-bearing, still less that they can yield any significant amount, but until they have been checked some uncertainties remain.

The prospects in the Former Soviet Union and China have not been so well known in the West, which has tempted some to attribute a large undiscovered potential to these areas. I think that the Soviet explorers were certainly as intelligent as their western counterparts, and the systematic exploration of the Soviet system was efficient, being spared the commercial constraints of the West. It is said that oil is found in the head of the geologist, and I think that Soviet heads were no thicker than ours. Their technology may not have been as advanced, but most of the oil in the West was found long ago, when technology was even less advanced. I therefore think that all the large productive basins have been found as well as most of the giant fields within them.

Most deep and very deep water areas are non-prospective for geological reasons, and the main prospective tracts have now been found. They depend on divergent plate tectonic settings, where the continents pulled apart, as occur in Gulf of Mexico and along the margins of the South Atlantic. Elsewhere, deltas may locally extend into deepwater but have to rely on the source rocks within the deltas themselves which are commonly lean and gas-prone.

Figure 15-1 provides the essential data on discovery, showing that 944 billion barrels of *Regular Conventional Oil* have been produced and that estimated reserves from known fields stand at 760 billion barrels. Together, that adds to a total discovery of 1.7 trillion barrels.

About sixty percent of what has been discovered lies in just over three hundred giant fields, many in the Middle East. The peak of giant discovery was in the 1960s, and the discovery rate has fallen dramatically in recent years.

Produced through 2004	944 Gb
Reserves	760
Discovered	1704
Yet-to-Find	146
Yet-to Produce	906
Ultimate	1850
Depletion Midpoint	**2003**
Depletion Rate	2.6%
5 yr-Av. Discovery	12.5

Fig. 15-1. The World's Regular Conventional oil endowment

Discovery as a whole also peaked in the 1960s being heavily influenced by the contribution of giant fields. Fewer and fewer fields are being found and the average size is falling. Consumption exceeded discovery in 1981 and the gap is widening, see Figure 15-2. Exploration drilling has been in decline for many years because there have been fewer and fewer valid prospects left to test. Modern technology and advances in knowledge have made it easier to identify which are valid.

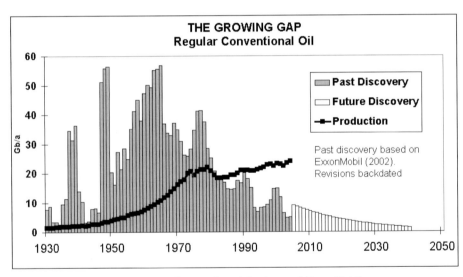

Fig 15-2 The Growing Gap (after Longwell, ExxonMobil)

HOW MUCH REMAINS

The sum of the *Reserves* and the *Yet-to-Find* gives how much remains to produce.

Estimating how much is *yet-to-find* is not easy, but nor is it quite as difficult as it was. The World has been so extensively explored that we can be sure that almost all, if not all, of its prolific basins have now been identified. The bulk of what remains to be found lies in ever smaller fields within the established provinces.

We can estimate this amount by old-fashioned geological judgment relating the maturity of exploration with the underlying geology, and there is now sufficient data to use statistical approaches.

When the first well is drilled in a basin, nothing is known about the ultimate distribution of field size, but when the last well is drilled everything will be known. As we get close to the finishing line, we can begin to see it clearly. Jean Laherrère has discovered a law of distribution stating that objects in a natural domain plot as a parabola when their size is compared with their rank on log-log scales. For example, the populations of the larger towns in a country can be plotted to yield the population down to the smallest settlement. It means that when the larger oilfields in a basin have been found, their size distribution can be used to predict what the *Ultimate* recovery will be. The difference between this and what has been discovered gives the *Yet-to-Find*.

Another approach is to plot cumulative discovery against the *wildcats* drilled. The plot is generally hyperbolic with the larger fields found first, and the asymptote equates with *Ultimate* recovery, subject to an economic cut-off for very small fields. Cumulative discovery may also be plotted over time. Lastly, production trends can be correlated with their related discovery trends and extrapolated to zero, giving an indication of how much is yet to produce.

It is a case of using all of these techniques, as well as judgment, to come up with the best estimate, remembering always to distinguish *Regular Conventional* oil from *Non-conventional* oil.

My best estimate computes that there are almost 150 billion barrels *yet-to-find*. With reserves of 760 billion barrels, it means that there are about 900 barrels *yet-to-produce*.

The distribution of the *yet-to-produce* is most uneven: about half of it lies in just five Middle East countries. The ten largest countries hold three-quarters of it.

PRODUCING WHAT REMAINS

The production of any finite commodity starts at zero, rises to one of more peaks and ends at zero. Think of your life-time spending pattern: you spend little in the cradle or the coffin, but have several peaks around middle age. It is the same with oil production in a country: peak comes around the midpoint of depletion. It could come a little before midpoint if there are a lot of giant fields found early; or it could come after midpoint if peak production were artificially restricted by, for example, pro-rationing or OPEC quota. But the general coincidence of peak and depletion midpoint is valid for both theoretical and empirical reasons.

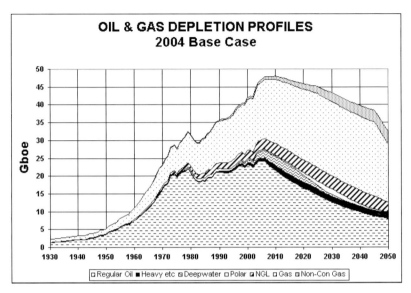

Fig. 15-3. Production forecast

Most countries are now either past midpoint or close to it, except the United States, which is far past it, and the five Middle East countries, which are far from it.

It means that the producing countries of the World can be divided into those past midpoint where production is set to continue to decline, and those which have not yet reached it where production can be held at current levels or even increase. We may further consider the five Middle East countries as an extreme category of the pre-midpoint group, because their *yet-to-produce* is so large and the depletion rate so low. They have behaved as swing producers, making up the difference between world demand and what the others can produce, although this role has effectively ended as they reach their practical productive capacities.

SCENARIOS

World demand for oil rose rapidly until the 1970s, when it briefly declined before again beginning to increase at a slow rate. Most forecasters now predict rising demand, driven by the expanding economies of the East and the growing World population. It has been rising at around 2% over the past few years, and many forecasts, probably wrongly, assume that that will continue to do so at around this level.

One can envisage a large number of alternative scenarios of supply and demand, but here we will be content with one even though there will surely be departures in the real world. It seems fair to assume that the World enters an epoch of vicious circles. They start with a price shock, coming when capacity limits are breached, which triggers a recession leading to a fall in demand, which reduces pressure on price. The economy then recovers, causing demand to build until it again hits the capacity ceiling, itself falling, which results in to a new price shock. The capacity limits are declining gradually from natural depletion, suggesting ever more severe vicious circles, until some new equilibrium is reached.

SHOCK

In an ideal world, governments would properly study the resource base and understand the principles of depletion. They do not, and in democratic societies probably cannot because they are elected for short terms and are therefore motivated to deliver short-term benefits to their electors. As a consequence, it is most unlikely that the governments of either the United States or the European Union will adopt an energy policy with the aim of preparing for the inevitable peak in oil production and subsequent scarcity. It looks very much as if the United States prefers a policy of trying to secure oil supply by military means. Even if successful, it would bring only temporary relief because the higher the short-term production, the steeper the subsequent decline, making a bad situation worse.[1]

However, shrewd financiers are now waking up to the enormity of what unfolds. They recognise that the financial system of the past was built on the assumption that the expansion of tomorrow provides the collateral for the debt of today, but now question if expansion is any longer possible in the face of the decline of oil production which furnished the energy that made the expansion possible. They find themselves in a dilemma of trying to protect their wealth yet avoiding any steps that might trigger a general collapse in a market where so much depends on sentiment. The financial reports of virtually all companies quoted on the exchanges make the tacit assumption of a "business-as-usual" supply of the cheap oil-based energy on which their businesses depend, but this assumption is no longer valid.

It follows that we are in for a shock and a big one at that, but, by its very nature, a shock is something that defies prediction or description in detail. The invasion of Iraq, with its not so hidden oil agenda, sets the scene for what must follow in an epoch of growing tension and uncertainty.

POST-SHOCK

But gradually, the realities will filter through. The Third World will be hit first: its oil-based energy consumption will begin to falter. We already have an example of what happened to Cuba when cheap Russian imports ended with the collapse of the Soviets

"Cuba has become an undeveloped country. Bicycles are replacing

automobiles. Horse-drawn carts are replacing delivery trucks. Oxen are replacing tractors. Factories are shut down and urban industrial workers resettled in rural areas to engage in labour intensive agriculture. Food consumption is shifting from meat and processed products to potatoes, bananas and other staples".[2]

But ironically, Cuba may set an example of a sustainable community, living within its means, becoming the envy of other less prepared places and communities.

The world price of oil has already doubled to around $50 barrel, without sign of reprieve. It may yet go higher before yielding a major economic and political discontinuity in the way the World lives. It heralds the end of rampant and mindless consumerism in the more developed countries that brought great suffering to many, especially in the Third World.

Every effort will be made to find alternative and renewable sources of energy. The use of renewable energy will expand rapidly and successfully although by no means offering a substitute for the cheap oil-based energy known so-far. Coal mining will be stepped up with adverse environmental consequences, especially in places like China where the power stations lack adequate smoke filters. Nuclear energy may stage a come back if the risks and problems of waste disposal can be resolved, although high grade uranium is itself resource constrained.

The greatest progress will however have to be made in terms of using less. The World will become a very different place with a smaller population. The transition will be difficult, and for some catastrophic, but at the end of the day, it may be a better and more sustainable place.

That seems to be a logical interpretation, but is it the correct one? I don't know, but I hope that the discussion which you have so patiently read will prompt you to think about it. I further hope that, having thought about it, you will make some provisions to protect yourselves as well as you can.

NOTES
1. See Klare, *Blood for Oil.*
2. Falcoff, 1995, quoting Preeg & Levine, 1993.

BIBLIOGRAPHY

(Number [#   ] refers to Bibliographic Notes and archives of ASPO IRELAND; [B#] is a book reference)

A

AAPG, 1991, *The Arabian Plate – producing fields and undeveloped hydrocarbon discoveries*; map published by Amer. Assoc. Petro. Geol.[#31]

AAPG Explorer, 1993, *Future energy needs not being addressed*; AAPG Explorer June 1993 [#156]

AAPG, 2001, Discussions and Reply re*"Energy Resources – cornucopia or empty barrel"*; AAPG Bulletin, Vol. 85, No. 6, June 2001 [#1333]

AAPG, 2000, *World Oil Resource Forecast Increases*; AAPG Explorer, pp. 24/25, June, 2000 [#1556]

AAPG, 2001, *USGS Assessment Study Methods Endorsed*; AAPG Explorer, November, 2001 [#1711]

Abelson A., 1996, *Crude awakening*; Barrons 1.5.96 [#431]

Abernathy, V.D., 2001, *Carrying capacity: the tradition and policy implications of limits*; Science & Environmental Politics http://www.esep.de/articles/esep/2001/article1.pdf

Abi-Aad N., 1995, *Challenges facing the financing of oil production capacity in the Gulf*; Petroleum Review, Inst. Pet. London Feb 1995 83-86 [#321]

Aburish S.K., 1994, *The rise, corruption and coming fall of the House of Saud*; 226p Bloomsbury, London [#B15]

Abraham K S., 1994, *Low oil prices killed the USSR*; Pet. Eng. Int., Sept. 1994 [#278]

Abraham, S., 2001, *We Can Solve Our Energy Problems*; The Wall Street Journal, Friday, May 18, 2001 [#1591]

Adam P., 1997, *Modern asset pricing – a map of the future?* Petroleum Review Feb. [#545]

Adams T., 1991, *Middle East Reserves*; Oilfield Review pp.7–9 [#27]

Adelman M.A., 1989, *Problems in modeling world oil supply*; Energy Modeling Forum, MIT Energy Lab Working paper MITEL -89-010WP [#104]

Adelman M.A., 1995, *The genie is out of the bottle: world oil since 1970*; MIT ISBN 0-262-01151-4

Adelman M.A. & M.C. Lynch, 1997, *Fixed view of resource limits creates undue pessimism*; Oil & Gas Journ. April 7th 1997 [#615-2]

Adelman M.A. and M.C. Lynch, 1997, *More reserves growth*; Oil & Gas Journ. June 9th [629]

Adelman, M.A., 2001, *Oil-Use Predictions Wrong Since 1974*; The Wall Street Journal, March 6, 2001 [#1772]

Adelman M.A, 1993 *Economics of Petroleum Supply*

Adelman M.A., 2002, *World Oil production & prices 1947–2000*, The Quarterly Review of Economics and Finance 42 pp.169–191, 2002 [#2152]

Ahlbrandt, T.S. & Charpentier, R R. & Klett, T R. & Schmoker, J W. & Schenk, C.J. & Ulmishek, G F., 2000, *Future Oil and Gas Resources of the World*; Geotimes, June 2000 [#1503]

Ahlbrandt, T., 2000, *USGS World Petroleum Assessment 2000*, USGS [#1577]

Ahlbrandt T.S. & P.J. McCabe, 2002, *Global Petroleum Resources: a view of the future*; Geotimes Nov. 2002 [#2030] [#2036]

Ahmed N.M., 2002, *The war on freedom*; Media Messenger p.398

Akehurst J., 2002, World oil markets and the challenges for Australia: ABARE Outlook 2002, Canberra 6 March 2002.

Al-Jarri A.S. & R.A. Startzman, 1997, *Worldwide supply and demand of petroleum liquids*; SPE 38782 [#699]

Al-Husseini S.,2004, *Why higher oil prices are inevitable this year, rest of decade*; Oil & Gas Journal, August 2

Alahakkone R.R., 1990, *Convenional supply of oil will slow down*; Oil & Gas Journ., February 5 1990 [#81]

Alberta Energy & Utilities Board., 1998, *Non-conventional oil production for Alberta* [#708]

Aldinger C., 1995, *World dependence on Gulf oil will grow*; Pentagon press release [#356]

Aleklett K., 2001, *Brannpunkt*; Svenska Dagbladet 23 March

Aleklett K., 2001, *Oljebrist hotar var civilisation*; Brannpunkt [#1208]

Aleklett K. and C.J. Campbell, 2003, *The peak and decline of world oil and gas production*; Minerals & Energy v.18.no1

Alezard N., J.H. Laherrere & A. Perrodon, 1992, *Réserves et resources de pétrole et de gaz des pays Mediterranées*; Revue de l'Energie 441, September 1992 [#140]

Alhajji, A F., 2001, *Middle East: Investment levels rise higher*; World Oil, August 2001 [#1488]

Ali M., 1999, *Oil Crisis*; Al Arab, 19 November [1020]

Allen R.V., 1966, *The man who changed the game plan*; The National Interest 44 summer 1996.

Alternative Energy Inst., 2001, *Turning the Corner: energy solutions for the 21st Century*, ISBN- 0-9673118-2-9

Alvarez C.G., 2000, *Economia y politica petrolera*: published by Ecopetrol, Colombia

Amenson H., 1988, *Oil discovery principles, exploration strategy*; Oil & Gas Journ. Oct 17 1988 [#89]

Amiel, B., 2002, *Bush's victory is the voice of an angry America*; The Daily Telegraph, Nov.11, 2002 [#2018]

Amuzwgar J., 1999, *Managing the oil wealth*; Tauris, 266 pp.

ANA, 1999, *Rohoel: In Zukunft ein kostbares Gut*; ANA, January 30, 1999 [#1505]

Anderson, B., 2001, *The creation of a Palestinian state is essential for the war against terrorism*; The Spectator, October 20, 2001 [#1671]

Anderson Forest, S., 2000, *Industrial Management: Energy – "There Is Not Enough Gas Around"*; Business Week, September 18, 2000 [#1388]

Anderson Forest, S., 2000, *Unnatural Demand for Natural Gas*; Business Week, April 3, 2000 [#1644]

Anderson R.B., 1999, *Gas tanks on full; supplied heading for empty*; The Oregonian April 10 [#1046]

Andrews S., 1997, *Letter to Editor*; Oil & Gas Journ. [#582]

APPEA,2002, *A natural gas development strategy for Australia* [#1855]

APS., 1994, *Middle East strategy to the year 2007; Conference programme and participants*; APS, Cyprus [#271]

ARAMCO, 1980, *Aramco and its world: Arabia and the Middle East*; (Ed. Nawwab I.I., P.C.Speers, & P.F.Hoye) Aramco. ISBN 0-9601164-2-7.

Arbatov A.,1994, quoted in Abraham K.,1994, *Economists project mixed view to 2000*; Petrol. Eng. Int., June 1994 .

Arbatov A.,1994, *The foreign concerns of the Russian oil and gas complex*; preprint APS Conf. [#293]

Arnold R., G.A.Macready & T.W.Barrington, 1960, *The first big oil hunt: Venezuela 1911-1916*; Vantage Press, New York 353p [#B3]

Arnold, W., 2001, *A Gas Pipeline to World Outside*; The New York Times, October 26, 2001 [#1727]

Arthur, C., 2003, *Oil and gas running out much faster than expected, says study*; The Independent, October 2, 2003 [#1962]

Arthur C., 2003, *Fossil Fuels Exhausted-Study Predicts Depletion of Oil and Gas*; Triangle Free Pres, December 30, 2003 [#2310]

Ashby T., 2000, *Venezuela not interested in spare capacity*; Reuters [#1058]

Associated Press, 2000, *It could be a cold, expensive winter*; The New Mexican, Wed.,December 13, 2000 [#1593]

Associated Press, 2001, *Rising use of cleaner natural gas a concern*; The Register-Guard, December 25, 2001 [#1738]

ASTM, 1982, *Standard for metric practice*; ASTM E-380-82 [#341]

Attanasi, E.D. & Mast, R.F. & Root, D.H. & USGS, 1999, *Oil, gas field growth projections: wishful thinking or reality ?*; Oil & Gas Journal, April 5, 1999 [#1596]

Attarian J., 2000, *The Age of Impiety*; Culture Wars, Feb. 2002 [#2040]   Attarian J., 2001, *The Unsustainability of Economism: Economism is not feasible economically*; The Social Contract, Summer 2001[#2028]

Attarian J., 2002, *The Coming End of Cheap Oil*; The Social Contract Vol.XII, No.4, 2002 [#2026]

Attarian J., 2002, *Economism and the National Prospect*; The Social Contract, Winter 1999–2000 [#2027]

Attarian J., 2002, *Economism vs Earth*; The Social Contract, Fall 2002 [#2038]

Attarian J., 2002, *Malthus Revisited*; The World & 1 pp.257–271 [#2039]

Attarian J., 2003, *After The Bombing: To Hell With Iraq- and Immigration*; VDARE.com, August 28, 2003 [#2239]

Attarian J., 2003, *Blood for Oil- and Other Things: The coming struggles over resources*; The Social Contract, Summer 2003 [#2246]

Attarian J., 2003, *You can't find oil or natural gas if it's not there*; The Detroit News, August 12, 2003 [#2247]

Attarian J., 2003, *The Jig Is Up: Supplies of oil and gas are running out*; The Social Contract, Fall 2003 [#2283]

Australian Treaty , 1979, *International Energy Program* [#782]

Ayres R.U. & P. Frankl, 1998, *Toward a non-polluting energy system*; Environ. Sci. & Tech. September 1. [#831]

Azar C. & Lindgren K. & Andersson B.A., 2003, *Global energy scenarios meeting stringent CO2 constraints-cost effective fuel choices in the transportation sector*; Energypolicy, Elsevier, 2003 [#2311]

B

Bachtold D., 2003, *Britain to Cut CO2 Without Relying on Nuclear Power*; Wall St. Journ., February 29 2003 [#2130]

Baer R., 2003, *The Fall of the House of Saud*; The Atlantic Monthly; May 2003 [#2146]

Bahree B., 1998, *Oil demand policies distort policies in the Gulf*; Wall St Journ. February 23 [#677]

Bahree B., 1999, *Global demand for oil is set to grow in 2000 at faster pace*; Wall St. Journ August 11 [#1004]

Bahree, B. & Barrionuevo, A., *OPEC Warns of Price Collapse Next Year*; The Wall Street Journal, Tuesday, October 30, 2001 [#1714]

Bahree, B. & Fialka, J., 2002, *Oil industry ponders Iraq risks*; Wall Street Journal August 30 2002 [#1913]

Bahree B. & Herrick T., 2003, *Officials say OPEC can't keep a lid on rising crude-oil prices*; Wall Street Journal, Feruary 28, 2003 [#2112]

Baker P., 2002, *Russian Oil being seen as lifeline, pipe dream*; The Washington Post, Sept.9, 2002 [#2097]

Baker Institute Study, 2000, *Japanese Energy Security And Changing Global Energy Markets: An Analysis of Northeast Asian Energy Cooperation and Japan`s Evolving Leadership Role in the Region*; James Baker III Institute For Public Policy of Rice University, No.13, May 2000 [#1437]

Bakhtiari A.M.S., 1999, *The price of crude oil*; OPEC Review 13/1 March

Bakhtiari A.M.S., 2000, 1999, *IEA, OPEC oil supply forecasts challenged*: *Oil & Gas Journ*, April 30

Bakhtiari, A.M.S., Shahbudaghlou, F., 2001, *IEA, OPEC oil supply forecasts challenged*; Oil & Gas Journal, April 30, 2001 [#1295]

Bakhtiari, A.M.S., 1999, *The price of crude oil*; OPEC Review, Vol. 23, No. 1, March 1999 [#1296]

Bakhtiari, A.M.S., 2001, *2002 to see birth of New World Energy Order*; Oil & Gas Journ, January 7, 2002 [#1763]

Bakhtiari Oil Consultants, 2001, *OPEC Member Countries: Forecasts For Oil Production Capacities (2001–2020)*; October 2001 [#1680]

Bakhtiari, A.M.S., 2003, *Middle East production to peak with next decade*; Oil & Gas Journ., July 14

Bakhtiari, A.M.S., 2003, *North Sea oil reserves; half full or half empty?*; Oil & Gas Journal, August 25, 2003 [#1943]

Bakhtiari, A.M.S., 2003, *Middle East oil production to peak within next decade*; Oil & Gas Journal, July 7, 2003 [#2220]

Bakhtiari, A.M.S., 2003, *The World Oil Production Capacity Model*; presented at the International Oil Conference, Copenhagan, Denmark, December 10, 2003 [#2276]

Baldauf S., 1998 *World's oil supply may soon run low*: Christian Science Monitor 23 September [#840]

Ball, J., 2002, *Bush Will Shift Gears on Auto Research, Favoring Fuel Cells Over 80 MPG Sedan*; The Wall Street Journal, January 9, 2002 [#1747]

Ball, J. 2004, *As prices soar, doomsayers provoke debate on oil's future*; Wall Street Journal September 21

Banerjee, N. and Kapner S., 2001, *Drilling in Alaska presents hard choices to BP*; New York Times [#1197]

Banerjee, N., 2001, *Fears, Again, of Oil Supplies at Risk*; The New York Times, October 14, 2001 [#1723]

Banerjee, N., 2001, *The High, Hidden Cost of Saudi Arabian Oil*; The New York Times, Oct 21, 2001 [#1725]

Banks F.E., 2002, *The World Oil Market: From Myth to Meaning*; Uppsala, 2002 [#2005]

Banks F.E., 2003, *An Inexpensive Lesson on Oil and Economics*; meeting of the International Association for Energy Economics, Prague June 5–8, 2003 [#2228]

Banks H., 1998, *Cheap oil: enjoy it while it lasts*; Forbes June 15 [#748]

Barber B.R., 1995, *Jihad vs. McWorld*; ISBN 0-345-38304-4

Bardi U., 2003, *La Fine del Petroleo*; Ed. Reuniti ISBN 88-359-5425-8 244p

Bardi, U., *Forecasting oil prices and productions*; *A comparison of historical trends with predictions*; Universita di Firenze, Italy, [#1949]

Barker T., 1995, *Taxing pollution instead of employment: greenhouse gas abatement through fiscal policy in the UK*; Energy & Environment 6/1 [#471]

Barker, R., 2001, *The Oil Spigot May Be Closing*; Business Week, November 12, 2001 [#1715]

Barkeshli F., 1996, *Oil prospects in the Middle East and the future of the oil market*; Oxford Energy Forum, 26 August 1996 10–11 [#483-1, #493]

Barnes J., 2003, *This war was not worth a child's finger*; The Guardian, April 4, 2003 [#2257]

Barrett M.E., 1994, *Amoco Corporation*; American Graduate School of International Management. [#499]

Barrett M.E., 1991, *Global Marine Inc*; American Graduate School of International Management. [#500]

Barrett W.J., 1992, *Calling all bottom fishers*; Forbes May 11th 1992 [#57]

Barrioneuvo A, 2001, *Oil services firms take a bath on deep-water drilling*; Industry Focus [#1225]

Barrlett D.L. & Steele J.B., 2003, *Why America is running out of Gas*; Time Magazine July 13, 2003 [#2204]

Barron's, 1994, *What Next for oil prices*; Barron's Feb 28 1994 [#209]

Barry, D., 2000, *Americans are just a few dinosaurs short of a full tank*; The Register Guard, Sunday, April 16, 2000 [#1645]

Barry J., 2002, *Pipeline brigade*; Newsweek, April 8 [#1831]

Barry R.A., 1993, *The management of international oil operations*. PennWell Books,Tulsa.

Bartlett A.A., 1978, *Forgotten fundamentals of the energy crisis*; Am. J.Phys. 46/9 Sept [#681]

Bartlett A.A., 1986, *A management program for non-renewable resources*; Am. J. Physics 54/5 [#734]

Bartlett A.A., 1990, *A world full of oil*; The Physics Teacher Nov.[#696]

Bartlett A.A., 1994, *Reflections on sustainability, population growth and the environment*; Population & Environment 16/1 [#682]

Bartlett A.A., 1996, *The exponential function, XI: the new flat earth society*; The Physics Teacher 34 [#679]

Bartlett A.A., 1997, *Is there a population problem?* Wild Earth Fall 1997 [#714]

Bartlett A.A., 1997, *An analysis of US and world production patterns using Hubbert curves*; draft [#678]

Bartlett A.A., 1998, *Reflections on sustainability, population growth and the environment – revisited*; Renewable resources Journal Winter 1997/8 [# 697]

Bartlett A.A., 1998, *Reflections in 1998 on the twentieth anniversary of the publication of the paper: "Forgotten fundamentals of the energy crisis"*, Negative Population Growth [#1240]

Bartlett A.A., 2002, *Energy Crisis: Fact or Fiction?*; text of a talk at the National Defense University, December 2002 [#2089]

Battle J., 1997, *From waste to potential*; Petrol. Review [#581]

Bauquis P-R., 2001, *A Reappraisal of Energy Supply and Demand in 2050*; Oil & Gas Science and Technology, Vol.56, No.4, pp. 389–402, 2001 [#1754 & 1843]

Bauquis P.R., 2003, *A Reappraisal of energy supply-demand in 2050 shows big role for fossil fuels, nuclear but not for nonnuclear renewables*; Oil & Gas Journal February 17, 2003 [#2120]

Bauquis P.R., 2003, *What energies for transportation in the 21st century?*; ASPO Seminar 26–27 May 2003 [#2205]

Bayless F., 2000, *Sky high oil prices freeze the northeast*; USA Today 10 February [#1080]

BBC, 2000, *The last oil shock*: The Money Programme Nov.8, 2000

Beaumont P. and J. Hooper, 1998, *Energy apocalypse looms as the world runs out of oil*; Observer Newspaper 26.7.98 [#792]

Beaumont P. and J. Hooper, 1998, *Running on empty*; Bremerton Sun, October 23 [#985]

Beardsley T., 1994, *Turning Green; Shell International projects a renewable energy future*; Scientific American 271/3 September 1994 [#260]

Becker, B., 1991, *U.S. Conspiracy to Initiate the War Against Iraq*; The Commission of Inquiry for the International War Crimes Tribunal, 1992 [#1310]

Becker R., 2002, *The battle for Iraqi oil*; Nexus, December 2002

Beeby-Thompson A.,1961, *Oil pioneer*; Sidgwick & Jackson, London

Begin J-F, 2003, *La fin du Petrole*; l'actualite 1 April [#2067]

Bell W., 1998, *The promise of methanol fuel cell vehicles*; Pet. Review December [#880]

Beller M., A. Chauvel and P. Simandoux, 1999, *The Challenge of North Sea oil and gas*; Revue de l'IFP 54/1 [1022]

Benoit G., 2000, *Energizing America*; The New American March 27 [#1162]

Bentley R.W., 1997, *The future of oil*; Seminar briefing, Reading University [#604 & 1857& 1858]

Bentley R.W., 1997, *Oil shock imminent if heavies are slow or expensive to produce*; Energy World **250** 20-22 [# 617-3]

Bentley R.W. et al., 2000, *Perspectives on the future of oil*, Energy Exploration & Exploitation 18/2-3

Bentley R.W., 2001, *Global oil and gas depletion*; own paper for EU Conference: what energy options for Europe in 2020? [#1247]

Bentley, R.W. & Whitfield, G.R., 2001, *Advice on Nuclear Energy Issues Relevant to NERC Science*; The University of Reading, Version 2, August 28, 2001 [#1357]

Bentley, R W. & Booth, R H. & Burton, J D. & Coleman, M L. & Sellwood, B W. & Whitfield, G R., 2000, *World Oil Supply: Near & Medium Term-The Approach Peak In Conventional Oil Production*; The University of Reading, August 22, 2000 [#1413]

Bentley, R.W. & Booth, R.H. & Burton, J.D. & Coleman, M.L. & Sellwood, B.W. & Whitfield, G.R., 2000, *Perspectives on the Future Oil*; Reprinted from Energy Exploration & Exploitation, Vol.18, Nos.2 & 3, 2000 [#1414]

Bentley, R.W., 2000, *Letter to G.R Davis re: Topics related to the "Oil Depletion and reporting; Royal Society of Chemistry*; April 19, 2000 [#1639]

Bentley, R.W., 2001, *Submission to the Cabinet Office Energy Review*; Draft, The Oil Depletion Analysis Centre, September 6, 2001 [#1653]

Bentley, R.W., 2001, *Letter to The Economist re: "Sunset for the oil business?"* (#1756); The Economist, p 22, November 24, 2001 [#1757]

Bentley, R.W., 2002, *Global oil & gas depletion: an overview*; Energy Policy 30, p 189- 205, 2002 [#1789 & 1838]

Bentley, R.W., *2002 Oil forecasts, past and present*; ASPO Workshop, Uppsala [#1839] [#2004]

Bentley R.W., 2003, *What does the Economist know that oil companies don't?*; letter to the editor@economist.com, June 30, 2003 [#2199]

Berger W., 2000, *Ol sprudelt nicht mehr lang*; Wirtschaft [#1245]

Berthelsen, John, 2003, *Asia starts to gasp for energy*; Asia Times Online Ltd., 2003 [#1941]

Bianco, A., 2001, *Cover Story: Exxon Unleashed*; Business Week, April 9, 2001 [#1630]

Binney, G., 2001, *The Petro-Population Parallel*; The Ecologist, Vol. 31, No.6, July/August, 2001 [#1677]

Bunyard P., 2000, *Fiddling while the climate burns*; The Ecologist 30/2 [#1159]

BGR, 1995, *Reserven, Ressourcen und Verfügbarkeit von Energierohstoffen 1995*; BGR Hannover 498pp [#B19]

Bilkadi Z., 1996, *Babylon to Baku*; Stanhope-Seta ISBN 0952-881608 230p.

Billo S.M., 1990, *Complex geology discussed by noted Arabian scientist*: Oil & Gas Journ, January 29 199 [#82]

BIP, 1995, *Quelle politique pour l'union europeannne*; BIP 7836 [#347]

Bird K.J., F.Cole, D.G.Howell & B. Leslie, 1995, *The future of oil and gas in northern Alaska*; Amer. Assoc. Petrol. Geol., 79/4 p.579 [#390]

Bishop R.S., 1998, *"Out of Oil" question demands an answer*; AAPG Explorer Nov [#900]

Bjørlykke K., 1995; *From black shale to black gold*; Science Spectra 2, 1995 44-49

Blakey E.S., 1985, *Oil on their shoes*; Amer.Assoc. Petrol. Geol.192p [#B5]

Blakey E.S., 1991, *To the waters and the wild*; Amer.Assoc. Petrol. Geol. [#B6]

Blankley, T., 2002, *Death of the West...* ; Washington Times, National Weekly Edition, January 14–20, [#1769]

Bloomberg, L.P., 2000, *Crude Oil Inventories Fall to 24-Year Low in US; Prices Rise*; Bloomberg, L. P., August 8, 2000 [#1371]

Bloomberg, L.P., 2001, *Total Says Oil, Gas Flowing From Second Kashagan Test Well*; Bloomberg, L.P., May 3, 2001 [#1442]

Bloomberg, 2000, *Oil Demand Growth to Outpace New Output Capacity, BP Amoco Says*; Bloomberg, June 21, 2000 [#1518]

Blum, W., 2000, *Electric vehicles in Switzerland*; Clean Energy 2000 [#1116]

Boeckman, O.C. et al., 1990, *Agriculture and fertilizers*; Norsk Hydro 245pp

Bohnet M., 1996, *Promotion of conventional and renewable sources of energy in developing countries*; in Kürsten M.(Ed) *World Energy – Charging Scene*: ISBN 3-510-65170-7

Bookout, J.F., 1989, *Two centuries of fossil fuel energy*; Episodes 12/4 [#331]

Boostein, D.J., 1985, *The discoverers*; Vintage 745pp

Borovik, S., 2003, *Energy Resources*; Nov. 27, 2003 [#2286]

Bossel, U., Eliasson, B. & Taylor, G., 2003, *The Future of the Hydrogen Economy: Bright or Bleak?*; Lucerne Fuel Cell Forum 2003 [#2183]

Boulton, G., 1998, *World may have more oil than it will ever need*; Salt Lake City Tribune May 19 [#749]

Bourdaire, J.M. et al., 1985, *Reserves assessment under uncertainty – a new approach*; Oil & Gas Journ. June 10, pp. 135–140.

Bourdaire, J.M., 1993, *Le pertrole dans l'economie mondiale*; Energie Universale, September 1993 [#180]

Bourdaire, J.M., 1998, *World energy prospects to 2020*; Brit. Inst. Energy Economics, 2 July [#770]

Bourdaire, J.M., 2000, *Petrole -Les fondaments de l'economie petroliere*, Encyclopoedia Universalis, France [#1234]

Bourdaire, J.M., 2000, *Energy and Economics: "Moving Backwards To The Future"*; July 2000 [#1361]

Bowen, J.M., 1991, *25 years of UK North Sea exploration*; in Abbotts, J. United Kingdom Oil & Gas fields; Geol. Soc Mem14 [#913]

Boy de la Tour, X., 1994, *Technologies petrolieres : les nouvelles frontieres*; Revue de l'energie 456 [#217]

Boyes, R., 2001, *Russian power and Gazprom*; The Times (UK) [#1194]

BPA, 2001, *California Woes Loom Large etc.*; Journal- A Monthly Publication of the Bonneville Power Administration (BPA), Oregon, April 2001 [#1632]

BPAmoco, 2000, *Oil Resources: Policy Implications of "A Balanced Assessment"*;   Energy Forum, Rice University, Houston, TX. May 2000 [#1430]

Bradbury, J., 2000, *Reinventing the North Sea*; Hart's E&P [#1253]

Bradbury, J., 2003, *Pressure mounts for UK operators*; Harts E&P, Feb. 2003 [#2109]

Bradley, R.L., 1999, *The growing abundance of fossil fuels*; The Freeman [#1130]

Bradsher, K., 2002, *China pursues alternatives to limit dependence on oil*; The Sunday Oregonian, Sept. 8, 2002 [#2098]

Brammer, R., 1995, *Oil Change*: Barron's March 6th 1995 4-4 [#333]

Brauer, B., 2001, *A Waning Honeymoon in Kazakhstan*; New York Times [#1195]

Brethour, P., 2002, *Industry fears Kyoto will hurt oil sands-Emission cutting Protocol could kill new projects*; theglobeandmail.com, September 26, 2002 [#2044]

Brethour, P., 2003, *Soaring Costs to Slash Expansion of Alberta's Oil Sands*; The Globe and Mail, May 10, 2003 [#2164]

Breton, T.R. & J.C. Blaney, 1991, *Production rise, consumption fall may turn Soviet oil exports higher*; Oil & Gas Journ. November 18 1991 110-114 [#7]

British Petroleum Co., *BP Statistical Review of World Energy*; Published annually by BP, London.

British Petroleum Co., 1959, *Fifty years in pictures*; publ. BP, London 159 p.

British Petroleum Co., 1979, *Oil crisis – again*; BP paper [#590]

Bronner, M.E., *US pins high hopes on Caspian oil*; Washington Times 22.11.96 [#655]

Brown, D., 1977, *Some reminiscences of an industrialist*; Eastern Hive Publishing ISBN 0-87960-109-4

Brown, D., 2000, *So, Are We Running Out of Oil ?*; AAPG Explorer, pp. 46, 47, April, 2000 [#1540]

Brown, D., 2000, *Bulls and Bears Duel Over Supply*; AAPG Explorer, pp. 12–15, May, 2000 [#1541]

Brown, K. & Herrick, T., 2003, *Rising Cost of Oil Isn't Sending Stocks Higher*; Wall St. Journ. March 3, 2003 [#2129]

Brown, L.R., 2001, *The state of the world*; W.E.Norton & Co.,268pp

Brown, L.R., 2001, *Eco-economy*; W.E.Norton & Co.,333pp

Brown, L.R., 2003, *Deflating the Bubble Economy Before it Bursts*; Earth Policy Institute September 4, 2003 [#1979]

Brown, L.R., 2003, *Plan B* – ISBN 0-393-05859-x

Browne, E.J.P., 1991, *Upstream oil in the 1990s: the prospects for a new world order*; Oxford Energy Seminar, September 1991. Publ. The British Petroleum Company, London [#42]

Brown, R.A., 2003, *Critical Paths to the Post-Petroleum Age*; Argonne National Laboratory, 2003 [#1940]

Brown, T., 1991 *Mexico oil assets called inflated*; The Arizona Republic December 10 1991 [#13]

Buderi, R., 1992, *Oil's downhill skid may be ending*; Business Week Feb. 1992 [#8]

Bundesministerium für Wirtschaft, 1993, *Securing Germany's economic future*; Report 338 [#481-1]

Bundesministerium für Wirtschaft, 1995, *Sources of energy in 1995: reserves, resources and availability*; Report 383 [#483-1]

Bureau of Resource Sciences, 199?, *Sufficiency of Australia's petroleum resources* [#537]

Burns, C., 2000, *Rising gas prices fuel push for hydrogen-powered cars*; CNN.com , September 15, 2000 [#1666]

Burns, S., 2001, *World running out of cheap oil that fuels economies*; Houston Chronicle,, July 16, 2001 [#1322]

Burrows, Shauna C., 2002, *Earth Emergency- a call to action*; Positive News, August, 2002 [#1887]

Business Week, 2000, *Is Big Oil Getting Too Big?*; Business Week, October 30, 2000 [#1415]

Business Week, 2001, *King Fahd of Saudi Arabia*; Special Advertising Section,   Business Week, November 26, 2001 [#1720]

Business Week, 1993, *The scramble for oil's last frontier*; Business Week Jan 11 1993 [#150]

Business Week, 1997, *The new economics of oil* November 3 [#939]

Bryant, A., 2000, *The New War Over Oil*; Newsweek, October 2, 2000 [#1404]

Byman, D., 1996, *Let Iraq collapse!*; The National Interest, Vol.45, pp. 48–60.

Byrne, H.S., 2001, *Too Much Power ?*; Barron`s, pp. 21–23, August 6, 2001 [#1647]

C

Calamai P., 2003, *Energy crisis looms, experts warn*; Toronto Star, Nov.26, 2003 [#2272]

Cameron, N., Bate, R., Clure, V. & Benton, J., 1999, *Oil and Gas Habitats of the South Atlantic: Introduction*; In: The Oil and Gas Habitats of the South Atlantic, Cameron *et al.*, Special Publication, 153, The Geological Society, London, 1999 [#1526]

Campbell, C.J. 1965, *The Santa Marta wrench fault of Colombia and its regional setting*; 4th Carib. Geol. Conf

Campbell, C.J. & H.Bürgl, 1965, *Section through the Eastern Cordillera of Colombia, South America*; Geol. Soc. Amer. 76 567-590

Campbell, C.J., 1974 *Structural classification of northern South America*; verhnd. Naturf Ges. Basel

Campbell, C.J., 1974, *Colombian Andes*; in Spencer A.M. (Ed) *Mesozoic-Cenozoic Orogenic Belts*; Mem. Geol. Soc., London

Campbell, C.J., 1974, *Ecuadorian Andes*; in Spencer A.M. (Ed) *Mesozoic-Cenozoic Orogenic Belts*; Mem. Geol. Soc., London

Campbell, C.J., 1975, *Ecuador* in Fairbairn R.W. (Ed) *Encyclopedia of world regional geology*; Hutchinson & Ross

Campbell, C.J. and E.Ormaasen, 1987, *The discovery of oil and gas in Norway: an historical synopsis*; in Spencer A.M. (Ed) Geology of the Norwegian oil and gas fields, Norwegian Petrol. Soc.493p.

Campbell, C.J., 1989, *Oil price leap in the early nineties*; Noroil 17/12 pp. 35–38

Campbell, C.J., 1991, *The golden century of oil 1950-2050: the depletion of a resource*; Kluwer Academic Publishers, Dordrecht, Netherlands; 345pp.

Campbell, C.J., 1992, *The depletion of oil* ; Marine & Petrol. Geol., v.9 December 1992, pp. 666 –671

Campbell, C.J., 1993, *The Depletion of the world's oil*; Petrole et technique. No.383. 5-12 Paris

Campbell, C.J., 1994, *World oil – Reserves, production, politics and prices*; Proc. 1993 NPF Conference, Stavanger, Norway *Quantification and prediction of hydrocarbon resources.* [#495]

Campbell, C.J., 1994, *An Oil Depletion Model: a resource constrained yardstick for production forecasting*; Rept., Petroconsultants S.A., Geneva.

Campbell, C.J., 1994, *The end of an era. What now?*; Abs. Sp. Lecture. Petrol. Explor. Soc. Gt. Britain Newsletter. March 1994. Also Video [#195]

Campbell, C.J., 1994, *Scraping the barrel*; The Economist August 6th 1994 [#259]

Campbell, C.J., 1994, *Scrambling for oil*; Time Magazine July 11 [#265]

Campbell, C.J., 1994, *The imminent end of cheap oil-based energy*; SunWorld 18/4 17-19

Campbell, C.J., 1994, *Nytt prissjokk pa olje – og ny seilskuteetid*; Stavanger Aftenblad 9 September 1994

Campbell, C.J., 1995, *Taking stock*; SunWorld 19/1 16-19

Campbell, C.J., 1995, *The Coming Crisis*; SunWorld 19/2 16-19

Campbell, C.J., 1995, *The next oil price shock: the world's remaining oil and its depletion*; Energy Expl. & Exploit., 13/1 pp. 19–46

Campbell, C.J., 1995, *Proving the unprovable*; Petroleum Economist, May 1995

Campbell, C.J., 1995, *Cassandra or prophet*; Petroleum Economist Oct. 1995 [#362]

Campbell, C.J., 1995, *Spar ny gullalder for olje* – Stavanger; Rogalands Avis 1.12.95 [#610-2]

Campbell, C.J., 1996, *The resource constraints to oil production : the spectre of a pending chronic supply shortfall*; in Kürsten M. (Ed) *World energy – a changing scene*; Proc.7th Int. Symposium BGR, Hannover, E.Schweizerbart'sche Verlagsbuchhandlung, Stuttgart 227pp.

Campbell, C.J., 1996, *The status of world oil depletion at the end of 1995*; Energy Exploration and Exploitation; March 1996

Campbell, C.J., 1996, *Oil Shock*; Energy World, June 1996

Campbell, C.J., & J.H.Laherrère, 1995, *Gauging the North Sea*; Platt's Petroleum Insight April 10 1995.

Campbell, C.J., & J.H.Laherrère, 1995, *The world's supply of oil 1930-2050*; Report Petroconsultants S.A., Geneva

Campbell, C.J., 1997, *Better understanding urged for rapidly depleting reserves*: Oil & Gas Journ. April 7th 1997. [#615-2]

Campbell, C.J., 1997, *The Coming Oil Crisis*; Multi-Science Publishing Co. 210pp.

Campbell, C.J., 1997, *Oil shock — now the production peak*; Energy World 250 p.23 [#616-3]

Campbell, C.J., 1997, *How to calculate depletion of hydrocarbon reserves*; Petroleum Economist September pp. 111–112

Campbell, C.J., 1997, *As depletion increases, energy demand rises*; Petroleum Economist September pp. 113–116

Campbell, C.J., 1997, *Depletion survey: Syria and Oman*; Petroleum Economist, October.

Campbell, C.J., 1997, *Egypt's greater emphasis on gas*; Petroleum Economist December, p. 33.

Campbell, C.J., 1997, *Depletion patterns show change due for production of conventional oil*; Oil & Gas

Journ., December 29, pp. 33–37

Campbell, C.J., 1998, *Remote areas remain underexplored*; Petroleum Economist January, p. 44

Campbell, C.J., 1998, *How secure is our oil supply?*; Science Spectra 12 18-24 [#636]

Campbell, C.J. and J.H. Laherrère, 1998, *The end of cheap oil*; Scientific American March 80-86

Campbell, C.J., 1998, *Running out of gas: this time the wolf is coming*; The National Interest, spring [#661]

Campbell, C.J., 1998 *Oil in crisis*; Resource Investor Summer 4-6 [#662]

Campbell, C.J., 1998, *Have all the elephants been found?* Petroleum Review March 24–26 [#700]

Campbell, C.J., 1998, *Major exporter nearing peak*; Petroleum Economist April 7–8

Campbell, C.J., 1998, *Reserves controversy*; Oil & Gas Journal April 20

Campbell, C.J., 1998, *Russia: a net importer by 2008*; Petroleum Economist, May 31–32

Campbell, C.J., 1998. *The future of oil*; PESGB Newsletter

Campbell, C.J., 1998, *The future of oil*; Energy Exploration & Exploitation 16/ 2-3

Campbell, C.J., 1998, *Explorers turning to the deepwater Atlantic margin*; Petroleum Economist July [#794]

Campbell, C.J., 1998, *The enigma of oil prices in times of pending oil shortage*; World Oil Prices: oil supply/demand dynamics to the year 2020; Centre for Global Energy Studies, 208–225

Campbell, C.J., 1998, *Maybe a third peak for Nigeria*; Petroleum Economist Dec.[#883]

Campbell, C.J., 1988, *Oil: a case of short-sighted vision*; Energy Day, December 17 1998 [#885]

Campbell, C.J., 1998, *L'avenir de l'homme de l'age des hydrocarbures*; Geopolitique 63, October

Campbell, C.J. and J.H.Laherrère, 1998 *"La fin du petrole bon marche" Pour la Science* May pp. 30–36

Campbell, C.J., 1998, *How soon will oil production peak ?* Petroleum Review October [#890]

Campbell, C.J., 1999, *Oil madness*; Geopolitics of Energy; January

Campbell, C.J., 1999, *A new mission for geologists as oil production peaks*; Swiss Association of Petroleum Geologists and Engineers, 4/1 19-34

Campbell C.J., 1999, *The dating of reserve revisions*; Petroleum Review. July pp. 44–45

Campbell, C.J., 1999, *The extinction of Hydrocarbon Man ?* Times Higher Educat. Supplement, November 12

Campbell, C.J., 1999, *The imminent peak of oil production*; Presentation at House of Commons, oilcrisis.com [#1101]

Campbell, C.J., 1999, *Norway profile: looking for the peak*; Tomorrow's Oil 1/1 May

Campbell, C.J., 1999, *Definitions*; Tomorrow's Oil 1/2 August

Campbell, C.J., 1999, *Venezuela-Conventional Oil*; Tomorrow's Oil 2/1 December

Campbell, C.J., 1999, *Study – Non-conventional oil*; Tomorrow's Oil 1/2 August

Campbell, C.J., 2000, *Peru and Nigeria*; Tomorrow's Oil 2/2 February

Campbell, C.J., 2000, *World's oil endowment and its depletion* Tomorrow's Oil 2/2 February

Campbell, C.J., 2000, *Italy* Tomorrow's Oil 2/3 April

Campbell, C.J., 2000, *Vietnam* Tomorrow's Oil 2/3 April

Campbell, C.J., 2000, *Myth of spare capacity setting the stage for another oil shock*; Oil & Gas Journ. March 20 [#1144]

Campbell, C.J., 2000, *Global warming*; The National Interest 59. Spring

Campbell, C.J., 2000, *Oil depletion in the US Lower 48*: Tomorrow's Oil 2/6

Campbell, C.J., 2000, *Champion of national sovereignty in OPEC*; Tomorrow's Oil 2/6

Campbell, C.J., 2000, *What do the USGS numbers really mean ?* Tomorrow's Oil 2/7

Campbell, C.J., 2000, *Depletion: the Democratic and Popular Republic of Algeria*; Tomorrow's Oil 2/7

Campbell, C.J., 2000, *Deep water heroism keeps oil imports at bay*; Tomorrow's Oil 2/8

Campbell, C.J., 2000, *The myths of oil*; Petroleum Review, October

Campbell, C.J., 2000, *Opec's spare capacity and strategy*; Tomorrow's Oil 2/9.

Campbell, C.J., 2000, *Analysis: the challenge of reserve evaluation*; Tomorrow's Oil 2/10 October

Campbell, C.J., 2000 *Indonesia*; Tomorrow's Oil 2/12 December

Campbell, C.J., 2000 *Depletion and denial: the final years of oil supplies* USA-Today November

Campbell, C.J., 2000 *The imminent peak in world oil production*; Middle East Economic Databook 101-10 Campbell C.J., 2000, *Die Ara niedriger Olpreise ist zu Ende*; Gastkommentar [#1259]

Campbell, C.J., 2000 *The imminent oil crisis* in CMDC-WSEC Blueprint for the Clean Sustainable Energy Age, Proc. Millennium Conference, Geneva

Campbell, C.J., 2000, *Depletion And Denial- The Final Years Of Oil Supply*; ? , [#1375]

Campbell, C.J., 2000, *Myth of spare capacity setting the stage for another oil shock*; Oil & Gas Journal, p 20,21, March 20, 2000 [#1544]

Campbell, C.J., 2000, *Peak Oil: an Outlook on Crude Oil Depletion*; MBendi Information for Africa, Website, February, 2002 [#1773]

Campbell, C.J. and Zagar J, 2001, *Oilfields – Maintenance expenses*; Hubbert Center Newsletter 2001/2-2 [1184]

Campbell, C.J., 2001 *The four lives of Trinidad*; Tomorrow's Oil 3/1 January

Campbell, C.J., 2001 *The imminent oil crisis – the missing element in the environment debate*; Tomorrow's Oil 3/2 Feb

Campbell, C.J., 2001 *Peak Oil : a turning point for Mankind*; Hubbert Center Newsletter 2001/2-1

Campbell, C.J., 2001 *The Oil Peak Oil : turning point*; Solar Today, 15/4 July-August

Campbell, C.J., 2001 *Peak Oil : a turning point for Mankind*; Climate Change & Energy Conf. Evora, Portugal proc.

Campbell, C.J., 2001 *Oil, Gas and Make-believe*; Energy Exploration & Exploitation 19/2-3 117-133

Campbell, C J.& Zagar J J., 2001, *Peak Oil: A Turning For Mankind / Oilfields- Maintenance Expenses*; Hubbert Center Newsletter # 2001/ 2-1 and 2-2 [#1289]

Campbell, C J., 2001, *The Oil Peak: A Turning Point*; Solar Today, pp. 40–43, July/August 2001 [#1299]

Campbell, CJ., 2002, *Peak oil: A turning point for humankind*; Culture Change, Issue #19, www.culturechange.org/issue19/peakoil.htm ,March, 2002 [#1787]

Campbell, C.J., 2002, *Beer drinkers get a grip on people power*; The Times Higher-Economics, October 18, 2002 [#2002]

Campbell, C J., 2002, *The Fuel that Fires Political Hotspots*: The Times Higher Education Supplement, May 17th 2002 [#1861]

Campbell, C J., 2002, *Forecasting global oil supply 2000–2050*: M.King Hubbert Center for Petroleum Supply Studies 2002/3

Campbell, C J., 2002, *Facing up to the future*: Norwegian Petroleum Diary 2/2002 [#1866]

Campbell, C J., 2002, *Bell signals death knell for oil*: Times Higher Educational Supplement [#1865]

Campbell, C J., 2002, *Ekonom nonchalerar klotets begransningar*: Svenska Dagbladet 20 August [#1881]

Campbell, C.J., F.Liesenborghs, J.Schindler, & W. Zittel, 2002 *Oelwechsel*; Detscher Taschenbuch ISBN 3-423-24321-4 260pp

Campbell, C.J., 2002, *Petroleum and people*; Population and Environment 24/3 November

Campbell, C.J., 2002, *The imminent oil crisis*; in Grob. G, ed. Blueprint for the clean, sustainable energy age; Verlag eco-performance, Geneva

Campbell, C.J.,2003, *Iraq and oil – Scenarios for the future*; Energy World 203 March [#2117]

Campbell, C.J. ,2003, *The Essence of Oil & Gas Depletion*; Multi-Science, 342p

Campbell, C.J., 2003 *Industry asked to watch for regular oil production peaks, depletion signals*; Oil & Gas Journ. July 14 38-45

Campbell, C.J., 2003, *The ageing of oil*; Petroleum Review, November pp. 44–45

Campbell, C.J., 2003, *Managing Oil Depletion Fairly*; Conferance: L'economia del nobile sentiero, Rimini, 2003 [#2222]

Campbell, C.J., 2003, *The Essence of Oil and Gas Depletion*; The Journal of Energy Literature, 2003 [#2241]

Campbell, C.J., 2003, *Ireland Faces A Devastating Energy Crisis But its Not Alone*; Cork Environmental Forum, Deember 2003 [#2295]

Campbell, C.J., 2004, *The Urgent Need for an Oil Depletion Protocol*; in Centro Pio Manzu, Trans. Conf. The Economics of the Noble Path, Rimini, October 2003

Campbell, C.J., 2004, *Le declin des reserves est depuis longtemps une evidence aux yeux des specialiostes*; Le Temps 29 April 2004

Campbell, C.J., 2004, *The peak and decline of world oil supply*; Energy Verwertungsagentur 1/2004

Campbell, C.J., 2004, *The end of cheap oil: a critical turning point of historic proportions*; EUi Transport Report Vol. 2 No. 23

Campbell, C.J., 2004, *The truth about oil and the looming world energy crisis*; Eagle Print (info@eagleprint.net)

Campbell, S.J.D., 2004, *The Use and Abuse of Modern Methodology for the Calculation of Reserves and the Assigning of Risk to those Reserves, and its Use in the Decision Making Process*; 4017 Stavanger Norway, [#1465]

Campbell, S.J.D. & N. Gravdal, 1995, *Prediction and evaluation of high porosity chalks in the East Hod Field*; Pet. Geoscience 1/1

Carey, J., 2003, *Taming the oil beast*; Business Week, Feb. 24, 2003 [#2100]

Carlisle, T., 2001, *It's Gooey, Unruly, Mythic- and Priceless*; The Wall Street Journal,, August 9, 2001 [#1352]

Carmalt, S.W. and B. St. John, 1986, *Giant oil and gas fields*; in Future petroleum provinces of the world; Amer. Assoc.Petrol. Geol. Mem.40 [#43]

Carpenter, D., 2000, *Oil price rockets higher; America barely flinches*; The Register Guard 22 January [#1070]

Carter, C. (AP), 2001, *California dream grows nightmarish*; Register- Guard, March 19, 2001 [#1627]

Caruba, A., 2003, *Beyond Petroleum, The Truth*; MichNews.com , January 2003 [#2053]

Cattaneo, C., 2003, *Gas prices stay high as supplies fall*; National Post February 24, 2003 [#2110]

Cattaneo, C., 2003, *Natural gas in dangerous decline, says analyst*; National Post, June 11, 2003 [#2179] [#2213]

Cavallo, A.J., 1995, *High Capacity Factor Wind Energy Systems*; Journal of Solar Energy Engineering Vol. 117, May 1995 [#2065]

Cavallo, A.J., 2002, *Predicting the peak in World oil production*: Natural Resources Research 11/3, September 2002 [#1998] [#2009] [#2063] [#2151]

Cavallo, Alfred, 2003, *Spare Capacity and Peak Production in World Oil*, Natural Resources Research, October 2003[#1932]

Cavallo, Alfred, 2003, *Technical Synopsis: Large Scale Utilization of US Wind Energy Resources*; 2003 [#2062]

Cavallo, A., 2004, *The illusion of plenty*; Bulletin of the Atomic Scientists, January/Feruary 2004 [#2313] [#2322]

Cavallo, A., 2004, *Spare Capacity(2003) and Peak Production in World Oil*; Natural Resources Research, October 2003 [#2314]

Cazalot, C.P., 1996, *Texaco exploration and production*; Petrol. Engineer Int., May 1996 17-22 [#464]

Cazier, E.C., *et al.*, 1995, *Petroleum geology of the Cusiana Field, Llanos Basin foothills, Colombia*; Amer. Assoc. Petrol. Geol. 79/10 1444-1463 [#403]

CCN, 2001, *U.S. Food Supply Threatened By Energy Crisis And Population Growth*; CCN, [#1679]

Centre for Global Energy Studies, 1992, *The Gulf crisis* 228 pp

CGES, 1992, *The long-run price of oil*: CGES study Vol. 3 [#194]

CGES, 1992, *The costs of future North Sea oil production*; CGES Rept. Vol. 3[#205]

CGES, 1994, *Oil production capacity in the Gulf, Vol.II t*

CGES, 1993, *Oil production capacity in the Gulf Vol II Saudi Arabia*; Centre for Global Energy Studies [#284

CGES, 1998, *World oil prices*; 316p

CGES, 2001, *The future of oil: meeting the growth in world demand*; Global Oil Report, Volume 12, Issue 1 [#1190]

CGES, 2001, *Non-OPEC Production: The Going Is Getting Tougher*; Global Oil Report, Vol.12, Issue 6, November/December, 2001 [#1755]

CGES, 2003, *Global Oil Report-Market Watch-Oil Reserves gained and lost in 2002*; February 27, 2003 [#2298]

Charpentier R.R., 2002, *Locating the summit of the oil peak*; Science 295 [#1841]

Charrier B., 2000, *Energy – what future?* Clean Energy 2000 [#1109]

Chazan G., 2002, *Caspian summit fails to resolve energy claims*; Wall St Journ 25/April [#1826]

Chevron Corp., 1993, *Saudi memories*; Chevron World winter/spring 1993 pp. 14–18

Chevron Corp., 1996 *The great Arabian discovery*; http://chevron. Com/chevron root/arab 50 [#523]

Chew K., 1998, *Hydrocarbon Resource Forecasting*; Presentation to Goldman Sachs [#729]

Cheney R., 1999, Speech to Institute of Petroleum, London; http://web.archive.org/web/20010810115257 http://www.petrroleum.co.uk/speeches.htm

Chicago Tribune, 1999, *US hails deal for pipeline in Caspian region*, Nov 28 [#1105]

Clack R.W., 2000, *The world needs more oil*; Oil & Gas Journ February 21 [#1088]

Clark, P., P. Coene and D. Logan, 1981, *A comparison of ten U.S. oil and gas supply models*; Federal Reserves Institute, Washington [#103]

Cleveland, C.J., 1992, *Yield per effort for additions to crude oil reserves in the Lower 48 United States 1946-1989*; Amer. Assoc. Petrol. Geol. 76/7 948-958

Cleveland, C.J. & R.K.Kaufmann, *Forecasting ultimate oil recovery and its rate of production: incorporating economic forces into the models of M. King Hubbert*; The Energy Journal 12/2 [#26]

Clifford, M.L., 2000, *Can this giant fly ?* ; Business Week, Feb 7th [#1072]

Clifford, D., 2000, *The Oil Age*; First published S.A. Humanist Post, June 2000, www.users.on.net , March 2002 [#1786]

Clifford, M.L., 2001, *An Oil Giant Stirs (China)*; Business Week, May 7, 2001 [#1660]

Clover, C., 2001, *The Balance Of Power*; The Daily Telegraph, February 10, 2001 [#1624]

CNN Industry Watch, 2000, *Crude price shock risk high, says Goldman*; CNN Industry Watch, August 22, 2000 [#1367]

Cocks, D., 2001, *How to energise the clever county*; The Australian Financial Review, Friday, July 20, 2001 [#1702]

Cocks, D., 2001, *There will be life- but not as we know it*; The Australia Financial Review, Friday, July 27, 2001 [#1703]

Cocks, D., 2001, *Coming to terms with complexity*; The Australia Financial Review, Friday, August 3, 2001 [#1704]

Coghlan, A., 2003, *'Too little oil for global warming*; Newscintist.com, Oct.1, 2003 [#1969]

Cohn, L. & Crock, S., 2001, *What To Do About Oil*; Business Week, October 29, 2001 [#1693]

Coleman, J.L., 1995, *The American whale oil industry: a look back to the future of the American petroleum industry;* Non-renewable Resources 4/3 273-288 [#603]

Colvin, M., 1994, *Britain's Gulf War ally helped Saddam build nuclear bomb*; Sunday Times July 24, 1994 [#247]

Commonwealth of Australia, 1996, *Sustainable energy policy for Australia.* [#538]

Conant, M.A., 1999, *The Universe of Oil* – ISBN 1-896091-41-5

Conlan, M.D., 2003, *Summertime Gas Spikes Could Lie Ahead to Close 2-4 BCF/Day Gap;* conlan@weeden, May 1, 2003 [#2182]

Conrad, R., *Crushing Energy Crisis to hit early in 2003!;* Special Investor Bulletin, Fall 2002[#2093]

Cooke, R.T., 2004, *Oil, Jihad and Destiny*; Opportunity Analysis ISBN- 1-930847-62-9

Cook, W.J., 1993, *Why Opec doesn't matter anymore*; US News & World Report 13/12/93 [#299]

Cope, G., 1998, *Have all the elephants been found?* Pet. Review 52/614 24-26

Cope, G., 1998, *Will improved oil recovery avert an oil crisis*; Pet. Review June 52/616 22-23

Cooper, C., 2001, *A Wheezing Iraqi Oil Refinery Tells All*; The Wall Street Journal, May 14, 2001 [#1622]

Cooper, C. & Herrick, T., 2001, *Oil Giants Struggle to Spend Profits Amid Shortage of Exploration Sites*; The Wall Street Journal Online, July 30, 2001 [#1313]

Cooper, C. & Herrick, T., 2001, *Pumping Money- Major Oil Companies Struggle to Spend Huge Hoards of Cash*; The Wall Street Journal, July 30, 2001 [#1678]

Cooper, M., 1998, *Oil production in the 21st Century*; CQ Researcher 8/29 [#788]

Cooper, J.C. & K. Madigan, 2002, *Debt overseas stirs up trouble at home*; Business Week May 6 [#1813]

Copp, T., 2000, *Fuel costs set to pump up grocery bills*; timesunion.com , July 21, 2000 [#1537]

Cornelius, C.D., 1987, *Classification of natural bitumen: a physical and chemical approach*; in Meyer, R.F. (Ed) *Exploration for heavy crude oil and bitumen*; AAPG Studies in geology #25

Cornot-Gandolphe, S., 1994, *The main issues in gas processing, liquifaction LNG transport and storage*; preprint APS Conf. [#294]

Corzine R., 1993, *Warning over Saudi output*; Financial Times, 4/11/93 [#282]

Corzine, R., 1999, *Saudi diplomats lead first-aid effort;* Financial Times, April 15 [#970]

Coward, M.P., Purdy, E.G., Ries, A.C. & Smith, D G., 1999, *The distribution of petroleum reserves in basins of the South Atlantic margins*; In: The Oil and Gas Habitats of the South Atlantic, Cameron et al., Geological Society, London, Special Publications, 153, pp. 101–131, 1999 [#1524]

Coy, P., 2000, *Revenge of the Old Economy*; Business Week, September 25, 2000 [#1394]

Coy, P., 2000, *A Modest Proposal for The U.S.*; Business Week, October 9, 2000 [#1407]

Coy, P., 2001, *The Energy Forecast*; Business Week, August 27, 2001 [#1656]

CPN., 1994, *A forecast of China's petroleum resources*; China Pet. Newsletter 1/9 April 27 1994 [#253]

Crandell, J.D. et al., 1994, *Depends on oil prices*; World Oil, February 1994[#200]

Creswell, J., 1996, *Global demand fuels need for more drilling*; Press & Journal Aberdeen [#404]

Creswell, J., 1999, *How big oil is slashing*; Press & Journal Aberdeen [#920]

Crilley & Co., 1969, *Stock Market Action Report: Oil In The Alaskan Arctic- The Prudhoe Bay Discovery*; August, 1969 [#1770]

Crock, S., 2001, *Rogue States: Why Washington May Ease Sanctions*; Business Week, May 7 2001 [#1661]

Crow, P., 1991, *Oil and bullets*; Oil & Gas Journ., Oct 21. 1991[#183]

Cummins, C., 2002 *Natural Gas prices rebound*; Wall St. Journ. 6-5-02 [#1849]

Cummins, C., 2003, *Tapping Iraq's Experts on Oil;* Wall St. Journ. 27-1-03 [#2077]

Cummins, C., 2003, *BP Works to Improve Margins;* Wall St. Journ. 15-1-03 [#2108]

Cummins, C., Warren, S. & Schroeder, M., 2004, *Shell Cuts Reserve Estimate 20% As SEC Scrutinizes Oil Industry;* Wall St. Journ., Jan.11, 2004 [#2316]

Cummins C. & Schroeder M., 2004, *Shell's Watts Draws Fire;* Wall St. Journ., January 14, 2004 [#2318]

Curnow, J., 2000, *Australia's Carrying Capacity*; Sustainable Population Australia Inc., April 26, 2000 [#1636]

Curran, R., 1998 *Uncertainty clouds fundamental strength of Canada's industry*; World oil [#672]

D

Dabrowski, A., 2000, *Mexico energy chief criticised over rush to drop oil price*; S. Francisco Chronicle 22 February [#1084]

Dallas Morning News, 1993, *Arco says Alaskan fields could increase reserves 60%*; Dallas Morning News April 14 1993 [#185]

Daly, H., 2003, *Steady-state economics*; The Social Contract v8. n.3

Daly, M.C., M.S. Bell & P.J. Smith, 1996, *The remaining resource of the UK North Sea and its future development*; in Glennie K & A Hurst NW Europe Hydrocarbon Industry; Geol. Soc. [#912]

Dan, J., 2003, *The Ultimate Deception*; -ISBN 0-9539951-1-9

Darley, J., 2004, *High Noon for Natural Gas* – ISBN 1-931498-53-9

Davey, E., 2002, *A Plain Man's Questions Concerning Eco-Footprinting*; OPT April 2003 [#2091]

Davies, P., 1994, *Oil supply and demand in the 1990s*; World Petrol. Congr. Topic 16 [#235]

Davidson, J.D. & W.Rees-Mogg, 1988, *Blood in the streets – investment profits in a world gone mad*; Sidgwick & Jackson, London 385p

Davidson, J.D. & W.Rees-Mogg, 1994, *The great reckoning*; Pan Books 602pp

Davidson, K., 1998, *Some experts say oil demand will soon exceed supply again*; Sta Barbara News 2 September [#839]

Davidson, K., 1998, *Petrol Pessimists*; Rocky Mountain News 13 Sept [#841]

Davis, R.E., 1952 *A method of predicting availability of Natural Gas on average reservoir performance*; SPE [#1870]

Deffeyes, K.S., 2001, *Hubbert's peak – the impending world oil shortage*; Princeton University Press 208pp

Derakhshan, M., 1999, *Iran's oil and gas: background and policy challenges*; International Bureau for Energy Studies.

Deitzman, W.D. *et al.*, 1983, *The petroleum resources of the Middle East*; U.S.Dept. of Energy / Energy Information Administration Report. DOE/EIA-0395 May 19 83 169p.

Dell, P.R., 1994, *Global energy market : future supply potentials*; Energy Exploration and Exploitation 12/1 59-72 [#377]

Demaison, G. and B.J. Huizinga, 1991, *Genetic classification of petroleum systems*; Amer.Assoc. Petrol. Geol.. 75 1626-43

Department of Energy, 1989, *Federal oil research: a strategy for maximizing the producibility of known U.S. oil*; U.S.Dept of Energy DOE/FE 0139 [#389]

Department of Energy, 1995, *US crude oil, natural gas and natural gas liquids reserves*; 1994 Annual Rept. [#429]

Department of Primary Industries & Energy; 1996 *Australia's petroleum resources*; Statistical Review 1997 [#755]

Department of Trade and Industry; 1996, *"Brown book"* extract on reserves [#401]

Department of Trade and Industry; 1998 *"Brown book" extract on reserves* [#825]

DeSorcy, G.J., *et al.*, 1993, *Definitions and guidelines for classification of oil and gas reserves*; Journ. Canadian Petrol. Technology 32/5, 0-21 [#338]

Deutsche Bank, 2000, *Prescription for the OPEC Meeting: A Dose of Prozac*; Deutsche Bank Global Oil & Gas Research, Energy Wire, June 15, 2000 [#1498]

Deutsche Bank, 2001, *Oil Team Conference Call- Energy Rebound in Sight: Looking through the near-term murk*; Deutsche Bank Alex. Brown, ??? , [#1750]

Dieoff.com, *Economics error: Illusion that money can replace physically exhausted resources*; dieoff.com, p.185, [#1342]

Dismal Scientist, 2000, *Oil and Gas Inventories, Analysis and Business Implications*; www.dismal.com , Released April 26, 2000 [#1574]

Ditchfield, M., 1999, *New sources of funding in a period of limited cash liquidity*; CWC conf, [#1040]

Dixon, G.E., 1989, *The Russian/CIS position in today's and tomorrow's global petroleum industry*; presentation by Petroconsultants [#179]

Djurasek, S., 1998, *The coming oil crisis*; NAFTA 49/6 167-168 [#884]

Doliner, M., 2002, *The Motive for the Invasion*; antiwar.com, November22, 2002 [#1975]

Dorsey, J.M., 2001, *Saudi Leaders Warns U.S. of "Separate Interests"*; The Wall Street Journal, Monday, October 29, 2001 [#1708]

Douthwaite, R., 1998, *Only a warning or is this the real thing*; [#879]

Dowthwaite, R., 1999, *The ecology of money*; Green books 78pp

Douthwaite, R., 2001, *Essay: When should we have stopped?*; The Irish Times, Saturday, December 29, 2001 [#1782]

Douthwaite, R. (Ed), 2003, *Before the wells run dry*; Green Books

Dreyfuss, R., 2003, *The Thirty Year Itch*; 2003 [#2133]

Dromgoole, P. & R. Speers, 1997, *Geoscore: a method for quantifying uncertainty in field reserve estimates*; Petroleum Geoscientist 3 1-12.

Duffin, M., 2000, *The End Of Cheap Oil (Abstract)*; murrey`s@msn.com, July 2000 [#1527]

Dunbar, R., 2003, *Oil Supply Debate*; Canadian Energy Research Institute, December 6, 2003 [#2288]

Duncan, R.C., 1995, *The energy depletion arch: new constraints enhance the Hubbert model*: Public Interest Environmental Law Conf., Univ. of Oregon [#400]

Duncan, R.C. ,1996, *Mexico's petroleum exports: safe collateral for a $50 billion loan?* Sp. Report, Inst.

on Energy & Man, Seattle[#399]

Duncan, R.C., 1996, *The Olduvai Theory: sliding towards the post-industrial stone age*; Inst. Energy & Man, Seattle [#491]

Duncan, R.C., 1996, *Fossil fuel prospects for the Twenty-First Century*; Inst. Energy & Man [#506]

Duncan, R.C. and W. Youngquist, 1998, *Encircling the peak of world oil production*; draft [#730]

Duncan, R.C. and W. Youngquist, 1998, *The world petroleum life-cycle*; PTTC Workshop, Petroleum Engineering Program, Univ. of S. California 22 October 1998 [#875]

Duncan, R.C. and W. Youngquist, 1998 *Is oil running out?* Science 282 [#882]

Duncan, R.C. and W. Youngquist, 1998, *The world petroleum life-cycle – Issue 3*; Inst. Energy & Man [#928]

Duncan, R.C., 1999, *World energy production: kinds? amounts? per capita? who*; Inst Energy & Man [#969]

Duncan, R.C., 1999, *Oil production per capita*; Oil & Gas Journal 92/20 May 17 [#1009]

Duncan, R.C, 2000, *The peak of world oil production and the road to the Olduvai gorge*; Pardee Keynote Symposia Geological Society of America [#1256]

Duncan, R., 2000, *The Olduvai Theory, an illustrated guide*; [#1275]

Duncan, R.C., 2001, *World Energy Production, Population Growth, and the Road to the Olduvai Gorge*; Population and Environment, Vol. 22, No. 5, May 2001, [#1307]

Duncan, R.C., 2001, *A peek inside the United Nations*, 2001,; Discourse &Disclosure, Vol. 6, No. 3, July 2001 [#1325]

Duncan, R C., 2000, *Crude Oil Production And Prices: A Look Ahead At OPEC Decision Making Process*; Presented at the PTTC Workshop, September 22, 2000 [#1405]

Duncan, R.C., 2000, *Letters to President Clinton*; (9/29/2000 and 5/13/1997), [#1406]

Duncan, R.C. & Youngquist, W., 1999, *Encircling the Peak of World Oil Production*; Natural Resources Research, Vol. 8, No.3, pp. 219–232, 1999 [#1462]

Duncan, R.C., 2000, *The Peak Of World Oil Production And The Road To The Olduvai George*; Abstract to a paper to be presented at The Pardee Keynote Symposia GSA Summit 2000, Reno, Nevada, November 13, 2000 [#1561]

Duncan, R.C., 2000, *The Heuristic Oil Forecasting Method: User`s Guide And Forecast #4*; Institute on Energy and Man, July 24, 2000 [#1562]

Duncan, R.C., 2001, *Energy, Institutions, and Sustainability: What this Means to the Nations of the World*; Institute on Energy and Man, April 25, 2001 [#1635]

Duncan, R.C., 2003, *Three world oil forecasts predict peak oil production*; Oil & Gas Journal, May 26, 2003 [#2251]

E

Easterbrook, G., 1998, *The oil crisis really*; Los Angeles Times June 7 [#764]

Easterbrook, G., 2000, *Opportunity Cost*; The New Republic, May 15

Easterbrook, G., 2000, *Hooray for expensive oil !- Opportunity Cost*; The New Republic, pp. 21–25, May 15, 2000 [#1548]

Eberhardt, W., 2000, *Am Tropf der Scheichs*; Focus 200/9 [#1134]

Eberstadt, N., 1997, *World population implosion*; AEI Nov. 1997 [#718]

Economist, The, 1993, *Pollution*; The Economist June 13th 1993 [#177]

Economist, The, 1993, *A shocking speculation about the price of oil*; The Economist, September15 1993. 87–88. [#189]

Economist, The, 1994, *Power to the people – a survey of energy*; June 18th 1994 [#242]

Economist, The, 1995, *Illusion of change*; The Economist 12/8/95 [#347]

Economist, The, 1995, *The future of energy*; October 7 1995[#379]

Economist, The, 1996, *From major to minor*; May 18th [#437]

Economist, The, 1996, *Pipe dreams in central Asia*; May 4th [#472]

Economist, The, 1966, *Kurdistan, which one do you mean?* August 10th

Economist, The, 1996, *The oil buccaneer*; November [#521]

Economist, The, 1997, *Connections needed*; 15.3.97 [#563]

Economist, The, 1998, *Colombia on the brink*; 8/8/98 [#815]

Economist, The, 1998, *French dressing*; July 10 [#999]

Economist, The, 2000, *In praise of Big Oil*; The Economist, October 21, 2000 [#1425]

Economist, The, 1994, *Russia`s Caucasian cauldron*; The Economist, August 6,1994, [#1457]

Economist, The, 2001, *Solid-state physics – "Superduperconductivity"*; The Economist, March 17, 2001 [#1440]

Economist, The 2001, *Oil Depletion- Sunset for the oil business?*; The Economist, pp. 97–98, November

3, 2001 [#1756 & 1837]

Economist, The, 2003, *Still holding customers over a barrel*; The Economist, Ocober 23, 2003 [#1965]

Economist, The, 2003, *There's oil in them thar sands*; June 26, 2003 [#2187]

Economist, The, 2003, *The End of the Oil Age*; October 25, 2003 [#2303]

Economist, The, 2004, *We woz wrong*; February 9, 2004 [#2321]

Edwards Economic Research, 1999, *Into the Millennium*; HSBC Middle East Economic Bulletin, [#987]

Edwards Economic Research, 1999, *The imminent peak in world oil production*; draft Yearbook [#988]

Edwards, J.D., 1997, *Crude oil and alternate energy production forecasts for the twenty-first century: the end of the hydrocarbon era*; Amer. Assoc. Petrol .Geol. 81/8 1292-1305 [#631]

Edwards, J.D., 1998, *"Optimist" predicts world oil demand will outstrip production in 2020*; Univ of Colorado [#937]

Edwards, R.H., 1999, *The imminent peak in world oil production*; Gulf Business Economic Databook [#1148]

Edwards, R.H., 2002, *The prospects for sustainable development in the Middle East*. HSBC M.East Economic Bulletin 2 qtr [#1820]

Edwards, R., 1998, *High level risk*; New Scientist [#768]

Edwards, R., 1998, *Mea culpa*; New Scientist 12 September [#838]

Edwards, W.R., 1997, *Energy supply*; Oil& Gas Journ. July 9 [#629]

Ehrenfeld, D., 1999, *The coming collapse of the age of technology*; Tikkun 14/1 [#966]

Ehrenfeld, T., 2002, *Iraq: It's the oil, stupid – what if this is the beginning of an oil war?* Newsweek September 30 [#1992]

Ehrlich, P.R. & A.H.Ehrlich, 1990, *The population explosion*; Simon & Schuster

EIA, 2000, *Long Term World Oil Supply (A Resource Base / Production Path Analysis)*; www.eia.doe.gov, [#1768]

EIA, 2001, *Appendix D International Energy Outlook 2001*; EIA [#1304]

Einhorn, C.S., 1994, *Well oiled*; Barron's June 27 [#244]

Eisenberg, D., 2001, *Nuclear Summer*; Time, p 58- 60, May 28, 2001[#1297]

Eisenberg, D., 2000, *Oil`s New Boss*; Time, pp. 34–39, October 9, 2000 [#1416]

Ellis, P.A., 1991, *New technology for gas finding: how important has it been ?* Oil & Gas Journ. September 30, 1991 [#213]

Ellis-Jones, P., 1988, *Oil: a practical guide to the economics of world petroleum*; Woodhead-Faulkner 347 pp

Emerson, T., 2002 *The thirst for oil*; Newsweek, April 8 [#1831]

Energy Charter Treaty, 2003, *Energy Charters Missions and Objectives*; Autumn 2003 [#2289]

Energy Day, 2002, *Petrobras hails big Campos find*; Energy Day, August, 2002 [#1892]

Energy Economist, 1993, *The 1996 oil shock ?*; Energy Economists, May 1993, 139/17 [#339]

Energy Economist, 1993, *Saudi Arabian sands*; Energy Econ. 145/14 November 1993 [#281]

Energy Information Administration, 1985, International energy outlook 1985; DOE/EIA 0484(85) [#18]

Energy Information Administration, 1988, *Annual energy review*; DOE/EIA 0384 (88) [#19]

Energy Information Administration,1990, *U.S. oil and gas reserves by year of field discovery*; DOE/EIA 0534 [#611-2]

Energy Information Administration-,1993, *International oil and gas exploration and development 1991*; U.S.Department of Energy, Washington

Energy Information Administration, 2000, *First quarter 2000 oil production and capacity estimates*; [#1147]

Energy Information Administration, 1986, *International energy outlook 1986*; DOE/EIA 0484(96) [#609-2].

Energy Information Administration, 1999, *Projections of oil production capacity*; [#951]

Energy Information Administration, 2001, *World Oil Market and Oil Price Chronologies 1970-2000 (3 pages of 35)* [#1193]

Energy, 1996, *Production in FSU*, Energy [#566]

Engel, M., 2000, *Promoting best industry practices*; IRU [#1021]

English, A., 2003, *Half full, or half empty?*; Telegraph Motoring, Oct.11, 2003 [#1937]

Environmental Media Services, 2003, *End of Cheap Oil Poses Threat to World Economy, Experts Say*; January 6, 2003 [#2055]

Esau, I., 1992, *Azeri initiative*; Offshore Eng. Auust 1992 [#116]

Esser, R., 2000, *Discoveries of the 1990s -were they significant ?*; CERA [#1151]

Eugene Register Guard, 2002, *Oil Projections for Rockies Scaled Back*; 18 December 2002 [#2092]

Eureka, 1991, *Total- Colombian discovery*; Eureka July 1991 [#4]

European Commission, 1995, *For a European Union Energy Policy*; Green Paper [#605-1]

European Commission, 2000, *The European Union's oil supply*; Oct. 4th Report

European Commission, 1999, *European Union Energy Outlook To 2020*; Special Issue, November 1999 [#1511]

Ewing, T E., 2000, *Energy Minerals: Bridge to Future*; AAPG Explorer, pp. 42–43, May 2000 [#1541]

ExxonMobil, 1999, *Tomorrow's energy needs*; Wall Street Jounal [#1196]

Exxon Mobil, 2001, *Exxon Mobil Contribution To The Debate On The Green Paper – Towards a European Strategy for the Security of Energy Supply*; Report, September 28, 2001 [#1774]

F

Falcoff, M., 1995, *The Cuba in our mind*; National Interest Summer 1995 [#363]

Farago, L., 1973, *The Game of Foxes*; Bantam Books, New York 878p

Feldstein, M., 2001, *Vouchers Can Free Us From Foreign Oil*; The Wall Street Journal, December 27, 2001 [#1752]

Ferguson, A., 1999, *Judging the oil experts*; OPT 17 November [#1073]

Ferguson, A., 2000, *A key parameter for ecology*; OPT 28 January [#1075]

Ferguson, A., 2000, *The right price of oil*; OPT 27 February [#1091]

Ferguson, A., 2000, *The Lugano Report*; OPT 11 April [#1158]

Ferguson, A., 2000, *The empirical pricing of oil*; OPT 10 April [#1160]

Ferguson, A., 2003, *The Crucial Limit: 11 Cubic KM Of Carbon Per Decade*; OPT 6 January 2003 [#2056]

Ferguson, A.R.B., 2000, *The impact of globalization on the United States*,; Optimum Population Trust [#1207]

Ferguson, A.R.B., 2000 *The social and ecological consequences of globalization*; Optimum Population Trust [#1207]

Ferguson, A.R.B., 2001, *Planning for the demise of cheap energy*; Optimum Population Trust [#1209]

Ferguson, A.R.B., 2001, *Renewable Liquid Sunshine:its scope and limits*; The Optimum Population Trust (U.K.), 30 May, 2001 [#1301]

Ferguson, A.R.B., 2001, *Depletion of Oil and Gas- globally and in the US*; The Optimum Population Trust (U.K.), 8 July, 2001 [#1339]

Ferguson, A.R.B., 2001, *Perceiving the Population Bomb*; World Watch, July/August 2001 [#1340]

Ferguson, A.R.B., 1999, *Judging the Oil Experts*; Optimum Population Trust (U.K.), 1 December, 1999 [#1468]

Ferguson, A.R.B., 1999, *Shale Oil Prospects*; Optimum Population Trust (U.K.), 2 December, 1999 [#1470]

Ferguson, A.R.B., 1999, *Ice Age, Glacial and Interglacial*; Optimum Population Trust (U.K.), 9 November, 1999 [#1471]

Ferguson, A.R.B., 2001, *Verdict on the Hydrogen Experiment*; Letter to the Editor of "Worldwatch", March 18, 2001 [#1581]

Ferguson, A.R.B., 2001, *Natural gas resources and US population growth*; Optimum Population Trust (U.K.), 17 June, 2001 [#1663]

Ferguson, A.R.B., 2001, *The US Population Explosion*; Optimum Population Trust (U.K.), 29 July, 2001 [#1664]

Ferguson, Andrew, 2002, *Verdict on the Hydrogen Experiment*; The Pherologist, Vol.5 No.3, August 2002 [#1911]

Ferrier, R.W., 1982, *The history of the British Petroleum Company: Volume 1, the developing years 1901–1932*; Cambridge University Press 801pp [#B7]

Ferriter, J.P.,1994, *International energy cooperation*; Petrole et Technique 389 [#264]

Fesharaki, F. & Varzi, M., 2000, *Investment Opportunities Starting To Open Up In Iran`s Petroleum Sector*; Oil & Gas Journal, February 14, 2000 [#1474]

Fialka, J.J., 2002, *"Greens" and Churches Join Hands in Environmental Mission*, The Wall Street Journal, Tuesday, March 26, 2002 [#1806]

Financial Times, 1998, *Realignment in the Gulf*; 26.2.98 [#684]

Financial Times, 2001, *Expecting the improbable: Middle East capacity and global oil demand to 2020*;

Financial Times, *Expecting the Improbable – Middle East capacity and global demand top 2020*; Energy Economist Briefings [#1186]

Fisher D, 2001, *BP Amoco the hottest prospect in the oil patch*; Forbes [#1198]

Fisher, W.L., 1991, *Future supply potential of US oil and natural gas*; Geophysics: The leading edge of exploration, December 1991 [#22]

Fisher, W.L., 1993, *Oil and gas prices are headed higher*; Amer.Oil & Gas Reporter, January 1993 [#146]

Fisher, W.L., 2002, *Domestic Natural Gas: The coming methane*; Geotimes, November 2002 [#2035]

Fisk, R., 2003, *This Looming War Isn't About Chemical Warheads or Human Rights: It's About Oil*; The

Independant, Jan. 18, 2003 [#2078]

Fisker, J.L., 2002, *Oil Depletion: One of the most important problems of the world*; June 8, 2002 [#2052]

Fitzpatrick, M., 2003, *Shell Oil Find: rig moves in to drill off shore*; Sunday Independant, April 27, 2003 [#2162]

Flanigan, J., 2001, *An Oilman`s Dream / New Rules, Industry Changes to Push Gas Prices Higher*; Los Angeles Times, Tuesday, May 8, 2001 [#1303]

Flavin, C., 2000, *Energy for a new century*; Worldwatch March/April [#1129]

Fleay, B., 1995, *The decline in the age of oil*; Pluto Press 152p.

Fleay, B., & J.H. Laherrère, 1997, *Sustainable energy policy for Australia; submission to the Department of Primary Industry and Energy Green Paper 1996*; Paper 1/97 Institute for Science and Technology Policy, Murdoch University, W.Australia [#606-1]

Fleay, B., 1998, *Chemical dependency grows*; West Australian 26/6/98 [#814]

Fleay, B., 1998, *Its time to refuel debate on oil* ; West Australian 18/4/98 [#815]

Fleay, B., 1998, *The Greens WA – Energy Platform*; [#835]

Fleay, B., 1998, *Climaxing oil: how will transport adapt*; Chartered Inst.,Transport Symposium Tasmania 6–7 November [#868]

Fleay, B., 1999, *Beyond oil*; Chartered Inst of Transport in Australia conf. [#977]

Fleay, B., 2000, *Oil supply, the crunch has arrived*; pers. comm. [#1125]

Fleming, D., 1999, *The spectre of OPEC*; Sunday Telegraph March 21 [#953]

Fleming, D., 1999, *The next oil shock?* Prospect April [#954-F1]

Fleming, D., 1999, *The Next Oil Shock ?*; Prospect, pp. 12–15, April, 1999 [#1549]

Fleming, D., 2003, *The Wages of Denial*; The Ecologist April 2003 [#2144]

Fletcher, S., 2003, *Energy futures markets remain mixed*; Oil & Gas Journal, July 16, 20003 [#2210]

Fletcher, S., 2003, *Analysts see Angola as future major oil producer*; Oil & Gas Journal, aug.6, 2003 [#2263]

Flint, J., 2002, *Hydrogen Bomb*; Forbes, p 100, March 4, 2002 [#1802]

Flittie, C.G. & B.M.Robertson, 199?, *William Elliott Humphrey*; Amer.Assoc. Petrol. Geol. [#132]

Flores, G., 1987, *Arc of the sun*; Lion & Thorne, Tulsa. 229p [#B4]

Flower, A.R., 1978, *World oil production*; Scientific American 238/3 [#720]

Foley, G. and A. van Buren, 1978, *Nuclear of Not ?* Heinemann 208pp

Folinsbee, R.,1977, *World view from Alph to Zipf*; Geol. Soc. Amer. 88 897-907

Forbes, 1992, *The next oil crisis?* Forbes Feb 17 1992 [#159]

Forbes, 1968, *$5 000 000 000 000 US oil shale treasure*; March 1 [#995]

Forbes, S., 1999, *Unending oil?* Forbes 14 June [#993]

Foreman, N.E., 1996, *Opec influence grows with world output in next decade*; World Oil February 1996 [#412]

Foreman, N.E., 1997, *The bear awakens: resurgence of FSU oil and gas*; World Oil February [#557]

Forero, J., 2003, *Venezuelan Oilman: Rebel with a new cause*; The New York Times International, Feb.9 2003 [#2115]

Foss, B., 2001, *Search for gas becoming difficult*; Evansville Courier, August 10, 2001 [#1351]

Fowler, R.M., 2000, *World Conventional Hydrocarbon Resources: How Much Remains To Be Discovered And Where Is It ?*; Robertson Research International Limited, ???, [#1382]

Fowler, R.M. & Burgess, C J. & Otto, S C. & Harris, J P. & Bastow, M A., 2000, *World Conventional Hydrocarbon Resources: How Much Remains To Be Discovered, Where Is It?*; Robertson Research International Limited, 2000, [#1384]

Franco A., 1997, *Latin America activity soars to new heights*; Hart's Int. Petrol. Eng., July [#625]

Frank, R. & Barrionuevo, A., 2001, *Phillips Petroleum Agrees to Buy Conoco*; The Wall Street Journal, November 19, 2001 [#1732]

Freedman, A.M. & Bahree, B., 2003, *Iraq is shipping large cargoes of crude, violating UN rules*; Wall Street Journal, February 21, 2003 [#2116]

Friedman, A., 1990, *Opec needs $60 bn extra for productivity increase*; Financial Times Feb 8th 1990 [#76]

Friedman, T.L., 1994, *Opec's lonely at the tap, but China's getting thirsty*; New York Times Apr 3 1994 [#222]

From the Wilderness, 2003, *Chinas rising grain prices could signal global food crisis*; From the Wilderness Publications, Nov.19, 2003 [#

Fromkin, D., 1989, *A peace to end all peace*; Avon Books 635p ISBN .0-380-71300-4.

Fuerbringer, J., 2000, *Market Place- With oil prices increasing and the election season at hand, the cry goes up for OPEC to raise output*; The New York Times, August 16, 2000 [#1396]

Fuller, J.G.C., 1971, *The geological attitude*; Amer. Assoc. Petrol. Geol 55/11 [#141]

Fuller, J.G.C., 1993, *The oil industry today*; Brit. Assoc. Advancement of Sci. Published as *The British Association Lectures 1993* by The Geological Society, London [#53]

Futurist, 1997, *Permanent oil shortage impossible*; May–June [#583]
Futurist, 1997, *The world is not running out of oil*; May–June [#1812]

G

Galinier, P., 2001, *Quand le petrole disparaitra…* ; Le Monde, December 6, 2001 [#1734]
Gallon Environment Letter, The, 2002, *Canadian Tar Sands*; The Gallon Environment Letter, Vol.6, No.23, Oct.8, 2002 [#1997]
Garb, F.A., 1985, *Oil and gas reserve classification, estimation and evaluation*; Journ. Petrol. Technol. March 373-390
Gardner, T., 2000, *US ups estimate of non-US recoverable oil by 20 pc*; Reuters [#1132]
Garza, M.M., 2001, *Power woes put nuclear in new light*; The Chicago Tribune, Sunday, April 1, 2001 [#1658]
Gauthier, D.L. *et al.*, 1995, *1995 national assessment of United States oil and gas resources – results, methodology and supporting data*: U.S. Geological Survey CD-ROM
Gelbard, A. and C. Haub, 1998, *Population explosion not over for half the world*; Population Today 26–3 March [# 712]
George, A., 1996, *Reality and risk in the Middle East*; preprint Oil & Gas Project Finance in the Middle East Conference, Dubai May 12–13 [#448]
George, D., 1993 *Caspian Sea reserves could rival those of Saudi Arabia*; Offshore/Oilman July 1993 [#172]
George, R., 1997, *Canada's oil sands: the unconventional alternatives*; Hart's Pet. Eng. Int., 29/5/97 [#597]
Georgescu-Roegen, N., 1995, *La decroissance*; Sang de Terre, Paris 254pp
Geoscience Australia, 2000, *Undiscovered resources*; Oil and Gas Resources of Australia, pp. 29–35 2000 [#2132]
Geoscience Horizons, 2003, *The Rise and Fall of the Hubbert Curve: Its Origins and Current Perceptions*; Geoscience Horizons, 2003 [#1978]
Geoscientist, 1996, *World oil supplies*, 6/1 [#440]
Gerth, J., 2003, *Oil experts see long-term risks to Iraq reserves*; New York Times, November 30, 2003 [#1981]
Gerth, J., 2003, *Iraq's oil reserves*; New York Times, April 11, 2003 [#2249]
Gever, J., R. Kaufmann & D. Skole, 1997, *Beyond Oil*; http://dieoff.org.[#644]
Ghadhban, T.A. *et al.*, 1995, *Iraq oil industry: present conditions and future prospects*; Report Iraq Oil Ministry [#344]
Ghorban, N., 1994, *The evaluation of recent gas export pipeline proposals in the Middle East*; preprint, APS Conf [#290]
Giampietro, M., Ulgiati, S. & Pimentel, D., 1997, *Feasibility of Large-Scale Biofuel Production*; BioScience, Vol. 47, No.9, October, 1997 [#1429]
Gibson, W.R. and C.F. Garvey, 1992, *3D seismic has renewed the search for stratigraphic traps*; World Oil Sept. 1992 [#65]
Giddens, P.H., 1955, *Standard Oil Company (Indiana) – oil pioneer in the Middle West*; Appleton-Century-Crofts, New York
Gidron, G., 1999, *Mergers, acquisitions and alliances: imperatives for success*: CWC conf.[#1038]
Giraud, A., 1993, *1973-1993 – L'eclairage du passe, l'approche du futur*; Profils IFP 94/1 [#216]
Glain, S.J., 1999, *Iraq says it can exceed pre-embargo oil output*; Wall St Journ. 29 November [#1068]
Glain, S J., 1999, *Iraq Says It Can Far Exceed Pre-Embargo Oil Output*; The Wall Street Journal, p. 20, November 29, 1999 [#1641]
Globe International, 1998, *Global equity and climate change*; Globe Int [#940]
Gold, D.H. 2000, *Oil price surge brings fear of crisis, but also hope for more crude output*; Investor's Business Daily [#1219]
Gold, R. & Smith, R., 2003, *effects of Gas Shortage Rip Through Economy*; Wall Street Journal, Feb. 27, 2003 [#2113]
Gold, T., 1988, *Origin of petroleum; two opposing theories and a test in Sweden*; Geojournal Library 9 85–92.
Goldberg, J., 1998, *Getting crude in Baku*; New York Times 4/Oct. [#851]
Goldman, I., 2000, *Natural Gas: The Five Stages to Market Panic*; SolarQuest Net NewsService, August 10, 2000 [#1377]
Goldsmith, E. and N. Hildyard, 1990, *Earth Report 2*; Mitchell Beazley 176pp
Goldsmith, J., 1994, *The trap*; Macmillan 216p [#B17]
Goodman, M. & N.C. Chriss, 1979, *Mexico oil estimates inflated, experts say*; Los Angeles Times |May

18 1979 [#14]

Goodstein, D., 2002, *US must face reality that world's oil supply is dwindling*; Oregonian 2 April [#1814]

Goodstein, D., 2003, *Energy, Technology and Climate: Running out of gas*; New Dimensions In Bioethics, 2003 [#2267]

Goodstein, D., 2004, *Out of Gas*; ISBN-0-393-05857-3

Goodwin, S., 1980, *Hubbert's Curve*; Country Journal, November 1980 [#164]

Gore, A., 1998, *Finding a third way*; Newsweek, 23 Nov. [#858]

Gould, D., 2004, *Here comes the sun to heat your home*; Irish Examiner, January 31, 2004 [#2319]

Gowan, P., 1999 *The Twilight of the European Project*; CounterPunch [1027]

Grace, J.D., R.H. Caldwell and D.I. Heather, 1993, *Comparative reserve definitions: U.S.A., Europe and the former Soviet Union*; Journ. Petrol. Technol. September 1993 866-872 [#174]

Graham, H., 1990, *US oil plot fuelled Saddam*; Observer. 21 Oct 1990 [#721]

Graham, H., 1991, *The wolf bites back*; South [#736]

Grant, L., 1997, *In support of a revolution*, NPG Forum [#747]

Gray, D., 1989, *North Sea outlook and its sensitivity to price*; Petrol. Revue. January 1989 [#35]

Greenhouse, S., 1992, *Chevron taps ex-Soviet reserves*; Ft. Worth Star Telegram, May 19 1992 [#45]

Greenwald, J., 1994, *Black Gold*, Time June 20 [#239]

Gray, J., 2001, *Wars of want*; The Guardian, August 21, 2001 [#1354]

Grob, G., 1997, *Driving forces – energy in ISO and IEC*; ISO Bull. April [#964-f1]

Grob, G., 1999, *New total approach to energy statistics and forecasting*; CMDC-WSEC [#1043][1819]

Groeneveld, R., 1966, *Auto LPG: global review and criteria for success*; Pet. Review May [#445]

Grollman, N.G., 1997, *Environmentally sustainable energy for the East Asia/Pacific Region*; APPEA Journ 722 [#752]

Grollman, N.G., 1998, *Pipelines, politics and prosperity*; APPEA Journ. 815 [#t53]

Gruy, H.J., 1998, *Natural gas hydrates and the mystery of the Bermuda triangle*; Hart's Petrol. Eng. Int. March [#717]

Guardian, 1998, *A nuclear white elephant*; World Press Review Dec. [#854]

Guenther, F., 2000, *Vulnerability In Agriculture: Energy Use, Structure And Energy Futures*; Presentation for the INES conference, KTH Stockholm, June 15, 2000 [#1502]

Gulbenkian, N., 1965, *Portrait in oil*; Simon & Schuster, New York [#B9]

Gunther, F., 2000, *Vulnerability in agriculture: energy use, structure and energy futures*: INES Conference, Stockholm

Gurney, J., 1997, *Migration or replenishment in the Gulf*; Petrol. Review May [#593]

H

Haggett, Scott, 2003, *Oh, How the patch has changed: Technology, finances revamp landscape*; Calgary Herald, August 2, 2003 [#1933]

Haggett, S., 2003, *Oilpatch: Gigantic natural gas finds now rare*; Calgary Herald, August 2, 2003 [#1934]

Haines, L., 1994, *Oil price recovery : for real or a dream*; NewsWell[#272]

Halbouty, M.T., 1970, *Geology of giant petroleum fields*; Amer. Assoc. Petrol. Geol. Mem.14

Haldorsen, H.H., 1996, *Choosing between rocks, hard places and a lot more: the economic interface*; in Doré A.G and R.Sinding Larsen (Eds), *Quantification and prediction of hydrocarbon resources*; NPF Sp. pub. 6., Elsevier ISBN 0-444-82496-0.

Hall, C.A.S., Cleveland, C .J. & Kaufman, R., 1992, *Oil And Gas Availability: A History Of Federal Government Overestimation*; In: Energy And Resource Quality, by all *et al.*, pp. 343–349, 1992, dieoff.com, [#1552]

Hall, C., Tharakan, P., Hallock, J., Cleveland, C. & Jefferson, M.; *Hydrocarbons and the evolution of human culture*; Nature, Vol. 426, November 20, 2003 [#2275]

Hallock, J.J. Jr *et al.*, 2004, *Forecasting the limits to the availability and diversity of global conventional oil supply*; Energy v. 29 2004.

Hamer, Mick, 2002, *Horse power beats diesel*; New Scientist July 2002 [#1897]

Hamilton-Bergin, S., 2004, *The No 19 Bus* – ISBN0-9545318-0-9

Hammer, A., 1988, *Hammer, witness to history*; Coronet Books 752pp

Hanson, J., 1997, *Fossilgate – the biggest coverup in history*; http://dieoff.org [#645]

Hanson, J., 2000, *A lethal education*; http://dieoff.org [#1168]

Hanson, J., 2000, *Fossilgate -as soon as the year 2000*; http://dieoff.org [#1168]

Hanson, J., 2000, *Energy Synopsis*; dieoff.com/synopsis.htm , December 30, 2000 [#1579]

Hardman, R.F.P., 2000, *The oil industry – a whipping boy for a world out of control*; draft [#1126]

Hardman, R.F.P., 1998, *The future of Britain's oil and gas industry*; Inst.Mining Eng. March [#711]

Hardman, R.F.P., 2001, *New Petroleum Provinces of the 21st Century?*; 20 July, 2001 [#1341]

Harigal, G.G., 1998, *Energy in a changing world*; Pugwash Meeting 243 [#946]

Harigal, G.G., 1998, *Nuclear energy for developing countries*; CIPRI [#947 F-1]

Harper, F.G., 1999, *Ultimate Hydrocarbon Resources in the 21st Century*; BP Amoco, AAPG Birmingham 1999 [#1399]

Harris, R., 1994, *Production Analysis shows growth in worldwide oil and gas reserves*: World Watch, Petroconsultants [#308]

Harris, R, 2001, *Drilling and discoveries in the year 2000 worldwide outside of North America*; IHS Energy Group [#1233]

Harts Petroleum Engineer International, 1996, *New paradigm: mining for oil*; Sept. [#514]

Hart's, *OPEC's ability to manage oil prices shaky, but group's long-term prospects still bright*; OPEC and oil prices Hart's 12/10/00 [#1257]

Harvey, H., 1993, *Innovative policies to promote renewable energy*; Advances in Solar Energy, Amer. Solar Energy Soc[#220]

Hatfield, C.B., 1995, *Will an oil shortage return soon?* Geotimes Nov. 1995 [#406]

Hatfield, C.B., 1997, *Oil back on the global agenda*; Nature 367 121 [#589]

Hatfield, C.B., 1997, *A permanent decline in oil production*; Nature 388 [#632]

Hatley, A.G., 1995, *The oil finders*; Centex 267pp

Haun, J.D. (Ed), 1975, *Methods of estimating the volume of undiscovered oil and gas resources*; Amer. Assoc. Petrol. Geol. Studies in Geology No.1.

Hawken, P., 1993, *The Ecology of Commerce – a declaration of sustainability*; Publ. Harper Business, New York 250p.

Hawkes, N., 1999, *Huge reserve of gas will fuel 21st Century*, The Times [#897]

Healion, K, 2001, *Energy Pie*; FEASTA [#1246]

healthandenergy.com, 2002, *Sources of Information about Oil Crisis (incl. C J Campbell)*; www.healthandenergy.com/oil_crisis.htm, March 2002 [#1785]

Hebert, H.J., 2000, *Clinton preaches patience on oil*; Assoc. Press. [#1141]

Hecht, Jeff, 2002, *You can squeeze oil out of a stone*; New Scientist, August 2002 [#1894]

Heinberg, R., 2002, *Behold Caesar*; October, 2002 [#1999]

Heinberg, R., 2003, *The Party's Over*; New Society Publishers 274p

Henderson, Simon, 2002, *The Saudi Way*; Wall Street Journal, August 12, 2002 [#1885]

Henry, J.C., 1996, *Managing oil and gas companies*; Oil & Gas Journ. November 4. [#524]

Herald Sun, 1997, *Power Cut*; 31st. May [#618-3]

Herkstroter, C., 1997, *Contributing to a sustainable future — the Royal Dutch/ Shell Group in the global economy*; Shell paper [#588]

Herrick, T. & Bahree, B., 2001, *OPEC Won't Curb Oil Until Others Do*; The Wall Street Journal, Thursday, November 15, 2001 [#1730]

Herrick, T., 2002, *Oil prices get a lift from decline in inventories*; Wall St Journ 2 Apr [#1827]

Herrick, T. & Warren, S., 2003, *Exxon Mobil CEO Warns of Effects of High Oil Prices*; Wall St. Journ., May 3 2003 [#2138]

Herrick, T., 2003, *Gas Prices Rock Chemical Industry*; Wall St. Journ., June 18, 2003 [#2174]

Hersh, S.M., 2001, *King's Ransom – How vulnerable are the Saudi royals ?*; The New Yorker, www.newyorker.com , October 24, 2001 [#1682]

Hersh, S.M., 2001, *King's Ransom: exposing a right royal mess*; The Australian Financial Review, Friday, November 16, 2001 [#1699]

Herwijnen, T. van & M. Groeneveld, 1999, *Global energy perspectives for power generation*; 3 IERE Conf. Kobe, proc.[971]

Higgins, A. & Hutzler, C., 2001, *China Pursues a Great Game of Its Own*; The Wall Street Journal, June 14, 2001 [#1398]

Higgins, G.E., 1996, *A history of Trinidad Oil*; Trinidad Express Newspapers, 498pp

Hill, P., 2000, *Gulf states will likely agree to increase oil production*; Washington Times [#1081]

Hiller, K., 1997, *Future world oil supplies – possibilities and constraints*; Erdol Erdgas Kohle 113/9 349-352 [#572]

Hiller, K., 1997, *Kohnwasserstoff-projekte der BGR von 1970 bis 1995*; Z. angew. Geol. 43/1 [#573]

Hiller, K., 1998, *Depletion midpoint and the consequences for oil supplies*; WPC 15. [#780]

Hiller, K., 1999, *Verfugbarkeit von Erdol*; Erdol, Erdgas, Kohle Jahrgand 115/2 [#918]

Hoagland, J., 1994, *The US-Saudi line is off the hook*; Int. Herald Tribune 31.1.94 [#206]

Hobbs, G.W., 1995, *Oil, gas, coal, uranium, tar sand resource trends on rise*; O&GJ Sept 4 1995 [#374]

Hobson, G.D., 1980, *Musing on migration*; Journ. Petrol. Geol. 3/2 [#311]

Hobson, G.D., 1991, *Field size distribution – an exercise in doodling ?* Journ. Petrol Geol 14/1 [#312]

Hogan, L., *et al.*, *Australia's oil and gas resources*; ABARE report [# 534]

Hogarty, T.F., 1999, *Gasoline: still powering cars in 2050*; The Futurist March [#1000]

Holder, K., 1998, *Worldwide oil shortage will soon affect us*; Evansville Courier 31 May [#751]

Holder, K., 1999, *Oil depletion problem being ignored*; Evansville Courier & Press, December 16 [1029]

Holditch, S., 2003, *The increasing role of unconventional reservoirs in the future of the oil and gas business*; November 2003 [#2297]

Holhut, R.T., 2002, *On Native Ground: Big oil, Bin Laden, The Bush team and Sept.11*; The American Reporter, Vol.8, No. 1942, October 1, 2002 [#2045]

Holley, D., 1997, *China's thirst for oil fuels competition*; Los Angeles Times July 29 [#635]

Hollis, R., 1996, *Stability in the Middle East – three scenarios*; Pet. Review May p. 205 [# 442]

Holmes, B. & Jones N., 2003, *Brace yourself for the end of cheap oil*; New Scientist, August 2, 2003 [#2238]

Holmgren, D.E., 2002, *Permaculture* – ISBN 0-646-41844-0

Homer-Dixon, T., 2001, *The Ingenuity Gap*; ISBN 0-676-97296-9

Hopkins, N., 2002, *Worry over Shell input to exploration*; The Times, ??? , [#1767]

Hopson, C., 2000, *A burning issue*; Upstream *Special feature* [#1277]

Hornberger, J.G., 2002, *Iraq, Iran and September 11: A Chronology*; The Future of Freedom Foundation, 2002 [#2042]

Horton, S. & N. Mamedov, 1996, *Investment in Azerbaijan's upstream requires attention to legal details*, World Oil April 1996 [#441]

Hotellet, R., 1998, *Tangled web of an oil pipeline*; Christian Science Monitor May 1st [#740]

Hotelling, H., 1931, *The Economics of exhaustible Resources*; Journ. Politic. Economy 1931 [#190]

Hovey, H H., 2001, *DJ Matt Simmons: Energy Crisis is "Extremely Serious"*; Dow Jones   News, April 6 2001 [#1287]

Howard, K., 2000, *Running on Empty*; Autocar 23. February

Howe, J., 2003, *The End of Fossil Energy* – ISBN 0-9743404-0-5

Howes, J., 1997, *Petroleum systems, resources of Southeast Asia, Australia*; Oil & Gas Journal December 15, 1997 [#2119]

Hoyos, C., 2003, *Energy companies see a big future for gas. But will the West's increasing dependance imperil it's fuel security?*; Financial Times August 15, 2003 [#2242]

Hubback, A., 2000, *10 years on, Iraq and West still poles apart*; Hart`s E&P, August 2000 [#1368]

Huber, P. & M. Mills, 1998, *King Fahd and the tide of technology*; Forbes November 16 [#942]

Hubbert, M.K., 1949, *Energy from fossil fuels*; Science 109, 103–109

Hubbert, M.K., 1956, *Nuclear energy and the fossil fuels*; Amer. Petrol. Inst. Drilling & Production Practice. Proc. Spring Meeting, San Antonio, Texas. 7-25.[#187]

Hubbert, M.K., 1962, *Energy resources, a report to the Committee on Natural Resources*; Nat. Acad. Sci. Publ. 1000D

Hubbert, M.K., 1969, *Energy resources*; in Cloud P. (Ed) *Resources and Man*; W.H. Freeman

Hubbert, M.K., 1971, *Energy resources of the Earth*; in *Energy and Power*, W.H. Freeman

Hubbert, M.K., 1976, *Exponential growth as a transient phenomenon in human history*; in Strom Ed., Scientific Viewpoints, American Inst. of Physics 1976 [#952]

Hubbert, M.K., 1980, *Oil and gas supply modeling*; in Gass S.I., ed. proceedings of symposium, U.S.Dept. of Commerce June 18–20,1980 [#492]

Hubbert, M.K., 1981, *The world's evolving energy system*; Amer. J. of Physics 49/11 1007-1029

Hubbert, M.K., 1982, *Technique of prediction as applied to the production of oil & gas*; in NBS Special Publication 631. U.S. Dept.Commerce/ National Bureau of Standards, 16–141.

Hubbert, M.K., 199? *Senate committee hearings of efforts to suppress study*; [#849]

Hubbert, M.K., 1998 *Exponential growth as a transient phenomenon in human history*; Focus [#902][#1909]

Huber, P. & M. Mills, 1998, *King Faisal and the tide of technology*; Forbes November 16 [#878]

Huddleston, B.P., 1998, *The availability of reserve estimates depending on the purpose*; Pet. Eng. Int. Sept [#842]

Humphrys, J., 2002, *We're planning a war, but don't mention the oil*; The Sunday Times December 8, 2002 [#2034] [#2068]

Hupe, R., 2000, *Entzug der Öldroge*; Die Woche, 22.12.00

Huxley, A., *Falling Off Huberts Peak*; NZ Listener, March 8, 2003 [#2307]

I

Ignatius, D., 2003, *Check That Oil*; Washington Post, November 14, 2003 [#2279]

IHS Energy Group, 2000, *1999 A Star Year For Oil And Gas Discoveries…*; Press Release, June 13, 2000 [#1514]

Imbert, P., J.L. Pittion and A.K. Yeates, 1996, *Heavier hydrocarbons, cooler environment found in deepwater*; Offshore April [#446]

Imperial Oil Ltd., 2002, *Imperial Oil 2002 Annual Report (Canadian Exxonmobil subsidiary)*; 2002

[#2149]

Independent Petroleum Association of America, 1993, *The promise of Oil and Gas in America*; final report of IPAA Potential Resources Task Force [#225]

Industrie & Environnement, 2003, *A Challenge Yet To Be Taken Up; The End Of Oil,* Industrie & Environnment, No. 275, June 12, 2003 [#1931]

Ibrahim, Y.M., 1990, *Widespread unrest threatens world oil supply, experts say*; Int. Herald Tribune 6/3/90 [#73]

Institute of Petroleum, 1993, *Valuable Saudi upstream data published*; Pet. Review December 1993 [#283]

Institute of Petroleum, 1995, *The UK continental shelf in 2010: is this the shape of the future?* Report [#481-1]

Interface News, 2002, *Buzzard oil discovery: biggest for 25 years?*; Interface, August, 2002 [#1888]

International Center for Technology Assessment, 2000, *The real price of gas* [#1136]

International Energy Agency, 1993, *Oil Market Report,* 7 Sept. 1993 [#175]

International Energy Agency, 1994, *World Energy Outlook* 1994 Edition.[#241]

International Energy Agency, 1995, *World Energy Outlook* 1995 Edition [#357]

International Energy Agency, 1998, *World Energy Prospects to 2020*; Report to G8 Energy Ministers, March 31 (www.iea.org/g8/world/oilsup.htm)

International Energy Agency, 1998, *World Energy Outlook* 1998 Edition

International Energy Agency, 1999, *World Energy Outlook and impact of Economic turmoil in Asia on oil prospects*; Report June [#998]

International Energy Agency, 1999, *World Energy Outlook* 224pp

International Energy Agency, 1999, *Meeting of the governing board*; Press Release [#1094]

International Energy Agency, 2001, *World Energy Outlook* 226pp

International Energy Agency, 2003, *Supply*; Monthly Oil Market Report, July 11, 2003 [#2227]

International Energy Agency, 2004, *World Energy Outlook* 226pp

International Herald Tribune, 2002, *Interview with Felix Zulauf, president of Zulauf Asset Management AG*; International Herald Tribune, March 2-3, 2002 [#1791]

Investors Chronicle, 1994, *Driven by supply*; Investors Chronicle 23/12/94 [#326]

Ion, D.C., 1980, *The availability of world energy resources*; Graham & Trotman 345pp

Ismail, I.A.H., 1994, *Untapped reserves, world demand spur production expansion*; Oil & Gas Journ. May 2, 1994 95-102 [#224]

Ismail, I.A.H., 1994, *The world oil production perspective; the future role of Opec and Non-Opec*; Preprint, APS Conf. [#285]

Ismail, I.A.H., 1994, *Future growth in OPEC oil production capacity and the impact of environmental measures*; Energy Exploration and Exploitation 12/1 17-58 [#378]

Ivanhoe, L.F., 1976, *Evaluating prospective basins*; in three parts – Oil & Gas Journ. December 13. 1976 [#108]

Ivanhoe, L.F, 1980, *World's prospective petroleum areas*; Oil & Gas Journ. April 28 1980 [#91]

Ivanhoe, L.F, 1984, *Oil discovery indices and projected discoveries*; Oil & Gas Journ. 11/19/84

Ivanhoe, L.F, 1985, *Potential of world's significant oil provinces*; Oil & Gas Journ. 18/11/85 [#144]

Ivanhoe, L.F., 1886, *Oil discovery index rates and projected discoveries of the free world*; in Oil & Gas Assessment. Amer. Assoc. Petrol. Geol. Studies in Geology #21, 159-178

Ivanhoe, L.F, 1987, *The decline of giant oilfield discoveries*; draft [#19]

Ivanhoe, L.F., 1987, *Permanent oil shock*; AAPG 71/5 [#309]

Ivanhoe, L.F., 1988, *Future crude oil supply and prices*, Oil & Gas Journ. July 25 111-11? [#97]

Ivanhoe, L.F., 1990, *Liquid fuels fill vital part of US Economy*; Oil & Gas Journ. Apr.23 106-109 [#15]

Ivanhoe ,L.F., 1990, *Competition increases to obtain oil imports*; Oil & Gas Journ Oct 29. 1990 [#49]

Ivanhoe, L.F., 1991, *Oil, gas dominant sources of energy in U.S.*; Oil & Gas Journ. Sept 30 1991 #387]

Ivanhoe, L.F., 1995, *Future world oil supplies: there is a finite limit*; World Oil Oct.1995 [#381]

Ivanhoe, L.F & G.G.Leckie, 1991, *Data on field size useful to supply planners*; Oil & Gas Journ, April 29 1991 [#30]

Ivanhoe, L.F., 1995, *Oil reserves and semantics*; M.King Hubbert Center for Petrol. Supply studies [#405]

Ivanhoe, L.F., 1996, *World oil supply*; personal comm. [#439]

Ivanhoe, L.F., 1997, *Updated Hubbert curves analyze world oil supply*; World Oil November [#508]

Ivanhoe, L.F., 1997, *Get ready for another oil shock*; Futurist Jan-Feb 1997

Ivanhoe, L.F., 2000, *Petroleum positions of the United Kingdom and Norway*; M.King Hubbert Center for Petrol. Supply studies 200/1 [#1166]

Ivanhoe, L.F., 2000, *Oil reserve revisions: Major OPEC and Communist countries 1979-99*; M.King Hubbert Center for Petrol. Supply studies 200/1 [#1167]

Ivanhoe, L.F., 2001, *Hubbert Center Newsletter 2001/1, Petroleum positions of Saudi Arabia, Iran, Iraq,*

Kuwait, UAE Middle East Region [#1232]

Ivanhoe, L.F., 2001, *Petroleum Positions Of Egypt, Libya, Algeria, Nigeria, Angola, Africa Region*; Hubbert Center Newsletter # 2001/3 [#1326]

Ivanhoe, L.F., 2002 *Canada's future oil production 2000-2050*; Hubbert Center Newsletter 2002/2 [#1818]

Ivanhoe, L.F. & Riva, J.P., 2000, *Exports- The Critical Part Of Global Oil Supplies /Petrophobia*; Hubbert Center Newsletter # 2000 / 4-1 and 4-2, October 2000 [#1412]

Ivanhoe, L.F., 2000, *World Oil Supply- Production, Reserves, And EOR*; Hubbert Center Newsletter # 2000 / 1-1, January 2000 [#1427]

Ivanhoe, L.F., 2001, *Petroleum Positions Of China, India, Indonesia, Malaysia, Australia, Far East & Oceania Region*; Hubbert Center Newsletter # 2001 / 4, October, 2001 [#1694]

Ivanhoe, L.F., 2002, *Canada's future oil production*; Hubbert Center Newslertter 2002/2 [#1818]

Ivanhoe, L.F., 2002, *Petroleum Positions of United States(US/48 & Alaska),Canada,Mexico North American Region*; Hubbert Center Newsletter #2002/4 [#2029]

Ivanovich, D., 1999, *World may learn to wean itself from oil*; Houston Chronicle, Sunday, October 24, 1999 [#1546]

Ivans, M., 2003, *There's gold in them thar war torn countries*; The Davis Enterprise Forum, June 22, 2003 [#2175]

J

Jacque, M., 1994, *Reserves mondiales de petrole*; geochronique 49. [#257]

Jaffe, A.M. and R.A. Manning, 1999, *The Shocks of a World of Cheap Oil*; Foreign Affairs 79/1[1031]

Jaffe, A.M. and R.A. Manning, 1999, *The myth of the Caspian great game: the real geopolitics of energy*; [#1123]

Jaffe, A., 2001, *Exxon Mobil, Others Say Needed Refineries Unlikely to Be Built*; Bloomberg.com: energy news, June 21, 2001 [#1315]

Jaffe, A., 2000, *The Outlook For OPEC and International Oil Markets*; James A. Baker III Institute For Public Policy Energy Forum, [#1438]

Jaggi, R., 2004, *Britain feared oil crisis could spark US military retaliation*; Financial Times, January 3, 2004 [#2306]

James, M., 1953, *The Texaco Story – the first fifty years 1902-1952*: Publ. The Texas Company . 115p

Jefferson, M., 1994, *World energy prospects to 2010*; Petrole et Techn. 389 [# 262]

Jenkins, D.A.L., 1987, *An undetected major province is unlikely*; Petrol. Revue Dec 1987 p 16 [#336]

Jenkins, S., 1997, *Exploding the myth*; The Times 12.Nov [#653]

Jenkins, S., 1999, *Will they never learn*; Times 9 April [#949]

Jenkins, S., 1999, *Weep for poor Orissa*; [#1025]

Jenkins, S., 2002, *The Bush White House is at war-with itself*; The Times, November 22, 2002 [#2020]

Jenkins, G., 1994, *World oil trade and tankers to 2007*; preprint APS Conf. [#291]

Jennings, J.S., 1996, *The millennium and beyond*; Energy World 240 June.

Jochen, V.A, & J.P. Spivey, 1997 *Using the bootstrap method to obtain probabilistic reserves estimates from production data*; Petroleum Engineer, Sept. [#651]

Johnston, D., 1998, *Oil and tax*; Encycl. Human Ecology, draft [#867]

Johnston, D., 1994, *International petroleum fiscal systems and production sharing contracts*; PennWell 325pp

Johnson, K., 2003, *Blasts Hinder Resumption of Iraqi Oil Exports*; Wall St. Journ., June 24, 2003 [#2171]

Johnson, K., 2003, *Is a great Iraqi oil field fading away?*; Wall St. Journ., June 23, 2003 [#2172]

Johnson, K., 2003, *Key Iraqi Pipeline Won't Reopen Before Years End*; Wall St. Journ., June 18, 2003 [#2173]

Jones, C., 2001, *Nature's system saves us billions*; The Australian Financial Review, p 20, September 7, 2001 [#1700]

Jones, G., 2003, *World oil and gas 'running out'*; CNN.com, October 2, 2003 [#1968]

Jones, N., 2002, *Strike it rich*; New Scientist, February 2, 2002 [#1765]

Journal de Geneve, 1998, *Un eminent geologue predit la fin du petrole a bon marche des 2008*; April 2 [#705]

Jung-Hüttl, A., 2000, *Kassandra-Rufe in Öl*: Suddeutsche Zeitung -online

Jungels, P., 1999 *How the predictability of profits and cash flow can be improved in a period of oil price uncertainty*; Presentation to Oil & Money Conf. [#989]

K

Kahn, J., 2000, *Surging prices show that oil's far from irrelevant in US economy;* Minneapolis Star Tribune February 21 [#1085]

Kahn, J., 2000, *Surge in oil prices is raising specter of inflation spike;* N.Y. Times 21 February [#1090]

Kaletsky, A., 1996, *Time has come to review the demand side disaster;* Times 25.1.96 [#411]

Karmon, Y. & D. Peretz, 1998, *Isreal;* Grolier [#929]

Kassler, P., 1994, *Two global energy scenarios for the next thirty years and beyond;* World Petrol. Congr. Stavanager.[#233]

Katz, D. & H. Payne, 2000, *Think again about eco-cars;* Wall St Journ. 20 January [#1067]

Kaufmann, R., 1991, *Oil production in the Lower 48 States:* Res. & Energy 13 [#96]

Kaufmann, R. & C.J. Cleveland, 1991, *Policies to increase US oil production likely to fail, damage the economy, and damage the environment;* Ann.Rev.Energy Environ. 1991 [#210]

Kaufmann, R., W. Gruen & R. Montesi, 199?, *Drilling Rates and expected oil prices: the own price elasticity of US oil supply;* Centre for Energy & Environmental studies, paper [#211]

Keegan, W., 1985, *Britain without oil;* Penguin 128pp

Kelley, A., 2000, USA:Experts duel over future oil supply scenarios; Reuters [#1178]

Kelsey, T., 1995, *Can 'mother' stop the sons of Islam?* Sunday Times, 31.12.95 [#418]

Kemp, A.G & L. Stephen, 1996, *UKCS future beyond 2000 depends on oil price and reserve trends;* World Oil October [#540].

Kempf, H., 2000, *Le petrole et la planete;* Le Monde Interactif, 4 Septembre 2000 [#1381]

Kennedy, P., 1994, *Preparing for the 21st Century;* Fontana 428pp

Kenney, J.F., 1996, *Impeding shortages of petroleum re-evaluated;* Energy World **250** June 1992.

Kergin, M., 2002, *Trust the market (and Canada);* Wall Street Journ. 15.6.02 [#1832]

Kerr, J., 1998, *The next oil crisis looms large – and perhaps close;* Science 281 [#790]

Kerr, J., 1998, *Warming's unpleasant surprise: shivering in the greenhouse;* Science 281 [#800]

Kerr, J., 1998, *West Antarctica weak underbelly giving way;* Science 281 July 24 [#812]

Khalimov, E.M., 1993, *Classification of oil reserves and resources in the former Soviet Union;* Amer. Assoc. Petrol. Geol. 77/9 1636 (abstract).

Khan, H.K., 1989, *Exploration promotion in India;* UN Seminar, Policy and Management of Petroleum Resources, Oslo [#101].

Kieburtz, G.B. *et al.*, 1997, *Strongest outlook in nine years;* World Oil Feb. [#559]

Kinzer, S., 1998, *On piping out Caspian oil;* New York Times 8/11/98 [#855]

Kinzer, S., 1998, *US bid to build a pipeline in Caucasus appears to fail;* New York Times 11/10/98 [#881]

Kinzer, S., 2003, *All the Shah's Men;* John Wiley & Sons 258p. ISBN 0-471-26517-9

Kjaergaard, T.,.1994, *The Danish revolution 1500-1800;* Cambridge 314pp

Klare, M.T., 2002, *Resource wars: the new landscape of global conflict;* Owl Books. pp289

Klare, M.T., 2002 *Bush's master plan;* www.alternet.org [#1822]

Klare, M.T., 2002, *Oiling the wheels of war;* The Nation, November 2002 [#2094]

Klare, M.T., 2003, *The Bush/Cheney Energy Strategy: Procuring (The Rest Of) The Worlds Oil;* draft [#2206]

Klare, M.T., 2004, *Blood and Oil: the Dangers and Consequences of America's Growing Petroleum Dependency;* Hamish Hamilton 265p.

Klebnikov, P., 1998, *Opec: the cowardly lion;* Forbes April 6 [#719]

Klemme, H.D., 1983, *Field size distribution related to basin characteristics;* Oil & Gas Journ. December 25. 1983 169–176

Klemme, H.D. & Ulmishek G.F., 1991, *Effective petroleum source rocks of the world: stratigraphic, distribution and controlling depositions factors;* Amer. Assoc. Petrol. Geol 75/12, 1908-185

Kleveman, L.C., 2003, *The New Great Game;* The Ecologist, April 2003 [#2145]

Knickerbocker, B., 1998, *West's balancing act: economy, nature;* Christian Science Monitor 10/8/98 [817]

Knoepfel, H., 1986, *Energy 2000;* Gordon & Breach 181pp

Knott, D., 1991, *Calm surface, deep currents: is industry storing up problems?;* Offshore, December 1991, 25–27 [#6]

Knott, D., 1994, *Opec, once all-powerful, faces a cloudy tomorrow;* Oil & Gas Journ. August 22 1994 [#255]

Knott, D., 1996, *Reserves debate;* Oil & Gas Journ. January 29. 40 [#526]

Knott, D., 1996, *BP sharpening focus on improved shareholder value, efficiency* ; Oil & Gas Journ. 8.7.96 [#501]

Knott, D., 1998, *Oil orthodoxies;* O&GJ 21 September [#837]

Koch, H.J., 2000, *IEA introductory address;* Clean Energy 2000. [#1106]

Koczot, S. & J. Orechowski, 2004, *Sekrety I klamstwa Shella;* Newsweek Polska 18/2004

Koen, A. 1996, *Day of reckoning*; Upstream news 23.12.96 [#529]

Kroenig, J., 1999, *Am Tropf der Scheichs*; Die Zeit, No.40, September 30, 1999 [#1452]

Krauss, C., 1997, *Mexican data suggest 30% overstatement of oil reserves*; New York Times 18th March [#577]

Krayushkin, V.A., *et al.*, 1994, *Recent application of the modern theory of abiogenic hydrocarbon origins: drilling and development of oil and gas frields in the Dneiper-Donets Basin*; 7th Int. Symposium on the observation of the continental crust through drilling., Sante Fe, New Mexico, proc. 1994 [#455]

Kronig, von J., 1999, *Am Tropf der Scheichs* ; DIE ZEIT 30 September [1023]

Kulke, H., I.Taner & A. Mayerhoff 1994, *China*; in Regional Petroleum Geology of the World. Borntrager, Stuttgart [#375]

Kulke, H., 1994, Sweden, in Regional Petroleum Geology of the World. Borntrager, Stuttgart [#376]

Kvint, V., 1990, *Eastern Siberia could become another Saudi Arabia*; Forbes September 17 1990 [#388]

L

Labibidi, M.M. *et al.*, 2000, *Depletion of petroleum resources*; Pres. Clean Energy 2000 Conf. [#1064]

Lafitte, R., 2000, *World Hydro potential*; Clean Energy 2000 [#1110]

LaGesse, D., 2001, *A low-gas warning*; U.S. News & World Report, September 17, 2001 [#1691]

Laherrère, J.H., 1990, *Hydrocarbon classification rules proposed*; Oil & Gas Journ. August 13. p.62

Laherrère, J.H., 1990, *Les Reserves d'hydrocarbures*; BIP 6629 [#167]

Laherrère, J.H., 1992 *Reserves mondiales restantes et a decouvrir* :ATFP Conf. Paris 18.4.91 Revue de Presse TEP No3 20.1.92 [#168]

Laherrère, J.H., 1993, *Le petrole, une ressource sure, des reserves incertaines*; Petrol et Technique, 383, October 1993 [#208]

Laherrère, J.H., A. Perrodon and G. Demaison, 1993, *Undiscovered petroleum potential: a new approach based on distribution of ultimate resources*; Rept. Petroconsultants S.A., Geneva

Laherrère, J.H., 1994, *Published figures and political reserves*; World Oil, Jan 1994 p. 33.[#207]

Laherrère, J.H., 1994, *Study charts US reserves yet to be discovered*; American Oil & Gas Reporter 37/9 99-104.

Laherrère, J.H., 1994, *Nouvelle approche des reserves ultimate – application aux reserves de gaz des Etas-Unis*; Petrole et Technique, Paris 392. 29-33

Laherrère, J.H., 1994, *Reverves mondiales de petrole: quel chiffre croire?*; Bull. Inform. Petrol. 7727,7728,7729

Laherrère, J.H., 1995, *World oil reserves: which number to believe?*; OPEC bull. Feb draft.[#256] Final [#346]

Laherrère, J.H., 1995, *An integrated deterministic/probabilistic approach to reserve estimations – discussion*; draft for Journ. Pet. Technol. [#332]

Laherrère, J.H., 1996, *Distributions de type "fractal parabolique" dans la Nature*: C.R.Acad. Sci. Paris 322 II

Laherrère, J.H., & A. Perrodon, 1996, *Technologie et réserves*; Draft Petrole et Technologie [#496]

Laherrère, J.H., 1996, *Upstream potential of the Middle East in a World context*; Proc. Oil& Gas project finance in the Middle East IBC Dubai Conf. May 1996 [#565]

Laherrère, J.H., A. Perrodon, and C.J.Campbell, 1996, *The world's gas potential*, Report, Petroconsultants

Laherrère, J.H., 1997, *Submission to the Green Paper "Sustainable energy policy for Australia"* [#535]

Laherrère, J.H., 1997, *Production decline and peak reveal true reserve figures*; World Oil Dec. 77- [#658]

Laherrère, J.H., 1997, *Evolution of development lag and development ratio*; IEA submission [#715]

Laherrère, J.H., 1997, *Distribution and evolution of recovery factor*; IEA submission [#1851]

Laherrère, J.H., 1998, *Development ratio evolves as true measure of exploitation*; World Oil, [#673]

Laherrère, J.H., 1998, *Modeles et realite*; Assoc. Nat. Dir. Finance, France (in press) [#810]

Laherrère, J.H. & D.Sornette, 1998, *Stretched exponential distributions in nature and economy*; The European Physical Journal B2 529-539 [#1829]

Laherrère, J.H., 1999, *Erratic reserve reporting*; Petroleum Review Feb [#914]

Laherrère, J.H., 1999, *World oil supply – what goes up must come down – but when will it peak ?*; Oil & Gas Journ. February 1 57-64 [#968]

Laherrère, J.H., 1999, *Reserve growth: technological progress, or bad reporting and bad arithmetic* Geopolitics of Energy [#1065]

Laherrère, J.H., 2000, *Oil reserves and potential of the FSU*, Tomorrow's Oil 2/7

Laherrère, J.H., Campbell, C.J.,Duncan, R.C. & McCabe, P.J., 2001, Discussions & Reply acc. to *"Energy Resourses– cornucopia or empty barrels?"* in AAPG Bulletin, V. 85, No. 6 (June 2001), pp.

1083–1097, 2001[#1306]

Laherrère, J.H , 2000, *Memoires et reflexions sur 45 ans de geophysique petroliere*;   For: Geologues, Union Francaises des Geologues, July, 2000 [#1533]

Laherrère, J.H., 2002, *Forecasting future oil production*; Presentation for the BGR, January, 2002 [#1764]

Laherrère, J.H., 2002, *Is FSU oil growth sustainable?*; Petroleum Review April, 2002 [#1810]

Laherrere, J.H., 2002, *The Future of Oil*; International Journal of Vehicle Design, 2002 [#2066]

Laherrere ,J.H., 2003, *Forecast of oil and gas supply to 2050*; Petrotech 2003 New Delhi [#2059]

Lamar, L., 1992, *World energy statistics*; Shale Shaker, May/June 1992 [#114]

Lanier, D., 1998, *Heavy Oil- A Major Energy Source for the 21ˢᵗ Century*; UNITAR Centre for Heavy Crude and Tar Sands, No. 1998.039, 1998 [#1697]

Lattice Group, 2002, *Initial response to the governments consultation on UK energy policy*; Lattice Group, June 2002 [#2049]

Laurier, D., *Le gaz: exploration, gisements, stockages*; ATFP Conference, Paris [#280]

Lavelle, M., 2003, *Living Without Oil*; US News & World Report, February 17, 2003 [#2101]

Laquer, W., 2002, *This isn't the time for peace*; Wall St. Journ. 27-03-2 [#1847]

Leach, G., 2001, *The coming decline of oil*; Tiempo, Issue 42, December, 2001 [#1766]

Leblond, B., 2003, *ASPO, ODAC see conventional oil production peaking by 2010*; Oil & Gas Journal, June 23, 2003 [#2191]

Leckie, G.G., 1993, *Hydrocarbon reserves and discoveries 1952 to 1991*; Energy Exploration & Exploitation, 11/1, 1993 [#214]

Lee, P.J. and P.R. Price, 1991, *Successes in 1980s bode well for W.Canada search*; Oil & Gas Journ. April 22 [#29]

Lee, P., 1992, *Shifting sands*; Fort Worth Star Telegram, January 15 [#10]

Legget, J., 1999, *The carbon war*; Penguin 338pp

Leggett, K., 2000, *China Is Likely to Establish A Strategic Oil Reserve*; The Wall Street Journal, August 28, 2000 [#1380]

Lehman Brothers, 2003, *Plummeting Supply to Drive Gas Demand Down, Prices Up*; Global Equity Research, Feb.21, 2003 [#2169]

Le Min, L. & Wisenthal, S. (Bloomberg News), 2000, *OPEC unlikely to boost oil output*; Houston Chronicle, Tuesday, May 30, 2000 [#1642]

Lencioni, L. *et al.*, 1996, *Breathing new life into an abandoned field*; Hart's Petroleum Engineer International, Dec. [#552]

Lenzner, R. & J.M. Clash, 1994, *Wrong again*; Forbes June 6 1994 [#231]

Lenzner, R., 1994, *The case for hard assets*; Forbes, June 20 1994 [#237]

Leonard, R.C., 1984, *Generation and migration of hydrocarbons on southern Norwegian shelf*; AAPG. 68 796 [#531]

Leonard, R.C.,1993, *Distribution of subsurface pressure in the Norwegian Central Graben and applications for exploration*; Petrol. Geol. of NW Europe Proc. 4th Conf. [#298]

Leonard, R.C., 1996, *Caspian Sea regional hydrocarbon development: opportunities and challenges*; 4ᵗʰ Kazakstan Int. Oil & Gas projects conf. [# 539]

Leonard, R., 2001, *The Current Oil Crisis And The Coming Age Of Natural Gas*; February 2001 [#1633]

Leonard, R., 2002, *Russian Oil And Gas: A Realistic Assessment*; Ray Leonard VP Exploration and News Ventures, YUKOS Exploration and Production, Uppsala, Sweden, May, 2002 [#1796]

Levitch, R.N., 1998, *Making sustainable energy available to all*; Geol. Soc. Amer. [#938]

Levy Forecast, 2002, *60% Risk of Recession within 12 Months*; The Levy Forecast, Vol.54, July, 2002 [#1876]

Lewis, B., 2002, *What Went Wrong?*; The Atlantic Monthly, January, 2002 [#1742] [#2096]

Lewis, C., 2001, *Natural Gas Hydrates- E&P Hazard or Large Business Opportunity?*; Hart`s E&P, 2001 [#1320]

Lia, A., 1966, *Elf fine tunes its strategy*; Pet. Engineer Int., May 1996 [#465]

Licking, E., 1998, *The world's next power surge*; Business Week, 14 December [#873]

Liesman, S., 2000, *OPEC oil cuts raise threat of shortage*; Wall St Journ., 20ᵗʰ January [#1065]

Liesman, S., & J.M. Schlesinger, 1999, *Blunted spike*; Wall ST Journ 15 December [#1069]

Liesman, S., 1999, *Texaco's strategy: produce less oil more profitably*; Wall St Journ., October 27 [#1019]

Lieven, A., 2001, *The Search For Strategy*, The Australian Financial Review, Friday, September 28, 2001 [#1698]

Lifsher, M., 2001, *Oil Producers Balk as Venezuela Tighten Terms On Investment*; The Wall Street Journal, November 15, 2001 [#1731]

Lifsher, M., 2003, *Oil Production In Venezuela Drops Off After Spring Surge;* Wall St. Journ. August 22, 2003 [#2234]

Lind, M., 2004, *A tragedy of errors;* The Nation February 23, 2004 [#1983]

Lippman, T.W., 1990. *Saudis come up with major oil find*; Washington Post October 15 [#52]

Lippman, T.W., 1992, *Saudis and US teamed on oil issues*; Fort Worth Star Telegram, July 24 1992 [#34]

List, F., 2000, *Die Ölförderung droht ab dem Jahr 2005 zu sinken; Technik & Wirtschaft*; 52 December 29 [#1243]

Littell, G.S., 1999, *World crude production: bad statistics produce poor conclusions*; World Oil June [#986]

Littell, G.S., 1999, *Bad data distorts industry, market perceptions*; World Oil April [#1172]

Lockhead, C., 1998, *IMF meets today, seeks answers*; San Francisco Chronicle 29 September [#848]

Loeb, P., 2001, *The money defense shield*; The Christian Science Monitor, [#1345]

Lomborg, B., 2001, *Running on Empty?*; The Guardian, 16 August, 2001 [#1350]

Longhurst, H., 1959, *Adventure in oil: the story of British Petroleum*; Sidgwick and Jackson, London 286p.[B#8]

Longwell, H., 2002, *The future of the oil and gas industry: past approaches ,new challenges*; World Energy 5/3 2002 [#2041]

Los Angeles Times, 1991, *Mexico lied about Proven Oil Reserves, report says*; December 10 1991 [#12]

Los Angeles Times, 2002, *Study says resource use is not sustainable*; June 25, 2002 [#1910]

Lotter, C. and S. Peters, 1996, *The changing European security environment*; Bohlau 335pp

Lovelock, J., 1988. *The ages of Gaia*; Oxford University Press 252pp.

Lübben, H. von & J. Leiner, 1988, *Öl: perspektiven im Upstream bereich*; Erdol,Erdgas,Kohle 104/5 May 1988 [#276]

Luhmann, J., 2003, *Turning Point of the oil age already in 2010?*; Neue Zuercher Zeiting, September 2003 p89 [#1951]

Lugar, R.G. and R.J. Woolsey, 1999, *The new petroleum*; Foreign Affairs January–February [#899]

LWV (League of Women Voters of Santa Cruz County), 2001, *The Oil Crash and You- Running on Empty!*; Santa Cruz Voter, February, 2001 [#1613]

Lynch, M.C., 1992, *The fog of commerce: the failure of long-term oil market forecasting*; MIT Center for Int. Studies September 92

Lynch, M.C., 1998, *Imminent peak challenged*; Oil & Gas Journal 28.1.98 p. 6 [#657]

Lynch, M.C., 1998, *Crying Wolf: warnings about oil supply*; MIT Center for International Studies [#728]

Lynch, M.C., 1998, *Farce this time*; Geopolitics of Energy; December-January [#925]

Lynch, M.C., 1999, *Crying wolf: warnings about oil supply*; Stanford website

Lynch, M.C., 1999, *The debate over oil supply: science or religion*; Geopolitics of Energy, August [#1133]

Lynch, M.C., 2001, *Oil prices enter a new era*; Oil and Gas Journal [#1214]

Lynch, M.C., 2001, *Forecasting Oil Supply: Theory and Practice*; DRI-WEFA, 2001 [#1311]

Lynch, M.C., 2001, *Closed Coffin: Ending the Debate on "The End of Cheap Oil" – A commentary*; M C Lynch, Chief Energy Economist, DRI-WEFA, Inc.,September, 2001 [#1675]

M

Mabro, R., 1996, *The world's oil supply 1930–2050 – a review article*; Journ. of Energy Literature II.1.96 [#469]

Mabro, R., 2002, *Russian versus Saudi oil*; Oxford Energy Forum [#1871]

MacArthur, C.E., 2002, *The adjustment*; 208pp

Macgregor, D.S., 1996, *Factors controlling the destruction or preservation of giant light oilfields*; Petroleum Geoscience 2. 197-217

Mack, T., 1994, *History is full of giants that failed to adapt*; Forbes 28 February 1994 [#195]

Mack, T., 1998 *Venezuela is changing the balance of power among the world's oil producers*; Forbes August1 [#695]

Mackay, N., 2003, *US: Saddam had no weapons of mass destruction*; Sunday Herald, May 4, 2003 [#2158]

Mackenzie, A.S., 2000, *Energy, petrochemicals and progress*; Clean Energy 2000 [#1124]

MacKenzie, J.J., 1994; *Transportation in the People's Republic of China: beginning the transition to sustainability*; preprint World Resources Inst. [#316]

MacKenzie, J.J., 1995, *Oil as a finite resource: the impending decline in global oil production*: World Resources Inst [# 394 436]

MacKenzie, J.J. and K. Courrier, 1996, *Cutting gas tax will make things worse*; Los Angeles Times May 8 [#438]

MacKenzie, J.J., 1996, *Oil as a finite resource: When is global production likely to peak?*; World Research Institute, Updated March 2000, [#1519]

MacKenzie, D., 1998, *Driven to destruction*; New Scientist 27 July [#805]

Mackenzie, W., 2001, *Centre Stage- Review of UK Central North Sea 2001 / Development Approvals / Production / Exploration Activity*; UK Upstream Report, No. 337, June 2001 [#1329]

Mackenzie, W., 1999, *Maturing Gracefully- Overview of the 1999 Probable Developments*; UK Upstream Report, No. 316, September 1999 [#1376]

Mackenzie, W., 2000, *Operating Like a Well Oiled Machine*; UK Upstream Report, No.329, October 2000 [#1422]

Mackenzie, W., 2000, *Production*; UK Upstream Report, No.329, October 2000 [#1423]

Mackenzie, W., 2000, *Make Hay While The Sun Shines?*; UK Upstream Report, No.328, September 2000 [#1426]

Mackenzie, W., 2001, *Midget Gems*; UK Upstream Report, No.340, September, 2001 [#1476]

Mackenzie, W., 2001, *2001 UK Oil Production Review etc.* ; UK Upstream Report, No.339, August, 2001 [#1477]

Mackenzie, W., 2002, *Key Events of 2001*; UK Upstream Report, No.344, January, 2002 [#1777]

Mackenzie, W., 2002, *Deepening Disappointments*; Latin America Upstream Report, Q2 2002 [#1987]

MacKillop, A., 1990, *On decoupling*; Int. Journ. Energy Research v14 38-105 [#1862]

Macleay, J., 1996, *Running on empty*; The Australian 24.10.96 [#510]

MacLeod, F., 1999, *The real motives for industry consolidation*; Deutsche Bank, November, 1999 [#1466]

Madron, R. & J. Jopling, 2003, *Gaian democracies*; Schumacher Briefings, Green Books 154p

Maegaard, P., 2000, *Wind power*; Clean Energy 2000 [#1111]

Maegaard, P., 2000, *Mobilizing the market for renewable energy*; Clean Energy 2000n [#1114]

Magoon, L.B., 2000, *Are We Running Out Of Oil?*; http://geopubs.wr.usgs/open-file/of00-320 [#1529]

Maiello, M., 2000, *Gas up and go*; Forbes, March 7 [#1083]

Main, A., 2000, *"Perspective" – Old drivers for the new oil crisis*; The Australian Financial Review, Saturday, September 9, 2000 [#1387]

Main, A., 2003, *Oil and the human factor*; Australian Financial Review, August 30, 2003 [#1938] [#2266]

Makansi, J., 2001, *The New Energy Value Chain*; Special Advertising Section, Business Week, August 13, 2001 [#1353]

Makhijani, A., 2001, *Magical Thinking- another go at transmutation*; Bulletin of the Atomic Scientists March/April 2001 [#1293]

Malone, A., 1996, *Revenge of the Saudi exile*; Sunday Times, 7 January 96 [#417]

Malsawma, Z.H., 1998, *A researcher's guide to population information websites*; Population today February 1998 [#713]

Manor, R. (Chicago Tribune), 2002, *Non-members may benefit as OPEC cuts*; San Jose Mercury News, Wednesday, January 2, 2002 [#1736]

Mansfield, P., 1992, *A history of the Middle East*; Penguin Books 373p.[#B15]

Marcel, V., 2002, *The Future of Oil in Iraq: Scenarios and Implications*; The Royal Institute of International Affairs, December 2002 [#2075]

Marine and Petroleum Geology, 1993, *Book Review of "The Golden Century of Oil"*; Marine and Petroleum Geology, Vol. 10, p. 182, April 1993 [#1455]

Martell, H., 1989, *Exploration and Resources, Venezuela*; UN Seminar Policy and Management of Petroleum Resources, Oslo [#98]

Martin, A.J., 1985, *Prediction of strategic reserves in prospect for the world oil industry*; Eds.T. Niblock & R.Lawless. Univ. of Durham 16-39

Martin, A.J. & P.B. Lapworth, 1998,*Norman Leslie Falcon*; R.Soc. Lond. 44 [#933]

Martin, P., 2001, *Population as weapon: Strategists see victory from the inside*; The Washington Times, National Weekly Edition, December 17–23, 2001 [#1771]

Martin, P, 2002, *Oil and "conspiracy theories" a reply to a liberal apologist for the US war in Afghanistan*; World Socialist Web Site [#1993]

Martinez, A.R. *et al.*, 1987, *Study group report: classification and nomenclature system for petroleum and petroleum reserves*; Wld Petrol. Congr. 325-342.

Martoccia, D, 1997, *Permanent oil shortage impossible*; Futurist May–June [#935]

Mast, R.F. *et al.*, 1989, *Estimates of undiscovered conventional oil and gas resources in the United States – a part of the nation's energy endowment*; U.S.Dept of Interior [#94]

Masters, C.D., 1987, *Global oil assessments and the search for non-OPEC oil*; OPEC Review, Summer 1987, 153-169 [#92][#1687]

Masters, C.D., 1991, *World resources of crude oil and natural gas*; Review and Forecast Paper, Topic 25, p.1-14. Proc. Wld. Petrol. Congr., Buenos Aires 1991 [#113]

Masters,C.D., D.H.Root & E.D. Attanasi, 1991, *Resource constraints in petroleum production potential*; Science 253.[#28]

Masters, C.D., 1993, *U.S.Geological Survey petroleum resource assessment procedures*; Amer. Assoc. Petrol. Geol. 77/3 452-453 (with other relevant references).

Masters, C.D., 1994, *World Petroleum analysis and assessment*; Wld. Petrol. Congr. Stavanger [#226]

Masters, C.D., 1994 *Bibliography of the world energy resources program*; USGS Open File 94-556 [#328]

Masters, C.D., D.H. Root and R.M. Turner, 1997, *World resource statistics geared for electronic access*;

Oil&Gas Journ, 13.October 98-104 [#660]

Masuda, T., 2000, *World Oil Supply Outlook to 2010*; For the 5th Annual Asia Oil & Gas Conference, May 28–30, 2000 [#1521]

Mathews, J., 1996, *World oil market faces instability and change*; ABQ Journ. 28.2.96 [#423]

Maxwell, C.T., 2000, *Integrated oil companies' liquids and natural gas (equivalent barrels) reserves ranking by size;* maxwell@weeden.com [#1268]

Maxwell, C.T., 2000, *This time, It's different;* maxwell@weeden.com [#1269]

Maxwell, C.T., 2000, *Crude oil prices may be peaking;* maxwell@weeden.com [#1272]

Maxwell, C.T., 2000, *The seven sisters reborn?;* maxwell@weeden.com [#1273]

Maxwell, C.T., *Oil Stocks Are Pressured By Crude Price Forecasts Into The Teens.But, Does The Evidence Support Such A Drop?;* maxwell@weeden, Vol. 2 / i.13, July 6, 2001 [#1312]

Maxwell, C.T., 2001, *Enough is Enough. Many Oil-Related Companies Are Now Under valued And Will Confirm This In Their Future Earnings Streams;* maxwell@weeden, vol. 2 / i.14, July 20, 2001[#1324]

Maxwell, C.T., 2000, *"Shape and Timing of the Next E & P Capital Cycle: Are We Doing Enough Now?";* Letter, August 9, 2000 [#1419]

Maxwell, C.T., 2000, *The Seven Sisters Reborn?;* maxwell@weeden, vol. 1 / i.25, October 17, 2000 [#1420]

Maxwell, C.T., 2001, *Integrated Oil Companies` Liquids And Natural Gas (Equivalent Barrels) Reserves Ranking By Size And Value To Shareholders;* maxwell@weeden , vol. 2 / i. 17, September 5, 2001 [#1481]

Maxwell, C.T., 2001, *America Will Strike Back;* maxwell@weeden , vol. 2 / I.19, September 17, 2001 [#1482]

Maxwell, C.T., 2001, *Heightened Terrorism: Potential Effects On The Oil Industry;* maxwell@weeden , vol. 2 / i.20, September 19, 2001 [#1483]

Maxwell, C.T., 2001, *La Creme de la Creme (Revised);* maxwell@weeden , September 4, 2001 [#1485]

Maxwell, C.T., 2000, *Natural Gas: Stage Five Commences;* maxwell@weeden , December 28, 2000 [#1585]

Maxwell, C.T., 2001, *OPEC`s Oil Pricing Leadership Will Bend, But Not Break;* maxwell@weeden, vol.2 / i.21, September 27, 2001 [#1690]

Maxwell, C.T., 2001, *U.S. Natural Gas Prices Have Fallen So Far, So Fast, Wildcatters Are Rejoicing;* maxwell@weeden, vol.2 / i.23, October 15, 2001 [#1718]

Maxwell, C.T., 2001, *Presentation on Oil and Gas Stocks;* ??? , October 31, 2001 [#1719]

Maxwell, C.T., 2001, *The oil industry outlook* ,Nov. 13 [#1845]

Maxwell, C.T., 2001, *The dawn of a new era in oil ricing and oil use,* Sept.7 [#1846]

Maxwell, C.T., 2001, *The Saudi concundrum,* Apr 26 [#1848]

Maxwell, C.T., 2002, *Suncor Energy (SU- $30);* maxwell@weeden, vol.3 / i.1, January 24, 2002 [#1780]

Maxwell, C.T., 2001, *OPEC vs. Non-OPEC Producers: The Outcome Will Be Lower Oil Prices For A While;* maxwell@weeden, vol.2 / i.25, November 15, 2001 [#1781]

Maxwell, C.T., 2002, *The Iraq Invasion Question;* Aug.23 2002 [#1923]

Maxwell, C.T., 2002, *Suncor Energy(SU-$17.90):Portrait of an Athabasca Tar Sands Giant in Short, Sweet Strokes;* weendenco.com Aug.26 2002 [#1925]

Maxwell, C.T., 2002, *Caught in the Energy Vise;*   GOTOBUTTON BM_2_ weendenco.com , Oct.7, 2002 [#2015]

Maxwell, C.T., 2002, *The Oil Industry Outlook;* revised Oct.9, 2002 [#2082]

Maxwell, C.T., 2002, *Suncor Energy: Third Quater Earnings Report is on Track;* maxwell@weedon, October 24, 2002 [#2081]

Maxwell, C.T., 2003, *Oil Investing When Prices Turn Down;* maxwell@weeden, Auust 9,2003 [#1930]

Maxwell, C.T., 2003, *Suncor Energy: Likely to Hit Aggressive 2002 Earning Target and Move Strongly Into 2003;* January 10, 2003 [#2058]

Maxwell, C.T., 2003, *The Reckoning;* maxwell@weedon, Feb.24, 2003 [#2139]

Maxwell, C.T., 2003, *My View: Stock Market Winner & Losers When Oil Prices Rise (Or Fall) Sharply;* maxweel@weeden, March 11 2003 [#2140]

Maxwell, C.T., 2003, *Suncor Energy (SU-$17.11 US)- Suncors historical earnings and outlook in (US Dollars);*   maxwell@weeden, March 18, 2003 [#2160]

Maxwell, C.T., 2003, *While Securing Iraqi Production is Early Military Objective, Tanker Deisruptions Could Temporarily Halt Persian Gulf Oil Outflows;*maxwell@weedem.com, March 21, 2003 [#2141]

Maxwell, C.T., 2003, *Peering Into The Future of Oil;* maxwell'weeden, June 3, 2003 [#2214]

Maybury, R., 2000, *Richard Maybury`s U.S. & World Early Warning Report For Investors;* August 2000 [#1379]

McCabe, P.J., et al., 1993, *The future of energy gases;* USGS Circ.1115 [#477-1]

McCabe, P.J., 1998, *Energy resources – cornucopia or empty barrel;* Amer.Assoc. Petrol. Geol. 82/11

2110-2134

McCarthy, T., 2001, *War Over Arctic Oil*; Time, February 19, 2001 [#1605]

McCartney, S. & M. Brannigan, 2000, *Airlines profit cut by higher fuel costs*; Wall St. Journ 19 January [#1066]

McCluney, W.R., 2004, *Humanity's Environmental Future*; ISBN 0-9744461-0-6

McCluney, W.R., 2004, *Getting to the Source* – ISBN 0-97844461-1-4

McClure, K., 2002, *An imminent peak in world oil production? The arguments for and against*; 13D Research July 2002 [#1908]

McClure, K., 2003, *The Case for Falling Production From Existing Oil Fields*; 13D Research, June 19, 2003 [#2217]

McCormack, M., 1999, *Twenty-first Century energy resources: avoiding crisis in electricity and transportation*; American Chemical Society address, [#967]

McCrone, A.W., 2001, *Looking beyond the petroleum age*; San Francisco Chronicle, Mar 4th [#1191]

McCutcheon, H., Osbon, R. & Mackenzie, W., 2001, *Risks temper Caspian rewards, potential*; Oil & Gas Journal, pp. 22–28, December 24, 2001 [#1761]

McKenzie, A., 2000, *The quality of life – a shared concern*; Presentation to Clean Energy 2000 [#1060]

McKibben, B., 1998, *A special moment in history*; Atlantic Monthly, May [#738]

McMahon, P., 2001, *Power suppliers run dry at worst time*; USA Today, Tuesday, March 27, 2001 [#1626]

McMullen, Porgam Murray, 1976, *Energy resources and supply*; Wiley [# 692]

McRae, H., 1994, *The world in 2020 : power, culture and prosperity, a vision of the future*; Haper Collins 302pp [#B16]

McRay, H., *Oil looks slippery*; Independent 7/5/93 [#161]

McRay, H., *No end of cheap oil*; Independent 11/11/93 [#191]

McRay, H., *The real question is why the price of oil is not even higher*; Independent 07/04/02 [#1821]

Meacher, M., 2003, *This war on terrorism is bogus*; The Guardian. London. Sept 6. [#2265]

Meacher, M., 2004, *Plan now for a world without oil*; Financial Times, January 5, 2004 [#2309]

Meadows, D.H. *et al.*, 1972. *The limits to growth*; Potomac 205 pp

Megill, R.E., 1993 *Discoveries lag oil consumption*; AAPG Explorer August 1993 [#171]

Megill, R.E., 1994, *Another look at finding cost*; AAPG Explorer July 1994 [#250]

Mellbye, P., 1994, *Norway's role, satisfying increasing demand for natural gas in Europe*; World Petrol Congr. Stavanger.[#234]

Meling, L.M., 2003, *How and for how long is it possible to secure a sustainable growth of oil supply*; Statoil ASA, Norway 2003 [#2294]

Menezes, F. Ramos de, 1994, *Worldwide criteria versus regional criteria*; Wld Petrol. Congr. Stavanger [#229]

Merrill Lynch, 1994, *Energy Monthly*; report [#324]

Merrill Lynch, 2002, *IEA Oil Market Report Analysis*; report [#2010]

Merrill Lynch, 2002, *Oil & Gas Production, Energy Monthly*; report [#1873]

Merrill Lynch, 2002, *Chevron Texaco Corp.-On track for improving returns*; report [#1875]

Meyercord, K., 2001, *The Ethic of Zero Growth*; www.zerogrowth.org

Middle East Economic Survey, 2003, *Middle East Oil Investment of $500bn needed Over Next 30 Years, says IEA*; Middle East Economic Survey, November 3, 2003 [#1970]

Miller, K.L., 2002, *Shells Game- its scenario spinners have made an art of prediction*; Newsweek September 16/23, 2002 [#1985]

Miller, R.G., 1992, *The global oil system: the relationship between oil generation, loss half-life and the world crude oil resource*; Amer. Assoc. Petrol. Geol. 76/4 489-500.[#302]

Miller, R., 2001, *Global Oil: Reserves, Discovery and Depletion*; Version 1.1, July, 2001 [#1674]

Miller, R., 2003, *Global Oil Supply Review 2003*; February 28, 2003 [#2105]

Milling, M.E., 2000, *Oil in the Caspian Sea*; Geotimes [#1173]

Mineral Management Service, 1995, *Gulf of Mexico OCS Region*; Internet[#955]

Mineral Management Service, 1995, *Gulf of Mexico*; Internet [#1076]

Mineral Management Service, 2000, *Outer Continental Shelf Petroleum Assessment 2000*; MMS2001-036 [#1828]

Mineral Resources of Canada, 1998, *Canada Energy Resources*; [907]

Minnear, M.P., 1998, *Forecasting the permanent decline in global petroleum production*; Thesis, Univ. of Toledo

Mitchell, A., 2003, *Non-OPEC Shortfall Bolsters OPEC Price Prospects*; Reuters, November 25, 2003 [#2271]

Mitchell, J., 1995, *The geopolitics of energy*; EU Report [#371]

Mitchell, J., 1996, *The new geopolitics of energy*; Royal. Inst. Int. Affairs April 1996 [#475-1]

Mitchell, J., 1997, *Renewing the geopolitics of energy*; Energy World 245 Jan. [#555]

Mohammed, A.H., 1989, *Crude oil production, refining, petrochemicals and research activities in Iraq*; UN

Seminar : Policy and Management of Petroleum Resources, Oslo [#102]

Mol Hirlap, 1996. *Hungarian oil history*; [#509]

Monastersky, R., *Geologists anticipate an oil crisis soon*; Science News 154/18 31 October [#865]

Monbiot, G., 2002, *In the Crocodiles Mouth*; The Guardian, November 5, 2002 [#2019]

Monbiot, G., 2003, *Bottom of the barrel*; The Guardian December 2, 2003 [#1980]

Monty, M., 1999, *Use data mining to discover odds of making a billion bbl discovery*; World Oil March [#1063]

Moody-Stuart, M., 1993, *Resources and resourcefulness*; AAPG Int. Conf., The Hague [#337]

Moody-Stuart, M., 2000, *Realising the value of scientific knowledge – geosciences in energy industries*; Sir Peter Kent lecture, Geological Society [#1278]

Moore, D.L., 1999, *The Pursuit of Strategic Value Enhancement*; CWC Conf. [1033]

Moran, M. & Johnson A., 2002, *Oil after Saddam: All bets are in- A great but quiet rush is on for a stake in Iraqs huge reserves*; MSNBC November 7, 2002 [#2031]

Morehouse, D.F., 1997, *The intricate puzzle of oil and gas "reserve growth"*; EIA, Nat. Gas Monthly [#1103]

Morrell, L., 2000, *Evaluation of the Southern Iran Oil Resource Base*; Iran Study, Draft- Not for publication [#1535]

Morgan, D. and Bustillo, M., 2001, *State approves bond sale to pay for power*; Los Angeles Times [#1220]

Morrison, D.R.O., 2000, *Energy in Europe -comparison with other regions*; Clean Energy 2000 [#1112]

Morse, E.L. & Richard, J., 2002, *The Battle for Energy Dominance*; Foreign Affairs Magazine, www.foreignaffairs.org , March/April, 2002 [#1792 & 1839]

Morse, E.L., 2003, *Is The Energy Map Next On The Neo-Conservative Cartography Agenda?*; Middle East Economic Survey, Vol. XLVI No.33, August 18, 2003 [#2259]

Mortished, C., 1995, *Shell thinks, then does the unthinkable*; Times 31/3/95 [#334]

Mortished, C., 1996, *Iraq oil coild be back on sale soon*; Times March 11[#421]

Mortished, C., 1997, *Tempus – BP*, 6.2.97 [#551]

Mortished, C., 1998, *BP shrugs off influence of oil prices with aid of self help*; The Times 11.2.98 [#666]

Mortished, C., 1998 *Set-back for Western fuel long term gains*; The Times 19.2.98 [#667]

Mortished, C., 1998 *BP to merge with Amoco in £67 bn deal*; The Times 12.8.98 [#786]

Mortished, C., 2003, *Oil fear takes gloss off BP deal*; Times Online February 12, 2003 [#2076]

Mosley, L., 1973, *Power play: oil in the Middle East*; Random House [#B10]

Mourik, M. & Shepard R., 2003, *Peak oil, money and markets*; 2020 Energy [#2264]

Mowlem, M., 2002, *The real goal is the seizure of Saudi oil*; The Guardian 5 Sept [#1901]

Mulgrew, I., 2002, *The truth is out there... right?*; The Vancouver Sun, Archives Story, canada.com network, Saturday, February 23, 2002 [#1788]

Murphy, D., 2002, *As China's car market takes off*; Wall St. Journ 3 July [#1868]

Myers, Jaffe, A. & Manning, R.A., 2000, *The Shocks of a World of Cheap Oil*; Foreign Affairs, Vol. 79, No.1, p 16- 29, January/February 2000 [#1434]

Myserson, A.R., 1998, *US splurging on after falling off diet*; New York Times October 22 [#877]

N

Nakicenonic, N. & Grubler A. & McDonald A., 1998, *Global Energy Perpectives*; 1998 [#2185]

Narimanov, A. and A. Palaz, 1994, *Baku region rich with oil, history*; AAPG Explorer October 1994 p.40 [#315]

Nasmith, J., 1996, *A story of oil price reporting 1945-1985*, Pipeline 13 May [#608-1]

Nation, L., 1993, *Delegates told oil prices must rise*; AAPG Explorer [#124]

Nation, L., 1995, *Hodel sees looming energy crisis*; AAPG Explorer May 1995 [#358]

Natural Gas Hydrates, 2002, *Future Supply Potential of Natural Gas Hydrates 2001-2002 Update*; Natural Gas Hydrates p81-87 2002 [#2167]

Naturhistorischen Museum, 1987, *Hans G. Kugler 1893-1986*; Memorial, Basel [#489]

Nehring, R., 1978, *Giant oil fields and world oil resources*; CIA report R-2284-CIA [#16]

Nehring, R., 1979, *The outlook for conventional petroleum resources*; Paper P-6413 Rand Corp.21p.

Nehring, R., 2000, *America's bigger resource*; Hart's E&P, Jan [#1071]

Newell, K.D. and J.C.Wong, 1992, *Re-exploration aided by computer graphics, mapping*; World Oil September 1992 39-42 [#64]

Newman, R.J., 1998, *A US victory at a cost of $5.5 trillion*; US News & world report July 13 [#801]

Newswell, 1991, *Higher oil prices to spur 1991 economic growth in southwest.* Newswell [#58]

New Scientist, 2003, *Energy Oil Special- Various Articles,* New Scientist, August 2, 2003 [#1929]

New Scientist, 2003, *Energy Special- Hydrogen*; August 16, 2003 [#2243]

New York Times, 2001, *Renewable Energy: today`s basics*; The New York Times, May 3 2001 [#1314]

New York Times, 2001, *Despite more drilling, gas production falls short*; The Register- Guard National, Sunday, July 22, 2001 [#1318]

New York Times, 2001, *Iran Is Accused of Threatening Research Vessel in Caspian Sea*; The New York Times, July 25, 2001 [#1344]

New York Times, 2001, *Reconsidering Saudi Arabia*; The New York Times, October 14, 2001 [#1722]

New York Times, 2001, *Osama bin Laden`s Wildfire Threatens Grip of the Saudi Royal Family*; The New York Times, November 6, 2001 [#1729]

New York Times, 2003, *War in the Ruins of Diplomacy*; Editorial, March 18, 2003 [#2130]

Nexon, M., 2001, *Petrole Chantages angolais*; Le Point, November 9, 2001 [#1709]

Nichols, J., 2003, *Spengler really understands*; Asia Times Online, 2003 [#1972]

Niiler, E., 2000, *Awash in Oil*; Scientific American, September 2000 [#1370]

Nikiforuk, A., 2000, *Running on empty, when Canada's natural gas reserves hit the crisis point, who will be left out in the cold?*; Canadian Business Magazine [#1255]

Norwegian Petroleum Directorate, 1993, *Improved oil recovery*; [#482-1]

Norwegian Petroleum Directorate, 1997, *The petroleum resources of the Norwegian continental shelf*; report [#620-3]

Norwegian Petroleum Directorate, 1997, *Discoveries on the Norwegian continental shelf*; report [#621-3]

Norwegian Petroleum Directorate, 1997, *Trends in petroleum resource estimates*; reprt [#659]

Norwegian Petroleum Directorate, 1997, *Classification of petroleum resources on the Norwegian continental shelf*; Report [#664]

Norwegian Petroleum Directorate, 1998, News Letter 12. [#777]

Norwegian Petroleum Directorate, 1998, News Letter 14. [#778]

Norwegian Petroleum Directorate, 1998, News Letter 15. [#886]

Norwegian Petroleum Directorate, 1999, *Undiscovered Petroleum Resources*; website [#974]

Norwegian Petroleum Directorate, 2000, *Two-thirds left to go*; NPD Diary [#1154]

Norwegian Petroleum Directorate, 2000, *Production Figures From The Norwegian Continental Shelf*; GOTOBUTTON BM_6_ www.npd.no/norsk/npetrres/prod_tal/produksjon.htm , Updated: February 3, 2000 [#1443]

Norwegian Petroleum Directorate, 2000, *Norwegian Production of Stabilized Oil and Gas*; 2000, [#1445]

Norwegian Petroleum Directorate, 2003, *Forecasts*; 2003 [#1959]

Nicolescu, N. & B. Popescu, 1994, *Romania*; in Kulke, H. (Ed.) *Regional petroleum geology of the world*; 21. Borntrager [#517]

Nuclear Issues, 2003, *When the oil runs out*; Nuclear Issues Vol.25 No.4 April 2003 [#2168]

Nuclear Issues, 2003, *In Denial*; Nuclear Issues Vol.25 No.5 May 2003 [#2180]

Nuclear Issues, 2003, *Atoms for peace*; Nuclear Issues Vol.25 No.11 November 2003 [#2285]

Nuttall N., 1998, *Arctic ice is now a third thinner than in 1976*; Times 23/11/98 [#874]

O

Obaid, N.E., 2000, *The oil kingdom at 100*; Washington Inst. For Near East Policy 136pp

Observer Worldwide, 2002, *Osama bin Ladens' "Message to the American People"*; Observer Worldwide November 24, 2002 [#2025]

O'Dell, S., 1994, *Prospects for non-opec oil supply*; 13th CERI Conf., [#314]

O'Connor, T.E., 199?, *The international development banking view of petroleum exploration and production in developing countries*; World Bank [#932]

Odell, P.R., 1994, *World Oil resources, reserves and production*; The Energy Journal v. 15 89-113

Odell, P.R., 1996, *Middle East domination or regionalisation*; Erdol, Erdgas,Kohle Heft4 [#435]

Odell, P.R., 1996, *Britain's North Sea oil and gas production – a critical review*; Energy Exploration & Exploitation 14/1/

Odell, P.R., 1997, *Oil shock- a rejoinder*; Energy World 245 [#554]

Odell, P.R., 1997 *Oil reserves: much more than meets the eye*; Petroleum Economist Nov. [#656]

Odell, P.R., 1998, *Fossil fuel resources in the 21st Century*; Presentation, Int. Atomic Energy Agency [#869]

Odell, P.R., 1999, *Oil and gas reserves: retrospect and prospect*; Geopolitics of Energy Jan [#956]

Odell, P.R., 1999, *Predicting the future: what lies ahead for fossil fuels?* Horizon June [#1096]

Odell, P.R., 1999, *Fossil fuel resources in the 21st Century*; Financial Times Energy [#1279]

Odum, H. and E. Odum, 1981, *Energy basis for man and nature*; McGraw Hill, New York

Oetliker, H., 2000, *Mobility and health*; Clean Energy 2000 [#1117]

Offshore, 1989, *Non-opec producers limited in ability to take advantage of rising oil demand*; Offshore June 1989 [#21]

Offshore, 1989, *Poor US performances underlie reserve buys, international move*; Offshore June 1989 [#22]

Offshore, 1996, *Improved recovery grow Norwegian reserves*; Offshore April 1996

Oil & Gas Journal, *World Production Reports*; December each year.

Oil & Gas Journal, 1996, *New life in US oil*; January 29 [#527]

Oil & Gas Journal, 2002, *Drop in Pemex's revised oil reserves figure 'significant'*; Oil & Gas Journal September 23, 2002 [#2001]

Oil & Gas Journal, 2003, *Peak-oil, global warming concerns opening new window of opportunity for alternative energy sources*; Oil & Gas Journal, August 18, 2003 [#1944]

Oil & Gas Journal, 2003, *Future energy supply*; Editorial, Oil & Gas Journal, August 18, 2003 [#1945]

Oil & Gas Journal, 2003, *Running out of oil*; Oil & Gas Journal, Sepember 15, 2003 [#1942]

Oil & Gas Journal, 2003, *Political Oil Supply*; Editorial June 23, 2003 [#2193]

Oil & Gas Journal, 2003, *ASPO sees conventional oil production peaking by 2010*; June 30, 2003 [#2195]

Oil & Gas Journal, 2003, *Majors replaced 101% of world oil and gas production in 2002*; August 1, 2003 [#2229]

Oilman The, 1996 ; *UK reserves and potential*; The Oilman 18 March pp. 3–5 [#430]

O'Mahoney, B., 2004, *Oil shortages to hit world economy in 2007*; Irish Examiner, February 7, 2004 [#2320]

OPEC Bulletin, 1997, *An early touch of millennium fever: the coming oil crisis?* October, p.3

Orphanos, A., 1995, *Looking for oil prices to gush*; Fortune 15/5/95 [#359]

Oswald, A., 2001, *Oil price puts skids under growth*; www.sunday-times.co.uk/news, September 2, 2001 [#1648]

P

Parade Magazine, 2002, *Chinese Oil Fills Foes With Fear*; Parade Magazine, p 14, January 27, 2002 [#1776]

Parent, L., 2003, *Natural Gas in North America- No one said it would be easy*; World Oil, February 2003 [#2142]

Parker, H.W., 2001, *After petroleum is gone, what then?*; World Oil September 2001 [#1830]

Parra, A.A., 1995, *Opec: the question of new oil*: 5th CGES Annual Conf. London

Patricelli, J.A. & C.L. McMichael, 1995, *An integrated deterministic/ probabilistic approach to reserve estimations*; Journ. Petrol. Technol. 47/1 49–53

Petterson, W.C.,1990, *The energy alternative*; Channel 4 Book 186pp.

Pauwels, J-P and F. Possemiers, 1996, *Oil supply and demand in the XXIst century*; Revue de l'energie, 477. [#458]

Pearl, D., 2001, *Suddenly, Pipeline Project Sparks Interest*; The Wall Street Journal, June 21, 2001 [#1332]

Pearce, F., 1999, *Dry Future*; New Scientist; 10 July [#1007]

Pearce, F., 1999, *Iceland's power game*; New Scientist 1 May [#1053]

Pearce, F., 2003, *Over a barrel?*; New Scientist, 29 Jan. 2003 [#2071]

Pearson, J.C., 1997, *Estimating oil reserves in Russia*, Petroleum Engineer Int., September [#650]

Pees, S.T., 1989, *Guidebook, history of the petroleum industry symposium*; Amer.Assoc. Petrol. Geol. [#613-2]

Perrin, F., 1994, *L'impact de la baisse des prix sur les strategies d'exploration-production*; 7th Seminaire petrolier et gazier Int., Paris [#368]

Perrodon, A., 1988, *Hydrocarbons in Beaumont E.A. and N.H.Foster (Eds) Geochemistry.* Treatise of petroleum geology, Amer. Assoc. Petrol. Geol. Reprint Series No.8 3-26

Perrodon, A. and J. Zabek, 1990, *Paris Basin: in Leighton M.W. et al. eds. Interior cratonic basins.* Amer. Assoc. Petrol. Geol. Mem 51 819p [#106]

Perrodon, A., 1991, *Vers les reserves ultimes*; Centres Rech.Explor.-Prod. Elf-Aquitaine 15/2 253-369. [#36]

Perrodon, A., 1992, *Petroleum systems, models and applications*; Journ. Petrol. Geol.15/3, 319-326.[#38]

Perrodon, A., 1993, *Historique des recherches petrolieres en Algerie*; 118 Congr. Nat. des soc. hist et scient, Pau, 323-340 [#395]

Perrodon, A., 1995, *Petroleum systems and global tectonics*: Journ. petrol. Geol. 18/4 471-476 [#3]

Perrodon, A., J.H.Laherrere and C.J.Campbell, 1998, *The world's non-conventional oil and gas*; Pet. Economist [#683]

Perrodon, A., 1998, *Production: les premices du declin*; Petrol. Int. 1735 [#870]

Perrodon, A., 1999, *Vers un changement de decor sur la scene petroliere*; Pet. Informations 1738 [#1095]

Perrodon, A., 1999, *Quel pétrole demain*, Technip, Paris 94p

Peters, K.E., 2000, *Review of the Deep Hot Biosphere by Thomas Gold*; AAPG bull. 84/1 Jan

Peters, S., 1999, *The West against the Rest: Geopolitics after the end of the cold war*; Frank Cass, Journal offprint from Geopolitics [#1235]

Peters, S., 2002, *Courting future resource conflict: the shortcomings of western response strategies to new energy vulnerabilities*; Political Science Dept., Giessen University [#2033]

Peterson, S., 1998, *Is Iran next power on nuclear stage*; Christian Science Monitor 8 July [#798]

Peterson, S., 1998, *Power struggle in Iran clouds view for US policymakers*; Christian Science Mon. 16/10/98 [#859]

Petrie Parkman, 1992, *Llanos foothills Trend/ Cusiana Field's potential reserves and production economics assessed*; Petroleum Research v IV EPPO1 Feb 22 1992 [#80]

Petro-Canada, 2003, *Petro-Canada Reviewing Oilsands Strategy*; Feruary 5, 2003 [#2157]

Petroconsultants S.A., 1993, *World Production & Reserve Statistics; oil and gas 1992*; Petroconsultants, London

Petroconsultants S.A., 1993, *Strategic petroleum insights*; Report [#192]

Petroconsultants S.A., 1994, *Oil production forecast*; Report [#340]

Petroconsultants S.A., 1996, *World petroleum trends 1996* [#512].

Petroconsultants S.A., 1996, *World exploration – key statistics 1995/1986*, [#525]

Petroconsultants S.A., 1997, *Ten-year decline in new field wildcat drilling reversed*; International oil letter June 28th [#623]

Petroconsultants, 1997, *Listing of heavy oil fields* [#637]

Petroconsultants, 1997, *Oil and Gas Reserves 1997* [#647]

Petroconsultants, 1997, *Venezuelan oil* [#675]

Petroconsultants, 1997, *World Petroleum Trends* [#676]

Petroconsultants, 1998, *Deepwater reserves*; report [#689]

Petroconsultants, 1998, *Russian reserves and wildcats* [#707]

Petroconsultants, 1998, *UK reserves* [#722]

Petroconsultants, 1998, *US production* [#731]

Petroconsultants, 1998 *Petroleum Trends* [#760]

Petroconsultants, 1998, *Oil production – discovery gap widens*: news release [#771]

Petroconsultants, 1998, *Oil & gas reserves added 1993–97*; Petroleum Review Oct [#899]

Petrodata, 1999, *The impact of oil price changes*; Upstream Economics Symposium [#1054]

Petrole & Gas, 2003, *The Day When Peak Will Come*; Petrole & Gas No.1765/66, October 2003 [#1966]

Petroleum Argus, 2001, *Upstream Pressure for change special report*; Petroleum Argus, June 27, 2001 [#1309]

Petroleum Economist, 1993, *Size and success win votes*; Pet. Econ. June 1993 [#342]

Petroleum Economist, 1998, *Africa deepwater*; Pet. Econ. Sept 1998 [#892]

Petroleum Economist, 1995, *Interview with C J Campbell- "Prophet or Cassandra"*, Petroleum Economist, September, 1995 [#1436]

Petroleum Engineer Int., 1993, *The Norwegian North Sea contains about 75 billion bbl*; Petrol. Eng. Int June 1993 [#162]

Petroleum Engineer Int., 1994, *Worldwide oil and gas activity forecast*; Petrol. Eng. Int January 1994 [#203]

Petroleum Engineer International, 1997, *SPE/WPC reserve definitions to provide more accurate consistent estimates*; September [#652]

Petroleum Engineer International, 1998, *Peru reels from Camisea bombshell*; September [#846]

Petroleum Review, 1996, *First stages in production of Australian synfuels gets underway*; Pet. Review August 1996 [#474]

Petroleum Review, 1997, *Shell/Texaco US downstream merger*; April [#578]

Petroleum Review, 1997, *Statoil/Sasoil gas conversion alliance*; Petrol. Rev. May [#595]

Petroleum Review, 1998, *Insights from the statistics*; Pet. Review September [#820]

Petroleum Review, 2000, *Discovery still lags production despite good 1999 results*; Petroleum Review, September 2000 [#1366]

Petroleum Review, 2000, *Different articles about OPEC, Nigeria, West Africa and the Balkans*; Petroleum Review, pp. 2 & 27–38, July, 2000 [#1539]

Petroleum Review, 2002, *Filling the global energy gap in the 21st century*; Petroleum Review August 2002 [#1905]

PetroMin, 2000, *Report: The looming crisis- Are you ready for this?*; PetroMin, November, 2000 [#1417]

Pettingill, H.S., 2001, *Giant field discoveries of the 1990s*; The Leading Edge, July, 2001 [#1657]

Petzet, A., 1999, *Decline in world crude reserves is first sense '92*; O&GJ December 22 [1026]

Petzet, A., 2000, *World resource estimate shows more liquids, slightly less gas*, April 17 & 10, 2000 [#1531]

Pfeiffer, D.A., 2002, *Oil Prices and Recession*; elizdale@juno.com, March, 2002 [#1795]

Pfeiffer, D.A., 2003, *US Intentions*; From The Wilderness Publications, March 7 2003 [#2122]

Pfeiffer, D.A., 2004, *The End of the Oil Age*; 264p www.lulu.com/allenadale

Phillips, P., 1998, *Project Censored Award* ; Sonoma State University [#864]

Phipps, S.C., 1993, *Declining oil giants, significant contributors to U.S. production*; Oil & Gas Journ. October 4. 1993 [#181]

Phipps, S.C., 1993, *Note on US field size distribution and known reserves*; unpublished note [#185]

Picerno, J., 2002, *If we really have the oil*; Bloomberg Wealth Manager Sept [#1899]

Pickens, T.B., 1987, *Boone*; Houghton Miffin, Boston [#B1]

Pickler, N., 2003, *Lieberman offers plan to reduce US reliance on foreign oil*; Star Tribune May 8, 2003 [#2192]

Pilger, J., 2001, *Inevitable ring to the unimaginable*; September 14, 2001 [#1673]

Pilger, J., 2003, *Bush and Blair Are In Trouble*; New Statesman, Feb.12, 2003 [#2305]

Pimental, D., 1994, *Implications of the limited potential of technology to increase the carrying capacity of our planet*; Human Ecology Review Summer [#904]

Pimental, D., 1998, *Energy and dollar cost of ethanol production with corn*; M.King Hubbert Center for Petroleum Supply Studies, Newsletter April [#727]

PIW, 1994, *PIW ranks the world's top 50 oil companies*; Petroleum Int. Weekly 12/12/94 [#335]

PIW, 1998, *The end of cheap oil*; PIW April 6th [#702]

PIW, 1998, *Crying wolf, warnings about oil supply*; PIW April 6th [#723]

PIW, 1998, *Cusiana News*; September 14 [#844]

PIW, 1999, *End of sight for North Sea output growth*; March 22 [#956]

PIW, 1999, *Mexico gains credibility by losing reserves*; March 29 [#975]

PIW, 2001, *US Majors` Hot Streak Ends In Second Quarter, But Profits Remain Robust*; PIW, July 30, 2001 [#1323]

PIW, 2000, *Do we have reasonable expectations from Opec?*; PIW #[1215]

Plassart, P., 1996, *Norvege: derniers puits de petrole avant le desert*; Le Nouvelle Economiste 1050 [#462]

Pochari, T.R., 2003, *The logic of the Arab-Israeli conflict; Oil, alliances, and the dollar*; World Affairs Monthly, Nov. 2003 [#1974]

Polak, A., 1997, *Amoco to sell of one-third of US energy properties*; Fort Worth Star Telegraph June 3 [#627]

Pooley, E., 2000, *Who`s Right About Oil ?*; Time, October 2, 2000 [#1403]

Pope H, 1997, *Oil and geopolitics in the Caucasus*; Wall St. Journ, April 25 [#598]

Popescu, B., 1994, *Algeria*; in Kulke H., (Ed) *Regional petroleum geology of the world*; Bd 21 Borntrager [#518]

Popescu, B., 1994, *Benin*; in Kulke, H. (Ed.) *Regional petroleum geology of the world*; Bd 21 Borntrager [#520]

Popescu, B., 1994, *The Guyanas*; in Kulke H., (Ed) *Regional pertroleum geology of the world*; Bd 21 Borntrager [#519]

Popescu, B., 1995, *Romania's petroleum systems and their remaining potential*; Petroleum Geoscience 1 337-350 [#507]

Popescu, B., 1996, *World oil suppliers at risk*; EEI Newsletter 9 [#426]

Popular Science, 2001, *Are We Really Running Out Of Oil? / What`s Next: Cars*; Popular Science Special Issue, Summer, 2001 [#1335]

Population Today, 1998, *Population, consumption trends call for new environmental policies*; Population Today 26/4 April [#732]

Porter, E., 1995, *Are we running out of oil?* API Discussion Paper 081 [#425 & 479-1]

Poruban, S., Bakhtiari, A.M.S. Emerson, S.A., 2001, *OPEC`s Evolving Role- Analysts discuss OPEC`s role*; Oil / Gas Journal, July 9, 2001 [#1327]

Potter, N., 1997, *Caspian production sharing*; Petroleum review. Feb [#546]

Powell, T.G., *Understanding Australia's Petroleum Resources, Future Production Trends and the Role of the Frontiers*; APPEA Journal, 2001 [#2127]

Power, M., 1992, *Lognormality in observed size distribuiion of oil and gas pools as a consequence of sampling bias*; Int. Assoc. of Mathematical Geology 24/8 [#806]

Power, M., 1992, *The effects of technology and basin specific learning on the discovery rate*; Journ. Canadian Pet. Technology 31/3 [#807]

Power, M. and J.D. Fuller, 1991, *Generating and using basin specific discovery and finding costs forecasts*; Energy Exploration & Exploitation 9/6 [#819]

Powers, B., 2003, *Mexico. Oil's next basket case*; Canadian Energy Viewpoint, September, 2003 [#1957

Pratt, W., 1952, *Toward a philosophy of oil finding*; Amer.Assoc. Petrol Geol., **26**/12 2231-36.

Presse Die, 2000, *OPEC- Monopol: Spaetestens... / Interview with Zittel*; Die Presse, March 10, 2000 [#1506]

Presse Die, 2000, *Physiker: Oelprodukte werden noch teurer*; Die Presse, May 29, 2000 [#1507]

Preusse, A., 1966, *Coalbed methane production – an additional utilization of hard coal deposits*; in Kürsten, M. (Ed.) *World Energy – a changing scene*; E. Schweizerbart'sche Verlagshandlung, Stuttgart ISBN 3-510-65170-7

Priddle, R., 2000, *Opinion: Oil market volatility – what can or should be done?*; June 6, 2000 [#1512]

Prins, J., 1999, *Capital formation and corporate transformation in the energy industry*: conf [#1036

Protti, G.J., 1994, *Canada's upstream petroleum industry*; Journ. Canada. Pet. Tech. May 1994 [#252]

Pulvertaft, C., 1999, *Verdens oliereserver*; GeologiskNyt [#1261]

Pursell, D., 1999, *Depletion: the forgotten factor in the supply demand equation, Gulf of Mexico analysis*: Simmons & Co rept. [#982]

Pursell, D.A. & Eades, C., 2001, *Crude Oil and Natural Gas Price Update – Tough Sledding Near Term … But Optimistic About 2003*, Simmons & Co Internat, Energy Industry Research, March 4, 2001 [#1790]

Purves, L., 2003, *There is no virtue or safety in a war like this*; The Times February.11, 2003 [#2084]

Q

Quinlan, M., 1999, *The oil price factor kicks in*; Pet. Review April [#1102]

R

Radford, T., 1997, *A history of Trinidad oil*; Petrol. Review April [#579]

Radler, M., 2001, *World crude, gas reserves expand as production shrinks*; Oil & Gas Journal, p. 125, December 24, 2001 [#1762]

Raeburn, P., 2001, *Commentary: This Clean Oil Deal Is Already Tainted*; Business Week, May 7, 2001 [#1662]

Ramares, K., 2003, *Review of: The Essence of Oil & Gas Depletion*; Online Journal, Dec.30, 2003[#2308]

Randall, K. & Grey, B., 2003, *New York Times' Thomas Friedman: "No Problem with a war for oil"*; World Socialist Web Site, 15 January 2003 [#2060]

Randol, W., 1995, *No gushers*; Barron's 6/2/95 [#318]

Rasmusen, H.J., 1996, *Bright future for natural gas*; Oil Gas European 2/1/96 [#467]

Rauch, J., 2001, *The New Old Economy: Oil, Computers, and the Reinvention of the Earth*; The Atlantic Monthly, pp. 35–49, January, 2001 [#1623]

Raymond, L.R., 2003, *Challenge, Opportunity and Change*; World Energy Vol.6 No.3 2003 [#2236]

Rechsteiner, R,2003, *Grun gewinnt: die letzte Olkrise und danach*; Orell Fussli, 215p

Read, R.D., 2001, *The North Sea: Oil Production Has Peaked! The GOM Model Must Come To The North Sea*; Simmons & Company International, October 18, 2001 [#1670]

Reed, S., 2000, *Energy*; Business Week, Jan 10 [#1128]

Reed, S., 2001, *A Test For The House Of Saud*; Business Week, November 26, 2001 [#1721]

Reed, S., 2001, *Can The Saudis Step On The Gas?*; Business Week, December 24, 2001 [#1740]

Reed, S., 2002, *Does OPEC Have Sand In Its Eyes?*; Business Week, July 1, 2002 [#1883]

Reed, S. & Bush, 2003, *The Oil Lord Strikes Again*; Business Week, Oct.27, 2003 [#1936]

Reed, S. & Palmeri, C., 2000, *The Alan Greenspan of OPEC*; Business Week [#1252]

Rees-Mogg, W., 1992, *Picnics on Vesuvius: steps towards the millennium*; Sidgwisk & Jackson 396p [#B17]

Rees-Mogg, W., 1999, *Troubled waters for oil*; The Times 30 Aug. [#991]

Rees-Mogg, W., 2001, *Paying the price for the triumph of illusion*; The Times, September 10, 2001 [#1668]

Reese, C., 1998, *We may get another great depression* ; Evansville Press 8/20/98 [#818]

Register Guard, 2001, *Production peak coming*; The Register-Guard, November 17, 2001[#1737]

Reifenberg, A., 1996, *April crude oil futures top $21-a-barrel*; Wall St. Journ March 15 [#422]

Reich, K., 2001, *Gauging the Global Fuel Tank`s Size*; Los Angeles Times, Monday, June 11, 2001 [#1308]

Reilly, D., 2003, *BP May Have Finally Won Forgiveness From Investors*; Wall St. Journ., February 12, 2003 [#2137]

Rempel, H., 2000, *Will the hydrocarbon era finish soon ?*; BGR, Presentation at the DGMK / BGR event "Geosciences in Exploration…"; Hanover, May 23, 2000 [#1608]

Reuters, 2003, *EU says oil could one day be priced in euros*; Brussels, June 16, 2003 [#2196]

Reuters, 2003, *OPEC may discuss trading oil in euros*; Dec.8, 2003 [#2290]

Reynolds, J., 2003, *Global warming to run out of gas*; The Scotsman, October 2, 2003 [#1967]

Rhodes, R. and D. Beller, 1999 *The need for Nuclear Power*; Foreign Affairs 79/1 [1030]

Riahi, M.L., 1998, *Deepwater exploration in the Gulf of Mexico*; Hart's Pet. Eng. Int., April [#737]

Richardson, C., 1996, *New, permanent world oil crisis by 2000* [#515]

Ridley, M., 1998, *Only hot air fuelled the petrol crisis*; Daily Telegraph 17 March [#710]

Rifkin, J., 2002, *The Hydrogen Economy: The creation of the worldwide energy web and the redistribution of power on earth*; Blackwell Publishers 2002 [#1907]

Riley, D. and M. McLaughlin, 2001, *Turning the corner: energy solutions for the 21st Century*; Alternative Energy Inst. 385pp

Rist, C., 1999, *Why We'll Never Run Out Of Oil*; Discover, pp. 80–87, June, 1999 [#1460]

Ritchie, W., 1998, *Fast track to success*; World Oil September [#833]

Ritson, N., 1998, *Maintaining production in the new millennium*; Petroleum Reviw, July [#714]

Riva, J.P., 1997, *U.S.conventional wisdom and natural gas*; M.King Hubbert Center for petroleum supply studies, Newletter 97/3 [#622]

Riva, J.P., 1999, *Is the world's oil barrel half full or half empty?*, M.King Hubbert Center 99/2 [#1052]

Riva, J.P., 2002, *Canadian gas, our ace in the hole?*, M.King Hubbert Center 99/2 [#1018]

Roach, J.W., 1997, *Reserves growth*; Oil & Gas Journ June 2 [#628]

Roadifer, R.E., 1984, *Size distribution of world's largest oil, tar accumulations*; in Mayer R.E. (ed) Exploration for heavy crude oil and natural bitumen; Amer. Assoc. Pet. Geol Studies in Geology #25

Roadifer, R.E., 1986, *Size distribution of world's largest oil, tar accumulations*; Oil & Gas Journ. Feb.26. 1986 93-98 [#17]

Roberts, J., 1992, *Saudi Ambitions*; Petrol. Revue. July 1992 [#39]

Roberts, J., 1995, *Visions & Mirages – the Middle East in a new era*; Mainstream ISBN 1-85158-429-3.

Roberts, J., 1996, *IEA studies Middle East*; Petrol. Review, Feb. 1996 [#416]

Roberts, J., 1999, *Caspian oil and gas flows west but doubts remain*; Petroleum Review February [#915].

Robertson, J., 1989, *Future Wealth*; Cassell 178p.

Robertson Research International Limited, 2000, *Extra 30 years oil and gas supply*; Press Release, June 12, 2000 [#1515]

Robinson, A.B. and Z.W.Robinson, 1997, *Science has spoken: global warming is a myth*; Wall St Journ December 4 [#746]

Robinson, B., 2001, *"The Big Rollover": World oil production decline predictions*; CSIRO Sustainability Network Australia, GOTOBUTTON BM_2_ www.bml.csiro.au/bigrol.htm, Updated October 5, 2001 [#1713]

Robinson, B., 2002, *Australia's growing oil vulnerability*; [#1859]

Robinson, J., 1988, *Yamani – the inside story*; Simon & Schuster 302pp

Robinson, D., 1992, *US oil firms find new lenders*; Petrol. Econ. December 1992 [#152]

Robinson, L. & Cary, P., 2002, *Princely Payments- Saudi royalty, it is claimed, make out like bandits on U.S. deals*; U.S. News & World Report, January 14, 2002 [#1741]

Robinson,, M., 2002, *Venezuela syncrude challenging Mideast oil in U.S.*; Reuters Limited, March, 2002 [#1794]

Robinson, W.C., 1998, *Global population trends*; Resources Spring [#750]

Rocky Mountain Institute, 1998, *Oil, oil everywhere ...*; Newsletter Spring [#733]

Rodenburg, E., *The decline of oil*; World Resources Inst., handbook [#343]

Roger J.V., 1994, *Use and implementation of SPE and WPC petroleum reserve definitions*; Wld.Petrol. Congr., Stavanger [#228]

Rohter, L., 2001, *Energy Crisis in Brazil Brings Dim Lights and Altered Lives*; The New York Times International, Wednesday, June 6, 2001 [#1302]

Roland, K., 1998, *Perceptions of future, often flawed, shape plans and policies*; Oil & Gas Journ February 23 [#701]

Root, D., E.Attenasi, and R.M. Turner, 1987, *Statistics of petroleum exploration in the non-communist world outside the United States and Canada*; U.S.G.S. Circ. 981 [#110]

Root, D., E. Attenasi and R.M. Turner-1989, *Data and assumptions for three possible production schedules for non-Opec countries*; Memorandum U.S.Dept. of Interior 26 September 1989 [#100]

Rosa, R., 2002, *Predicaments of an Economy Running Short on Energy*; University of Evora, January 2002 [#2085]

Rosenberger, C., 1997, *Moscow's multipolar mission*; Inst. Study of Conflict, Ideology & Policy 8/2 [#828]

Ross J., 1998, *Industry Outlook Report*; A.D.Little [#916]

Rossant, J. & P. Burrows, 1994, *Pain at the pump*; Business weekly July 4 [#246]

Rothenberg, M., 1994, *Risk factors in Azerbaijan*; World Oil, Feb 1994 [#202]

Rothschild, E.S., 1992, *The roots of Bush's oil policy*; The Texas Observer February 14 1992 [#33]

Rubin, D, 2001, *OPEC cuts seen as new strain on economies*; Seattle Times [#1227]

Rubin, J. & Buchanan, P., *Why Oil Prices Will Have To Go Higher*; CIBC World Markets Inc. Occasional Report #28, February 2, 2000 [#1401]

Rubin, J., Shenfeld, A. & Buchanan, P., 2000, *The Wall /How High Must Oil Prices Rise?*; CIBC World Markets Inc., Monthly Indicators, October 2000 [#1409]

Rubin, J., 2000, *Running On Empty*; CIBC World Markets, October 2000 [#1411]

Rudel, D., 2000, *Das Zeitalter des Erdoels geht zu Ende*; Yahoo! Schlagzeilen, Monday, June 5, 2000 [#1493]

Rudnitsky, S., 2003, *Turkmenistan's E&P projects achieve good pace*; World Oil January 2003 [#2070]

Ruppert, M.C., 2003, *Beyond Bush II*; From the Wilderness Publications, 2003 [#1976]

Ruppert, M.C., 2002, *The Unseen Conflict*; From the Wilderness Publications, October 18, 2002 [#2008]

Ruppert, M.C., 2003, *Dis-Intergration*; From the Wilderness Publications, February 27, 2003 [#2103]

Ruppert, M.C., 2004, *Crossing the Rubicon – the Decline of the American Empire at the End of the Age of Oil*; 675p New Society Publishers

Russian Information Agency; 1998, *Russia's E&P sector undergoes a slow transition*; World Oil June [#757]

S

Salameh, M.G., 2000, *Can the oil price remain high ?* ; Pet. Review April [#1078]

Salameh, M.G., 2002, *Filling the global energy gap in the 21ˢᵗ Century*; Pet. Review August [#1905]

Salameh, M.G., 2002, *Can Caspian oil challenge Middle East supremacy?*; Pet. Review December [#1905]

Salameh, M.G., 2004, *How realistic are OPEC's Proved Reserves*; Pet. Review August

Salomon, Smith Barney, 2001, *OPEC Analysis*; OPEC Monitor, No.103, July 30, 2001 [#1479]

Salpukas, A., 1999, *An oil outsider revives a cartel*; New York Times Oct 24 [#1132]

Samuelson, R.J., 2001, *The American energy fantasy*; Newsweek [#1228]

Samuelson, R.J., 2001, *The Energy War Within Us*; Newsweek, p 28/29, May 28, 2001 [#1298]

Sandrea, I. & O. al Buraiki, 2002, *Future of deepwater, Middle East hydrocarbon supplies*; Oil & Gas Journ [#1896]

Sampson, A. ,1988, *The seven sisters: the great oil companies and the world they created*; Coronet, London. [B#11]

Sardella, M. & Darley, J., 2003, *A Statement On Global Oil Peak (in the light of war in Iraq)*; COPAD, [#2258]

Sarkis, N., 2000, *Petrole, le troisieme choc*; Le Monde Diplomatique Mars 2000 [#1149]

Scheer, H., 2002, *The solar economy*; Earthscan Publications, 347p

Schindler, J. & Zittel, W., 2000, *Der Paradigmawechsel vom Oel zur Sonne*; Natur und Kultur, 1/1, pp. 48–69, 2000 [#1510]

Schoell, M. and R.M.K. Carlson, 1999, *Diamondoids and oil are not forever*; Nature 6 May [#1049]

Schollnberger, 1996, *A balanced scorecard for petroleum exploration*; Oil Gas European 2/1/96 [#466]

Schollnberger, 199?, *Energievorrate unde mineralische rogrstoffe : wie lange noch*; Osterichische Akad. Wissenshft. Bnd 12.[#834]

Schollnberger, 1996, *Projections of the world's hydrocarbon resources and reserve depletion in the 21 century*; Houston Geol. Soc. Bull November [#861]

Schrempp, J.E., 2000, *Energy for the Future*; Speech at the opening of the World Engineer's Convention, Hannover, Germany, June 19, 2000 [#1538]

Schroeder, W.W., 2002, *Clear Thinking about the Hydrogen Economy*; World Energy, Vol.5 No.3, 2002 [#2135]

Schuyler, J., 1999, *Probabilistic reserves definitions, practices need further refinement*; O&GJ May [#1042]

Schuler, G.H.M., 1991, *A history lesson: oil and munitions are an explosive mix*; Oil & Gas Journ. November 18, 1991[#182]

Schweizer, P., 1994 *Victory: the Reagan administration's secret strategy that hastened the collapse of the Soviet Union*; Atlantic Monthly Press, New York 284p ISBN 0-87113-567-1

Science et Vie, 1995, *Energie: un fantastique tresor cache au fond de mers*; Science et Vie April 1995 [#350]

Sciences et Avenir, 1997, *Petrole: la penurie a partir de 2015?* Aug.[#634]

Sciolino, E., 1998, *It's a sea! It's a lake! No Its a pool of oil*; New York Times 21 June [#761]

Sciolino, E., 2003, *Restless Saudi Youth: Young peoples rebelliousness reflects boredom, bigger social issues*;

The Oregonian, 2003 [#2114]

Scott, R.W., 1995, *Bloody fiasco*; World Oil February 1995 [#345]

Scott, R.W., 1997, *Points to ponder*; World Oil July [#626]

Scott R.W., 1998 *Saudi stuff*; World Oil November [#871]

Seago, D, 2001, *This energy crisis isn't going away*; The News Tribune [#1224]

Seal,   C.,   2002,   The   9/11   evidence   that   may   hang   George   W.   Bush; www.scoop.co.nz/mason/stories/HL0206/S00071 [#1860]

See, M., 1996, *Oil mining field test to start in East Texas*; Oil & Gas Journ November [#541]

Sell, G., 1938, *Statistics of petroleum and allied substances*; The Science of Petroleum v1 1938 [#145]

Selley, R., 2000, *World oil discoveries and production diagram*; Adapted from Selley, Changing Oil: Briefing Paper New Series, 10, January, 2000, The Royal Institute of International Affairs, London, [#1640]

Seskus, T., 2003, *Rocketing Gas Prices May Hinder Oilsands growth*; National Post, February 21, 2003 [#2111]

Shafranik, Y.K., 1993, *Fuel and energy complex of Russia: modern conditions and perspectives*: printed lecture to Univ. of Leiden May 1993

Shah, S., 2003, *Crude: The Story of Oil*; Jun.16 2003 [#2197]

Sheasby, W.C., 2003, *The Coming Panic over the end of oil – Coming to a ballot box near you*; ASPO USA Green Party, September, 2003 [#1954] [#2170]

Shell, 1995, *The evolution of the world's energy system 1860–2060*; Shell [#434]

Shell, 2001, *Energy needs, choices and possibilities*; Shell 69 pp

Shepherd, R.,2000, *All eyes on the deepwater prize*; Offshore engineer, [#1061]

Shilling, A.G., 1993, *The poor get poorer*; Forbes 8/11/93 [#300]

Shirley K., 1992, *Colombia finds wows explorers*; AAPG Explorer, Aug. 1992 [#121]

Shirley, K., 1997, *Russia's potential still unrealized*; AAPG Explorer August [#653]

Shirley, K., 2000, *Caspian ready to face major tests*; AAPG Explorer Feb [#1077]

Shirley, K., 2001, *Angola Hottest of the Hot Offshore*, Explorer [#1192]

Shirley, K., 2000, *Discoveries Are Getting Smaller*; AAPG Explorer, p 6,10, January, 2000 [#1545]

Sierra, J., 1994 *European energy supply security*; Petrole et Tech. 389 [#263]

Silvia, J.E.,2002, *US more dependent than ever on imported oil*; Wachovia Securities [#1817]

Simeoni, C., 1994, *Prospects for gas export pipelines from North Africa and Middle East*; Preprint APS Conf., [#288]

Simienski, A., 2000, *Energy Puzzler on Saudi upstream investment*; Deutsche Bank [#1089]

Simienski, A., 2000, *Energy Puzzler on $25-30 oil price sustainablility*; Deutsche Bank [#1139]

Simon, B., 1990, *Oil project approaches last hurdle*; Financial Times February 2 1990 [#75]

Simon, B., 2003, *Canada is losing ability to fill US natural gas needs*; The New York Times, June 26, 2003 [#2188]

Simon, B., 2003, *Canada Is Losing Ability to Fill US Gas Needs*; New York Times, June 26, 2003 [#2235]

Simmons, M.R., 1994, *It's not like '86*; World Oil February 1994 [#201]

Simmons, M.R., 1995, *Strong market indicators*; World Oil, February 1995 [#330]

Simmons, M.R., 1995, *Despite sloppy prices, fundamentals tighten*; Pet. Eng. Int. September 1995 [#392]

Simmons, M.R., 1995, *1995 global wellhead review and drilling review; a new era for the oil service industry*; Simmons & Co report.[#402]

Simmons, M.R., 1996, *Robust demand strengthens outlook*; World Oil February 1996 [#415]

Simmons, M.R., 1997, *Are our oil markets too tight?* World Oil Feb. [#558]

Simmons, M.R. et al., 1997, *Our hydrocarbon system in uncharted waters*; Simmons & Co [#602]

Simmons, M.R., 1998, *Facts don't support weakening market*; World Oil, February [#671]

Simmons, M.R., 1998, *The impact of Asia's economic crisis on oil and gas now and in the future*; SPE Conf. Perth, Australia 13 October [#857]

Simmons, M.R., 1998, *The perils of predicting supply and demand*; Energy conf. N.Orleans [#909]

Simmons, M.R., 1999, *1998: a year of infamy*; World Oil, Feb.

Simmons, M.R., 1998, *It is not 1986, but could it be worse*; A. Andersen 19th Ann Energy Symp. [#950].

Simmons, M.R., 1999, *Why oil prices need to rise: the illusion of oilfield cost and technology*: Oil, Gas & Energy Quarterly [#1034]

Simmons, M.R., 2000, *The Earth in balance: has energy capacity maxed out*; Proc. Conf. Bridgewater House, London November16

Simmons, M.R., 2001, *Digging Out of Our Energy Mess: The Need For An Energy Marshall Plan*; AAPG, June 5, 2001 [#1330]

Simmons, M.R., 2001, *Investing in Energy: An Exercise Not for the Faint-hearted*; Managed Funds Association Forum 2001, New York City, July 11, 2001 [#1336]

Simmons, M.R., 2000, *An Energy White Paper*; April 2000 [#1393]

Simmons, M.R., 2000, *Energy in the New Economy: The Limits to Growth*; Energy Institute of the

Americas, October 2, 2000 [#1402] and [#1418]

Simmons, M.R., 2002, *2001: In like a lion, out like a lamb*; World Oil.com- Online Magazine, February, 2002 [#1779]

Simmons, M.R., 2002, *Depletion and US Energy Policy*; ASPO Workshop, Uppsala [#1853]

Simmons, M.R., 2002, *The Growing Natural Gas Supply Imbalance: The Role that Public Lands Could Play in the Solution*; Subcommittee on Energy & Mineral Resources of the House of Representatives, July 2002, [#1916]

Simmons, M.R., 2002, *Analyzing the Intergrateds*; July 29 2002 [#2043]

Simmons, M.R., 2002, *It Is Not Hard To Fix Corrupt Investment Research*; October 27, 2002 [#2017]

Simmons, M.R., 2003, *Are Oil & Murphys Law about to meet?*; World Oil, February 2003 [#2131]

Simmons, M.R., 2003, *The Dawning of a new oil and gas era:Is a sea change ahead?*; World Energy Vol.6 No.3 2003 [#2237]

Simon, B., 2001, *Embracing Canadian Energy*; The New York Times, October 24, 2001 [#1726]

Singer, S.F., 2003, *Oil Depletion*; Oil & Gas Journal-letters to the editor, August 11, 2003 [#2244]

Singham, M., 2003, *Cheney's Oil Maps: Can the real reason for war be this crass?*; Counterpunch, July 19, 2003 [#2211]

Sitathan, T., 2001, *Far East: Natural gas use increasing*; World Oil, August, 2001 [#1489]

Skrebowski, C., 1996, *World production and reserves added by wildcatting 1986–1996*; pers. Comm. [#547]

Skrebowski, C., 1998, *Sisters to wed*; Pet. Review September [#822]

Skrebowski, C., 1998, *Is this the third oil shock?* Petroleum Review October [#887]

Skrebowski, C., 1998, *Iraqi production set to triple* Petroleum Review October [#888]

Skrebowski, C., 1999, *Reducing drilling costs -the key to further non-OPEC development*; Pet. Review June [#981]

Skrebowski, C., 1999, *Fossil fuel resources in the 21st Century*; Pet. Review [#992]

Skrebowski, C., 2000, *How much can OPEC actually produce*; Pet. Review April [#1069]

Skrebowski, C., 2000, *A silly game with no winners*; Pet. Review June [#1177]

Skrebowski, C., 2000, *The North Sea – a province heading for decline ?*; Pet. Review September

Skrebowski, C., 2000, *The perils of forecasting*; Pet. Review Sept

Skrebowski, C., 2000, *Discovery still lags production despite good 1999 results* Pet. Review September

Skrebowski, C., 2000, *Asking the wrong question about oil reserves*; Petroleum Review, September 2000 [#1364]

Skrebowski, C., 2000, *The perils of forecasting*; Petroleum Review, September 2000 [#1365]

Skrebowski, C., 2004, *Oil field mega- projects 2004*; Petroleum Review Jan. 2004 [#2273]

Slessor, M. and J. King, 2002, *Not by money alone*; Jon Carpenter 160 pp

Sludge, C.D., 2002, *Discussing oil with the beeb*; Scoop: Sludge Report #141 September 24., 2002 [#1994]

Smil, V., 2003, *Energy at the Crossroads: Global Perspectives and Uncertainties*; MIT Press, 2003 [#2304]

Smith, D., 2000, *Will the oil sheiks kill off Goldilocks ?*; The Sunday Times, September 10, 2000 [#1389]

Smith, M.R., 2002, *Oil energy security in the Asia-Pacific Region*; Petromin April [#1869]

Smith, M.R., 2002, *Energy security in Europe*; Pet. Review August [#1871]

Smith, M.R., 2001, *Environmentalists can relax. Oil supplies will decline sooner than most geoscientists are prepared to accept*; PESGB, Newsletter August–September

Smith, M.R., 2002, *Energy security in Europe*; Petroleum Review August [#1871]

Smith, M.R., 2002, *US oil supply vulnerability growing*; Offshore August [#1918]

Snow, N., 1995, *Nazar's dismissal does not mean change in Saudi oil policy*; Pet. Eng. Int. September 1995 [#393]

Socci, T., 1999, *Surface temperature changes and biospheric responses in the northern hemisphere during the last 1000 years*; US Global Change Serminar 17 May [#980]

Solomon, C., 1993. *The hunt for oil*; Wall Street Journ. August 25 1993 [#169]

Solomon, Smith Barney, 2000, *Opec monitor* [#1152]

Sonnen, Zeitung, 2000, *Bis zum letzten tropfen*; 4/00

Sørenson, J-E., 2000, *Trade and the environment in the WTO*; Clean Energy 2000,[#1108]

Soros, G., 1995, *Soros on Soros: staying ahead of the curve*; John Wiley & Sons, ISBN 0-471-12014-6. 326p.

Soros, G., 1998, *The Crisis of global capitalism*; Public Affairs 243p

Spencer, J.E. & Rauzi, S.L. (Arizona Geological Survey), 2001, *Crude Oil Supply and Demand: Long-Term Trends*; Arizona Geology, Vol.31, No.4, Winter, 2001 [#1753]

S.P.E. & W.P.C., 1999, *Petroleum resource definitions*; J. Petr. Geol [#1100]

Specter, M., 1998, *Population implosion worries a greying Europe*; New York Times 10/7/98 [#804]

Speight, R., 1998, *Hydrocarbon Man: a threatened subspecies*; Shell International [#930]

Spiegel, Der, 2000, *Auswege aus dem Energienotstand*; Der Spiegel, No. 23, 2000 [#1513]

Spring, C., 2001, *When the lights go on: understanding energy*; Emerald Resource Solutions 120pp

Spring, C., 2003, *Review of Dr. C.J. Campbells The Essence of Oil and Gas Depletion*; September 9, 2003 [#2260]

Standard, Der, 2000, *Biomasse als Ausweg bei Energiemangel*; Der Standard, May 27, 2000 [#1508]

Stabler, F.R., 1998, *The pump will never run dry*; Futurist November [#943]

Srodes, J., 1998, *No oil painting*; Spectator 20 Auust [#789]

Srodes, J., 1998, *Here we go again*; Barron's 19 October [#902]

Stabler, F.R., 1998, *The pump will never run dry*; Futurist November [#853]

Stanley, B., 2002, *Oil supply seen set to fall*; Washington Times 28 May [#1856]

Stanley, B., 2002, *Oil experts draw fire on warning*; The Detroit News [#1863]

Stanton, W., *The Human Population Explosion*; Wildlife Trusts / Somerset, ? [#1343]

Stark, P.H., 2003, *Global Petroleum Industry Perspective 2003- A look at the Year Ahead*; IHS Energy, Denver, 2003 [#2067]

Starke, L. (ed.), 2000, *The environmental trends that are shaping our future*; Vital Signs 2000 [#1283]

Starling, P., 1997, *Oil market outlook*; Petroleum review February [#544]

Starobin, P. & Crock, S., 2001, *Putin's Russia* and *From Evil Empire To Strategic Ally*; Business Week, November 12, 2001 [#1710]

Statoil, 1996, *Oil price scenarios towards 2020*; Statoil Report [#478-2]

Steakley, L., 2002, *In the Pipeline*; Wired, January, 2002 [#1748]

Steeg, H., 1994, *World energy outlook to the year 2010*; 7th Int. Oil & Gas Seminar, Paris [#251]

Steeg, H., 1997, *De nouveaux chocs pétroliers nous menacent ... mais notre insouciance est totale*; Le Temps Stratigique, Geneve [# 642]

Steele, E.J., 2003, *You Get What You Need*; Conspiracy Pen Pal, May 14, 2003 [#2163]

Stelzer, Irwin, 2002, *Time to end our reliance on Saudi oil*; The Sunday Times, August, 2002 [#1898]

Stone, R., 2002, *Caspian Ecology Teeters On the Brink*; Science, Vol. 295, January 18, 2002 [#1775]

Stoneley, R., 1993, *Book Review of "The Golden Century of Oil 1950-2050"*; Cretaceous Research, p. 250, No. 14, 1993 [#1453]

Stosur, G.J. & R.W.Luhning, 1994, *Worldwide EOR activity in the low price environment*; Pet.Eng.Int. August 1994 p.46.[#267]

Stott, P., 2001, *Hot Air + Flawed Science = Dangerous Emissions*; The Wall Street Journal, Monday, April 2, 2001 [#1629]

Stott, P., 2002, *Cold Comfort for "Global Warming"*; The Wall Street Journal, March 25, 2002 [#1807]

Stow, A.R., 1996, *Consequences of US oil dependence*; Energy 3 November [#649]

Stow, D., 1999, *Into the abyss*; Guardian 29 September [#1014]

Stratfor, 2000, *Russia tries to influence OPEC's next decision* [#1145]

Stratfor, 2000, *US forces realign in Persian Gulf*; Stratfor [#1152]

Stuermlinger, D., 2000, *Die grosse Explosion der Benzinpreise kommt erst*; Hamburger Abendblatt, May 29, 2000 [#1509]

Sub-Committee B; 2001, *Meeting: Security Of Energy Supplies In The EU*; European Union Committee, House Of Lords, Monday, October 29, 2001 [#1735]

Suddeutsche Zeitung, 2001, *Kassandra-rufe in Ol*; Suddeutsche Zeitung Online, Wissenschaft [#1244]

Sullivan, A., 1992, *Iraq isn't expected to resume oil exports soon despite progress in talks with UN*; Wall St. Journ., Janary 13 1992

Suncor Energy Inc., 2003, *Suncor Energy provides details of its next phase of growth- Company plans to boost oil production to 330,000 barrels per day*; April 3, 2003 [#2159]

Sunday Times, 1993, *King Fahd*; Sunday Times 10/10/93 [#310]

Sunday Times, 1994, *Uncle Sam gets heavy with Fahd*; Sunday Times 9/10/94 [#269]

Sunday Herald, 2003, *Five Israelis were seen filming as jet liners ploughed into the twin Towers ...*, Sunday Herald, November 2, 2003 [#1961]

Sustainable Population Australia Inc., 2001, *A Dangerous Gamble- The assessment of Australia's carrying capacity based on use of fossil fuel*; Sustainable Population Australia Inc.(Occasional paper), June 2001 [#1317]

Sydney Morning Herald, 1997, *Report claims not enough oil to meet growth in Asian demand* 12.2.97 [# 550]

Syncrude, 1998, *Mining black gold*; World Oil June 1998 [#758]

Szulc, E., 1998, *Will be run out of gas?* ; Evansville Courier 19.7.98 [#795]

T

Taber, J.J., F.D.Martin & R.S.Seright, 1997, *EOR screening criteria revisited*; SPE Reservoir Engineering

August [# 639]

Takin, M., 1972, *Iranian geology and continental drift in the Middle East*; Nature 235 [#303]

Takin, M., 1988, *Energy cycles: can they be avoided* ? Opec Bulletin Oct. 1988 [#305]

Taki,n M., 1989, *The high cost of misunderstanding Opec*; 14th Congr. Wld Energy Conf.[#306]

Takin, M., 1993, *OPEC, Japan and the Middle East*; OPEC Bull. 4/2 (March–April 1993) 17-34.[#158]

Takin, M., 1994, *How much gas is there in the Middle East*; Pet. Review July 1944 [#307]

Takin, M., 1996, *Many new ventures in the Middle East focus on old oil, gas fields*; Oil & Gas Journ. May 27 [#457]

Takin, M., 1996, *Future oil and gas: can Iran deliver?* World Oil Nov. [#516]

Takin, M., 1998, *Prospects after the Riyadh agreement*; Petroleum Review, July [#713]

Takin, M., 1998, *Oil*; Metals and Minerals Ann. Review [#457]

Takin, M., 2001, *OPEC-consumer cooperation needed to ensure adequate future oil supply*; Oil & Gas Journal, December 3, 2001 [#1760]

Takin, M., 2003, *OPEC and the expansion of Iran's Oil capacity*; MEES, November 24, 2003 [#2300]

Takin, M., 2003, *Sanctions, Trade Controls & Political Risk 03-Case Study: Iran*; CGES November 5 & 6, 2003 [#2301]

Tanner, J., 1992, *Iran, in need of revenue, lifts oil output*; Wall St. Journ 19/10/92 [#115]

Tanzer, A. & Ghosh, C., 2001, *Insatiable- China and the rest of developing Asia are driving the world oil market*; Forbes, July 23, 2001 [#1321]

Tavernise, S. & Brauer, B., 2001, *Russia Becoming an Oil Ally*; The New York Times, Oct 19, 2001 [#1724]

Tavernise, S., 2001, *Exxon Says Way Is Cleared for Development in Russia*; The New York Times, October 30, 2001 [#1728]

Taylor, A., 1999, *Oil Forever*; Fortune November 22 [1021]

Taylor, B.G.S., 1997, *Towards 2020: a study to assess the potential oil and gas production from the UK offshore*; Petroleum Review, February [#543]

Taylor, E., 1996, *Future Chill*; Tyler Morning Telegraph 12 Feb [#468]

Tchuruk, S., 1994, *Les relations entre les societes et les gouvernements dans l'industrie mondiale du petrole et du gas*; Petrole et Technique 389, July 1994, [#266]

Teitelbaum, R.S.,1995, *Your last big play in oil*; Fortune 30 October 1995 [#382]

Thackery, F., 1998, *Opec future rests on Asian tigers' return to growth*; Petroleum Review July [#772]

Thews, K., 2000, *Allmahlich lauft die Erde leer*, Die Stern 4.5.2000 [#1369]

Third World Traveller, 2000, *Global oil reserves alarmingly overstated*; 1999 Censored Foreign Policies News Stories [#1180]

Thomas, M., 2000, *Azerbaijan back on track*; Hart's E&P, August 2000 [#1369]

Thomas, S. & Zalbowitz M, ?, *Fuel Cells – Green Power*; Los Alamos National Laboratory [#1276]

Thomasson, M.R., 2000, *Petroleum geology: is there a future* ? AAPG Explorer May [#1172]

Thompson, A.S., 1997, *Nuclear energy proves neither clean nor safe*; The Register Guard Opinion December 2 [#709]

Thomson, B., 2001, *The Coming Oil Crash*; GOTOBUTTON BM_7_ www.altogetherenergy.com, 2001 [#1654]

Thurow, L.,1996, *The future of capitalism*; Nicholas Brealey, 385pp

Tibaijuka, A.K., *Human Rights, Social Ecology and The Family*; The World Family Policy Forum-USA, 15–17 July, 2002 [#2224]

Tibbs, II., 1997, *Global change: a context for transport planning*; Ecostructure Report [#670]

Tickell, C., 1996, *Climate & history*; Oxford today 8/2 [#428]

Tillerson, R.W., 2003, *The challenges ahead*; speech at Institute of Petroleums' IP 2003 [#2102]

Tinker, S.W. & Kim, E M., 2001, *Research: Energy Policy for the Future*; Geotimes, June 2001 [#1331]

Times, The, 1994, *State of terror*; Editorial, April 29 1994.[#219]

Tissot, B. and D.H. Welte, 1978, *Petroleum formation and occurrence*; Springer verlag, New York [#B12]

Toal, 1999, *The big picture*; Oil & Gas Investor 1/99 [#866]

Toman, M. & J. Darmstadter, 1998, *Is oil running oil?* Science 282 2 October [#882]

Townes, H.L., 1993, *The hydrocarbon era, world population growth and oil use – a continuing geological challenge*; Amer. Assoc. Petrol. Geol. 77/5, 723–730.[#157]

Tomitate, T., 1994, *World oil perspectives and outlook for supply-demand in Asia-Pacific region*; World Petrol. Congr. Topic 16. [#232]

Toniatti, G., 2003, *Brazil furthers its leadership in deepwater E&P*; World Oil, Nov.ember 2003 [#1971]

Total, 1994, *Gas in FSU and European security of Supply*; Company Presentation [#215]

Trainer, T., 1997, *The death of the oil economy*; Earth Island Journal Spring [#743]

Transport Energy Strategy Committee, 2003 *Interim Report*; Govt. W. Australia June 2003 [#2207]

Traynor, J.J., Sieminski, A. & Cook, C., 2000, *OPEC – Shortage? What Shortage?*; Deutsche Bank, Global Oil & Gas, November 9, 2000 [#1751]

Trout, R., 2001, *Global insight from oil expert- Interview with C J Campbell*; Hobbs News – Sun, August, 2001 [#1665]

Tugendhat, C. and A. Hamilton, 1968, *Oil -the biggest business*; Eyre Methuen. ISBN 0-413-33290-X.

Tull, S., 1997, *Habitat of oil and gas in the Former Soviet Union*; Geoscientist 7/1 [#549]

Tyler, N. and N.J. Banta, 1989, *Oil and gas resources remaining in the Permian Basin: targeted for additional hydrocarbon recovery*; Bureau of Economic Geology, Univ. of Texas. Circular 89-4. 20p.[#55]

Tyson, R. & Spencer S., 2001, *Production Profiles vary, but US independent all post big 1Qs*; Platts Oilgram News, Vol.79, No.80, 2001[#1288]

U

Udall, S.L., 1974, *The energy balloon*; 288pp McGraw Hill

Udall, S.L., 1980, *America's Trip in the Energy Swamp*; The Washington Post, January 6, 1980 [#1395]

Udall, S.L., 1998, *The Myths of August*; Rutgers University Press 397pp

UKOOA & Nat. History Museum, *Britain's offshore oil and gas*; 56pp

Ulmishek, G.F., R.R. Charpentier and C.C. Barton, 1993, *The global oil system: the relationship between oil generation, loss, half-life and the world crude oil resource: discussion*; Amer. Assoc. Petrol. Geol. 77/5 896-899.

Ulmishek, G.F- and C.D.Masters, 1993, *Oil, gas resources estimated in the former Soviet Union*; Oil & Gas Journ, December 13. 59-62 [#249]

United Financial Group, 2003, *Russia Oil & Gas Comment- Oil & Gas in the 21st Century*; Deutsche Bank December 16, 2003 [#2296]

Upstream Feature, 2000, *Plugging Into Untapped Reserves – Different Articles*; Upstream Feature, p 22-27, October 27, 2000 [#1446]

USGS, 1995, *1995 national assessment of United States oil and gas resources*; USGS circular 1118. [#319]

USGS, 1996, *Ranking of oil basins* Open File [#638]

USGS, 2000, *USGS reassesses potential world petroleum reserves*; Press release [#1135]

USGS, 2000, *World petroleum assessment – Description and results*; DDS-60

USGS, 2000, *USGS Petroleum Assessment 2000 – Description and Results*; ? [#1362]

USGS, 2002, *2002 petroleum resource assessment of the NPRA, Alaska*; Fact Sheet 045-02 [#1890]

V

Vahrenholt, F., 1999, *Global market potential for renewable energies*; Shell Solar [#963-F1]

Valdmanis, R., 2003, *Oil – The worlds largest addiction*; Reuters, July 17, 2003 [#2215]

ven Koevering and N.J. Sell, 1986, *Energy – a conceptual approach*; Prentice Hall 271pp

Vargo, J.M., *et al.*, 1993, *Gulf of Suez plays center on tilted fault blocks*; World Oil May 1993 [#154]

Verhovek, H., 2001, *Alaska's Arctic dilemma*; The Register-Guard West, Sunday, March 18, 2001 [#1590]

Verlegger, P.K., *Economics error: illusion that money can replace physically exhausted resources*; Dieoff.com [#1266]

Veziriglu, T.N., 2000, *Hydrogen energy system*; Clean Energy 2000 [#1115]

Vlierboom, F.W., B. Collini and J.E. Zumberge, 1986, *The occurrence of petroleum in sedimentary rocks of the meteor impact crater of Lake Siljan, Sweden*; 12th Europ. Assoc. Organic. Geochem. International meeting, Julich Sept 1985; Org. Geochem **10** 153–161

Vogel, T.T., 1998, *Oil's weakness sparks a sinking feeling in Venezuela*; Wall St Journ. [#802]

Volsett, J., A.Abrahamsen & K. Lindbo, 1994, *An enhanced resource classification – a tool for decisive exploration*; preprint, Wld Petrol, Congr. Stavanger [#227]

Von Rainer, H., 2000, *Entzug der Oldroge*; DieWoohe [#1242]

von Sternberg, B., 2000, *Energy independence is no longer en vogue*; Star Tribune, Minneapolis, Minnesota, Monday, April 3, 2000 [#1643]

W

Wall Street Journal, 2000, *Iraq may have filled tankers, but oil is still not flowing*; Wall Street Journal [#1254]

Wall Street Journal, 2001, *No Policy, No Win*; The Wall Street Journal, Thursday, August 9, 2001 [#1348]

Wall Street Journal, 2001, *World Demand for Oil To Rise a Million Barrels A Day Despite Slump*; The Wall Street Journal, ?, [#1349]

Wall Street Journal, 2002, *Opec warns of possible shortfall*; The Wall Street Journal, December 20, 2002 [#2087]

Wall Street Journal, 2003, *In Russia, BP Bets on a Partner That Burned It*; February 27, 2003 [#2136]

Wallop, M., 1998, *Unless stopped, the global warming movement could cause economic disaster*; World Oil Sept [#8

Walsh, J.H., 2003, *The coming three peaks in the world energy system and their relationship to climate change and poverty*; jhwalsh@ca.inter.net [#2123]

Wardt, J.P. de, 1996, *Operational realities in the '90s*; World Oil February 1996 [#413]

Warin, O.N., 1997, *Mineral exploration into the millennium*; Geoscientist 7/3 [#562]

Warman, H.R., 1972, *The future of oil*; Geographical Journ. 138/3 287-297

Warman, H.R., 1973, *The future availability of oil*; proc. Conf. World Energy Supplies, by Financial Times, London 11p [#1]

Warren, S. & McKay, P.A., 2004, *Methods for Citing oil Reserves Prove Unrefined*; Wall St. Journ., January 14, 2004 [#2317]

Washington Post, 2003, *Perilous natural gas shortage*; June 10, 2003 [#2177]

Washington Post, 2003, *Greenspan on the natural gas problem*; June 12, 2003 [#2178]

Watkins, T.H., 2000, *The young are riding high, so their fall may be hard*; Oregonian January 2 [#1057]

Watt, K.E.F., 2003, *A test of six hypotheses concerning the nature of the forces that govern historical processes*; University of California, 2003 [#2231]

Watt, K.E.F., 2003, *Nations as Evolving Thermodynamic Systems*; 2003 [#2292]

Wattenberg, R.A., 1994, *Oil production trends in the CIS*; World Oil June

Welch, D., 2002, *Fuel Cells Are Getting Hotter*; Business Week, April 15, 2002 [#1808]

Weiner, J., 1990, *The next hundred years*; Bantam 312pp

Weiner, T., 2003, *Pemex: Mexico's oil giant*; The Oregonian, January 22, 2003 [#2083]

Weiss, M., 1998, *Today's technology – can it take the heat?* Pet. Eng. Int. September [#845]

Wellmer, F-W, 1994, *Rerserven und reservenlebensdauer von energierohstoffen*; Energie Dialog, July 1994 [#274]

Wellmer, F-W., 1997, *Factors useful for predicting future mineral commodity supply trends*; Geol. Rundsch 86 311-321 [# 735]

Wetuski, J., 2002, *Oil Will Dominate 2030's Energy Picture, But Supply Vulnerability Grows*; Oil and Gas Online, December 2002 [#2047]

Weyant, J.P. and D.M.Kline, 1982, *Opec and the oil glut: outlook for oil export revenues in the 1980s and 1990s*; Opec Review Winter 1982 334–365 [#93]

Whalen, J., 2001, *Vote for governor of Russian zone excites a war of words over oil*; WSJ [#1226]

Wheelwright, T., 1991, *Oil and world politics*; Left Book Club 220pp

Whipple, D., 1998, *Expert: world faces new oil crisis in next 12 years*; Caspar Star Tribune 6/11/98 [#850]

Whipple, D., 2002, *Blue Planet: The end is near*; Science & Technology, September 27, 2002 [#1995]

Widdershoven, C., King, G.H.H., Pabst, M. & Fuller, D., 2001, *Africa: Major Discoveries abound*; World Oil, August, 2001 [#1486]

Will, G., 1996, *Heavy oil – the jewel in Western Canada's oil play*; Petroleum Review November [#513]

Will, G., 1997, *Promise from Canada's east coast*; Petrol. Rev. April [#580]

Williams, B., 2000, *Radical changes loom for work place, work flows for future energy industry*; IPE [#1281]

Williams, B., 2003, *Progress in IOR technology, economics deemed critical to staving off worlds oil production peak*; Oil & Gas Journal, August 4, 2003 [#2245]

Williams, B., 2003, *Debate grows over US gas supply crisis as harbinger of global gas production peak*; Oil & Gas Journal, July 21, 2003

Williams, B., 2003, *Heavy hydrocarbons playing key role in peak-oil debate, future energy supply*; Oil & Gas Journal, July 28, 2003 [#2253]

Williams, C.J., 2000 *Norway looks beyond oil boom*; Los Angeles Times February 26 [#1082]

Williams, P., 1998, *Half full or half empty*; Oil & Gas Investor 18/2 February [#680]

Williams, P., 1997 *Orinoco*; Oil & Gas Investor 17/11 [#686]

Williamson, P.E., 1997, *Potential seen in areas off Northwest, Southeast Australia*; Oil & Gas Journal, November 24 1997 [#2118]

Willingham, B.J., 1994, *Energy shortage looms again*; The Nat. Times Mag, October/November [#261]

Wind Energy Weekly, 1996, *New oil price shock seen looming as early as 2000*; Wind Energy Weekly 15 684 12 February [#895]

Winter, F. de, 1976, *Solar energy happenings in other countries*; US Energy Res. & Dev. Admin. [#221]

Winter, F. de, (?) *Economics and policy aspects of solar energy*; Altas Corp., Santa Cruz [#230]

Winter, F. de, 1998, *The start of much higher petroleum prices: some scenarios*; draft [#687]

Wittrup, A.F., 2001, *Vind, sol, og vand*; Ingenioraen [#1239]

Witwer, D., 2000, *Environmental benefits of heat pumps*; Clean Energy 2000 [#1120]

Wohl, M., 2000, *New collectors for solar air conditioning and architectural use*; Clean Energy 2000 [#1122]

Wonacott, P., 2002, *GM's China unit enters partnership*; Wall St Journ. May 24 [#1867]

Wood, Mackenzie, 1997, *UKCS technical reserves*; N.Sea Report 295 [#824]

Wood, Mackenzie, 1999, *Maturing gracefully*, UK Upstream Report 323 [#1174]

Wood, Mackenzie, 1999, *Something new in the pipeline?*, UK Upstream Report 316 [#1062]

Wood, Mackenzie, 2001, *UK Oil production forecast 2001*; UK Upstream Report, Feb.2001, Number 333 [#1188]

Wood, Mackenzie, 2003, *Latin America Upstream Insights: Venezuela-The Cost of recovery*; July 2003 [#2209]

World Energy Assessment, 2000, *Summary description* ; Clean Energy 2000 [#1107]

World Energy Assessment, 2000, *Energy and the challenge of sustainability*; United Nations Development Programme, 2000 [#1921]

World Energy Council, 2000, *Energy for Tomorrow's World* 175pp

World Energy Outlook, 2003, *Exploration and development*; chapter 4-oil, 2003 [#1973]

World Energy Outlook, 2002, *Methodology for Projecting Oil Production*; chapter 3- The energy market outlook, 2002 [#1996]

World Oil, 2001, *Western Europe: Exploration targets becoming smaller* and *FSU / Eastern Europe: A roaring comeback continues*; World Oil, August, 2001 [#1487]

World Oil, 2001, *OPEC retains control*; World Oil, August, 2001 [#1749]

World Oil, 2002, *Gulf of Mexico Deepwater Map* [#1815]

World Oil, 2002, *Norweigan oil minister predicts another half-century of production*; August 2002 [#2095]

World Resources Institute, 1996, *World Resources- A Guide To The Global Environments – The Urban Environment*; World Resources 1996–1997, Oxford University Press, 1996 [#1334]

World Tribune. Com, 2002, *Saudi royal family "in complete panic" during December riots*; Worldtribune.com , Thursday, January 3, 2002 [#1743]

Worldwatch, 1999, *New century to be marked by growing threats, opportunities*; [#927]

Wright, D., 1994, *Small investors can strike tax bonanza in oil industry*; Sunday Times 27/2/94 [#218]

Wright, S J., 2001, *Information Letter: To BPA Customers and Citizens of the Pacific Northwest*; Bonneville Power Administration, Oregon, March 29, 2001 [#1628]

Wright, T.R, 1991, *Whither oil prices*; World Oil May [984]

Wright, T.R., 1999, *Oil prices could rise still more*; World Oil Oct.[#1018]

Wright, T.R., 2001, *A guyser of oil, at Gladys City*; World Oil [#1223]

X

Xiaojie, X., 1997, *China reaches crossroads for strategic choices*; World Oil April [#568]

Y

Yaffe, B., 2003, *Guess what? We are running out of gas and oil*; Vancouver Sun, September 12, 2003 [#1947]

Yamani, A.Z., 1995, *Oil's global role -the outlook to 2005*; MEES 38/33 [#355]

Yamani, A.Z, 1997, *Containment is too risky*; Petrol Review May [#594]

Yamani, H.A.Z., 1997, *Sheik's son urges Saudi Arabia to flood world oil markets*; Reuter Info. Service 25/8/97 [#687]

Yasunaga, Y., 1994, *Japan's new energy policy and strategy*; Preprint APS Conf. [#279]

Yeomans, M., 2004, *Oil – Anatomy of an Industry* – ISBN 1-56584-885-3

Yergin, D.,1991, *The Prize: the epic quest for oil, money and power*; Simon & Schuster, New York, 877p [#B2].

Yergin, D, 1988, *Energy security in the 1990s*; Petroleum Review November 1988 [#88]

Yergin, D, and J. Stanislaw 1989, *The commanding heights*; Simon & Schuster p. 457

Yergin, D, 2002, *US energy security lies in diversity of supply*; Oil&Gas Journ, 2 September 2002 [#1904]

Yergin, D., 2002, *A crude view of the crisis in Iraq*; Washington Post December 6, 2002 [#2074]

Yergin, D., 2002, *The Oil Equation;* The Register Guard, December 15, 2002 [#2088]

Yergin, D., 2003, *Gulf oil- how important is it anyway?;* Financial Times, March 22, 2003 [#2134] [#2143]

Yihe, X., 2003, *China finds More oil, gas in Tibet;* Wall St. Journ., June 24, 2003 [#2198]

Yonas, G., 1998, *Fusion and the Z pinch;* Sci Amer. August [#779]

Youngquist, W., 1997, *Geodestinies: the inevitable control of earth resources over nations and individuals;* Nat. Book Co., Portland 500p.

Youngquist, W., 1998, *Spending our great inheritance – then what ?* Geotimes 43 7 24-27 [#791]

Youngquist, W., 1998, *Shale Oil – the elusive energy;* Hubbert Center Newsletter 98/4 1-7

Youngquist, W., 2000, *Alternative Energy Resources;* Pre-Conference Papers- Kansas Geological Survey Open-File Report 2000-51, October 2000 [#1421]

Youngquist, W., 1999, *Comments about "The Coming Oil Crisis";* Journal of Geoscience Education, v. 47, p.295, 1999 [#1432]

Youngquist, W., 1998, *Shale Oil- The Elusive Energy;* Hubbert Center Newsletter #98 / 4 October, 1998 [#1463]

Youngquist, W., 1999, *The Post-Petroleum Paradigm- and Population;* Reprinted from Population and Environment: A Journal of Interdisciplinary Studies, Vol.20, No.4, 1999 [#1706]

Youngquist, W., 2002, *Alternative Energy Sources;* Kansas Geological Survey [#1850]

Youngquist, W., 2002, *The Hydrogen Economy-Book Review;* Natural Resources Research, Vol. 11, No. 4, December 2002 [#2150]

Youngquist, W. & Duncan, R.C., 2003, *North American Natural Gas: Data Show Supply Problems;* Natural Resources Research, Vol.12 No.4, December 2003 [#2315]

Yozwiak, S., 1997, *Solar electric plant rising near Flagstaff;* Arizona Republic May 10 [#706]

Z

Zach, B.A., 1998, *Canada's petroleum industry* [#898]

Zagar, J.J., 1999, *World oil depletion: the crisis on our threshold;* IADC Conf. [#958]

Zakaria, F., 1996, *Thank goodness for a villain;* Newsweek September 18. [#494]

Zeldin, T., 1994, *An intimate history of humanity;* Minerva 488pp

Zellner, W., 1994, *Steamed about natural gas;* Business Week 10/10/94 [#273]

Ziegler, P.A., 1993, *Plate-moving mechanisms: their relative importance;* Journ. Geol. Soc.,150/5

Ziegler, W.H., 1990, *World oil and gas reserves;* unpublished report to EU [#147]

Zischka, A., 1933, *La guerre secrète pour le pétrole;* Payot, Paris

Zittel, W., 2000, *Fossile Energiereserven (nur Erdol und Erdgas) und mogliche Versorgungpasse aus Europaische Perspektive;* LB-Systemtechnik

Zittel, W., 2001, *Analysis of the UK oil production, contribution to ASPO;* own paper [#1218]

Zittel, W. & Schindler J., 2001, *Natural Gas- Assessment of the Term Supply Situation in Europe;* L-B-Systemtechnik Gmbh, May 28, 2001 [#1294]

Zittel, W., 2000, *Comment on EIA Presentation: Long Term world oil supply;* 2000 [#1386]

UNITS, ABBREVIATIONS, CONVERSION FACTORS AND GLOSSARY

UNITS

S.I. (Système International) units are the recommended system replacing the Imperial system, which is however still entrenched in the oil industry.

ABBREVIATIONS

b = barrel (US) = 159 litres = 42 US gals (Also B)
boe = barrel of oil equivalent (see below for conversion)
cf = cubic feet = 0.0283 m³
Ma = million years
toe = ton of oil equivalent
Giant oilfield = >500 Mb initial reserves

k (kilo) =	10^3	= thousand	(Also M)
M (mega) =	10^6	= million	(Also MM and m)
G (giga) = 10^9	= billion (US)	(Also MMM, B, bn)	
T (tera) = 10^{12}	= trillion (US)		

CONVERSION FACTORS

sq. mile	sq. km	acre	hectare
1	2.59	640	259
0.386	1	247	100

barrel	cu. m	tonne
1	0.159	0.136
6.29	1	0.855
7.33	1.17	1

(Depends on oil gravity: above applies to Arabian Light 33.5° API)

Gas
1 m³ = 35.3 cf
1 cf = 0.0283 m³

Energy equivalents
1000 m³ gas = 1 toe = 7.168 boe (in calorific terms)
10 000 cf gas = 1 boe (in value @$1.8/kcf, $18/b)
1 Tcf = 0.1 Gboe
1 b/d = 50 t/a (32° API oil)

A SHORT GLOSSARY OF OIL AND TECHNICAL TERMS USED IN THIS BOOK

Acreage: the amount of land held under concession.

Alidade: an instrument for measuring distances, used with a planetable for surveying.

Anticline: an arch-like geological structure at the top of a fold in which petroleum tends to accumulate.

Appraisal (or delineation) wells: boreholes drilled to confirm the extent of a discovery.

Basement: non-prospective rocks beneath a sedimentary basin: often granite or metamorphic rocks.

Basin: a geological depression filled with sedimentary rocks, often containing petroleum.

Backwardation: a term in trading when the price of oil for future delivery is below present delivery.

Block: a commonly rectangular licence area or concession.

Cap-rock: an impermeable rock the prevents the escape of oil from a trap (also seal).

Choke: a restriction placed in a well to reduce flow. Also "choking-back" production.

Condensate: a liquid hydrocarbon that condenses from gas at surface conditions

Contango: the converse of backwardation.

Dip: the angle at which strata are inclined from the horizontal.

Downstream: the marketing and refining end of the oil business.

Evaporites: salt and similar rocks deposited by the evaporation of sea-water, commonly forming effective seals preventing the escape of oil from a reservour.

Farm-in: the converse of farm-out

Farm-out: a transfer of ownership in a petroleum concession to a new company, which will commonly drill one well at its expense to earn a 50% undivided interest. Oil rights change hands frequently, and many oilfields have been found by the incoming company.

Fault: a geological fracture such that the rocks on one side are offset relative to those on the other. Petroleum often accumulates against faults.

Flaring: the practice of setting light to unwanted gas, for which no current market exists.

Flysch: a coarse grained sediment resulting from the rapid erosion of a mountain chain, as found in front of the Alps.

Independent company: a general term for oil exploration and production companies other than the large integrated companies.

Licence: used largely in Europe, meaning concession.

Migration: the process by which petroleum moves through the earth from the rock (source-rock) in which it was generated.

Nappe: a large horizontally displaced mass of rocks, typical of Alpine geology.

NGL: Natural Gas Liquid, a liquid obtained from gas naturally or by processing.

Oil window: the depth at which oil is generated in a particular area.

Operator: a member of a group of companies in a joint venture charged with managing the operations. The other companies are known as Non-operators.

Palaeontology: the study of fossils, which are used to date and correlate geological formations.

Plane-table: a table mounted on a tripod, used in the traditional method of geological survey.

Play: an oil play is a group of prospects having similar geological characteristics.

Pore-throat: the communication between pores in a reservoir rock.

Prospect: an undrilled geological structure which is perceived to have the characteristics necessary for it to contain oil or gas.

Round: an offering at a particular time of concessions by governments.

Salt-plug: salt, being mobile and less dense than other rocks, may flow and penetrate the overlying sequence of rocks. Oil and gas sometimes accumulate around salt-plugs, also known as diapirs.

Sedimentary environment: the physical environment in which the rocks were laid down, such as shallow-water, deltaic, fluvial etc.

Seismic surveys: a survey procedure that involves the release of energy at the surface and recording the reflections from buried structures. It is now the most widely used exploration tool. The results are called seismic data. The surveys are conducted on a grid, each being known as a seismic-line. The surveys are said to be "shot".

Semi-submersible rig: a rig mounted on two submerged pontoons that float in calmer waters beneath the wave base.

Signature bonus: a cash payment on signature of a concession.

Stratigraphy: the study of the sequence and nature of sedimentary rocks.

Stratigraphic trap (strat-trap): a trap for oil or gas formed by stratigraphic factors alone, such as where a sandstone occurs as a lenticular body.

Stripper well: a well producing a few barrels a day from a nearly depleted oilfield, mainly in the United States.

Sub-sea completion: a technique of installing the wellhead and production equipment on the seabed as opposed to on a platform above the waves.

Tank-farm: an area containing large oil tanks.

Thrust: a low-angled fault.

Transcurrent fault: a fault on which movement has been mainly horizontal.

Upstream: the exploration and production end of the oil business.

Wildcat: an exploration borehole, which if successful finds a new oil or gas field.

Appendix II
WORLD OIL ENDOWMENT

WORLD — Unit:Gb (billion barrels) — **REGULAR CONVENTIONAL OIL PRODUCTION** — To 2100 — 2004 — 17/03/2005

Country	Region Code	Present kb/d 2004	Present Gb/a 2004	Past Total	Past 5yr Trend	Reported Reserves World Oil	Reported Reserves O&GJ	Deductions Static	Deductions Other	% Rept'd	Future	Past & Future	NEW FIELDS	ALL FUTURE	TOTAL	Revised Depletion Rate	Revised Depletion Mid-Point	Peak Disc	Peak Prod
Saudi Arabia	A	8750	3.19	100	2%	259.3	259.4	-39.64	0	180%	144	245	15.4	160	260	2.0%	2013	1948	2013
Russia	B	8850	3.26	130	8%	65.4	60.0	-6.27	-37	80%	75	205	14.6	90	220	3.5%	1996	1960	1987
US-48	C	3559	1.29	173	-4%	22.7	21.9	0	-9	90%	24	198	2.3	27	200	4.6%	1971	1930	1971
Iran	A	3940	1.43	57	1%	105.0	125.8	-20.74	0	180%	70	127	7.9	78	135	1.8%	2007	1961	1974
Iraq	A	2070	0.75	29	-4%	115.0	115.0	-8.72	0	220%	52	81	19.1	71	100	1.0%	2025	1928	2025
Kuwait	A	2050	0.74	32	3%	96.5	99.0	0	0	180%	55	87	2.7	58	90	1.3%	2020	1938	2015
Venezuela	D	1878	0.68	47	-5%	52.5	77.2	0	-30	225%	34	82	5.7	40	87	3.2%	1999	1941	1970
Abu Dhabi	A	1955	0.71	19	1%	65.0	92.2	-10.54	0	200%	46	65	4.5	51	70	1.4%	2025	1964	2011
China	B	3494	1.27	31	2%	15.5	18.3	-2.52	0	75%	24	55	4.6	29	60	4.2%	2003	1959	2003
Mexico	D	3410	1.24	32	3%	14.6	14.6	0	0	70%	21	53	2.7	24	56	5.0%	2000	1977	2004
Libya	E	1550	0.56	24	2%	30.5	39.0	0	0	190%	21	44	5.5	26	50	2.1%	2005	1961	1970
Nigeria	E	2350	0.85	24	3%	33.0	35.3	0	-5.6	175%	20	44	3.8	24	48	3.1%	2004	1967	2004
Kazakhstan	B	996	0.35	7	9%	-	9.0	-0.68	0	30%	30	37	8.3	38	45	0.9%	2036	2000	2030
Norway	F	2940	1.07	19	-2%	9.4	8.5	0	0	75%	11.3	30	3.2	14	33	6.9%	2002	1979	2001
UK	F	1830	0.66	21	-5%	4.3	4.5	0	0	60%	7.5	29	2.4	10	31	6.3%	1997	1974	1999
Indonesia	G	973	0.35	21	-5%	5.5	4.7	-0.35	0	60%	7.8	28	2.1	10	31	3.4%	1992	1945	1977
Algeria	E	1205	0.40	13	10%	14.0	11.8	0	0	95%	12.4	25	2.6	15	28	2.8%	2006	1956	1978
Canada	C	1100	0.01	20	0%	5.0	178.9	-0.40	-174.8	3100%	5.8	25	0.7	6.4	26	5.9%	1987	1958	1973
Azerbaijan	B	298	0.10	8.3	2%	-	7.0	-0.22	0	60%	11.7	20	2.5	14.2	22.5	0.8%	2014	1871	2009
N.Zone	A	597	0.21	7.1	-1%	4.75	5.0	-2.41	0	95%	5.3	12.3	1.7	6.9	14	3.0%	2004	1951	2003
Argentina	D	80	0.25	8.5	-2%	2.7	2.7	0	0	75%	3.6	12.1	0.9	4.5	13	5.5%	1996	1960	1998
Oman	H	767	0.28	7.6	-18%	5.7	5.5	-1.32	0	110%	5.0	12.6	0.4	5.4	13	4.9%	2001	1962	2001
Egypt	E	712	0.25	9.2	-2%	2.3	3.7	0	0	120%	3.1	12.3	0.7	3.8	13	6.4%	1995	1965	1995
India	G	685	0.25	6.1	1%	4.0	5.4	-0.25	0	120%	4.5	10.6	0.9	5.4	11.5	4.4%	2003	1974	2004
Qatar	H	782	0.28	7.3	3%	27.4	15.2	-0.78	-25	375%	4.1	11.4	0.1	4.2	11.5	6.4%	1998	1940	2004
Malaysia	G	855	0.31	5.9	5%	3.1	3.0	-0.88	0	75%	4.0	9.9	0.6	4.6	10.5	6.4%	2002	1973	2004
Colombia	D	530	0.19	5.9	-5%	1.5	1.5	-0.39	0	50%	3.1	9.0	1.0	4.1	10	4.8%	1999	1992	1999
Australia	G	430	0.15	6.1	-8%	4.0	1.5	0	-1	80%	1.9	8.0	2.0	3.9	10	3.9%	1999	1967	2000
Angola	E	480	0.17	5.0	-7%	8.9	5.4	-2.02	-9.9	140%	3.9	8.8	0.7	4.5	9.5	3.7%	2004	1971	1998
Ecuador	D	518	0.18	3.6	6%	5.0	4.6	-0.34	0	110%	4.2	7.8	0.2	4.4	8	4.1%	2006	1969	2004
Brasil	D	400	0.14	5.0	2%	9.8	8.5	0	0	425%	2.0	7.0	0.0	2.0	7	6.7%	1995	1975	1986
Romania	B	102	0.03	5.8	-3%	0.5	1.0	-0.12	-12	110%	0.9	6.7	0.3	1.2	7	3.1%	1970	1857	1976
Syria	H	504	0.18	4.2	-1%	2.4	2.5	-2.19	0	100%	2.5	6.7	0.3	2.8	7	6.1%	2000	1966	1995
Turkmenistan	B	216	0.07	3.1	10%	-	0.5	-0.33	0	50%	1.1	4.2	1.3	2.4	5.5	3.2%	1998	1964	1973
Dubai	H	350	0.12	4.0	5%	1.2	4.0	-2.23	0	500%	0.8	4.8	0.2	1.0	5	11.3%	1991	1970	1991
Trinidad	D	130	0.04	3.3	2%	0.8	1.0	-0.14	0	85%	1.2	4.5	0.3	1.5	4.7	3.2%	1985	959	1978

Country																			
Trinidad	D	130	0.04	3.3	2%	0.8	1.0	-0.14	0	85%	1.2	4.5	0.3	1.5	4.7	3.2%	1978	959	1985
Brunei	G	190	0.06	3.1	1%	1.1	1.4	-0.85	0	110%	1.2	4.4	0.1	1.4	4.5	4.8%	1978	1929	1989
Gabon	E	235	0.08	3.0	-6%	2.3	2.5	-0.74	0	170%	1.5	4.5	0.0	1.5	4.5	5.5%	1996	1985	1997
Ukraine	B	80	0.02	2.7	2%	-	0.4	-0.09	0	40%	1.0	3.7	0.3	1.3	4	2.2%	1970	1962	1984
Denmark	F	393	0.14	1.6	5%	1.3	1.3	0	0	120%	1.3	2.9	0.6	1.9	3.5	7.1%	2004	1971	2005
Yemen	H	350	0.12	1.5	0%	2.9	4.0	-1.52	-0.6	340%	1.2	3.0	0.5	1.6	3.5	7.3%	1999	1978	2003
Peru	D	81	0.02	2.4	-3%	0.9	1.0	0	0	110%	0.9	3.3	0.2	1.1	3.5	2.7%	1983	1861	1988
Vietnam	G	340	0.12	1.1	2%	2.3	0.6	-0.76	0	30%	2.0	3.1	0.4	2.4	3.5	5.0%	2005	1975	2009
Uzbekistan	B	134	0.04	1.2	-3%	-	0.6	-0.26	0	50%	1.2	2.3	0.4	1.6	2.7	3.0%	1998	1992	2008
Congo	E	240	0.08	1.7	-2%	1.4	1.5	-0.87	-0.5	210%	0.7	2.4	0.3	1.1	2.7	7.7%	2001	1984	2000
Germany	F	69	0.02	2.0	2%	0.3	0.4	0	0	120%	0.3	2.3	0.2	0.5	2.5	4.6%	1966	1952	1977
Italy	F	115	0.04	1.0	5%	0.5	0.6	-0.07	-0.3	80%	0.8	1.7	0.3	1.0	2	3.9%	2004	1981	2005
Sudan	E	287	0.10	0.4	5%	6.3	0.6	-0.27	0	50%	1.1	1.6	0.6	1.8	2.2	5.6%	2005	1980	2009
Tunisia	E	70	0.02	1.2	11%	0.5	0.3	-0.15	0	75%	0.4	1.7	0.3	0.8	2	3.3%	1981	1971	1998
Chad	E	247	0.09	0.1	-2%	-	-	0	0	-	1.2	1.3	0.7	1.9	2	4.6%	2008	1977	2014
Thailand	G	154	0.05	0.5	-	0.5	0.6	-0.11	0	80%	0.7	1.3	0.3	1.1	1.6	5.0%	2005	1981	2008
Cameroon	E	70	0.02	1.1	8%	-	0.4	-0.70	0	110%	0.4	1.4	0.1	0.4	1.5	5.1%	1986	1977	1994
Bolivia	D	35	0.01	0.4	-5%	0.5	0.4	-0.03	0	80%	0.6	1.0	0.3	0.8	1.2	1.6%	2010	1966	2016
Bahrain	H	34	0.01	1.0	5%	-	0.1	-0.03	0	60%	0.2	1.2	0.0	0.2	1.2	5.0%	1970	1932	1977
Netherlands	F	44	0.01	0.8568	2%	0.1	0.3	-0.03	0	40%	0.3	1.1	0.1	0.3	1.2	4.5%	1989	1980	1991
Turkey	H	42	0.01	0.8574	10%	0.3	0.3	-0.03	0	150%	0.2	1.1	0.1	0.3	1.2	4.3%	1991	1969	1992
Croatia	B	19	0.01	0.5055	-5%	0.1	0.1	-0.03	0	24%	0.3	0.8	0.1	0.5	1	1.4%	1988	1950	2003
Hungary	B	22	0.01	0.6937	-3%	0.1	0.1	0	0	70%	0.1	0.8	0.2	0.3	1	2.6%	1987	1964	1987
France	B	23	0.01	0.7415	-5%	0.2	0.1	0	0	95%	0.2	0.9	0.1	0.2	0.95	3.9%	1988	1958	1987
Pakistan	F	62	0.02	0.5047	-4%	0.3	0.3	-0.02	0	100%	0.3	0.8	0.4	0.4	0.9	5.4%	1992	1983	2001
Austria	G	18	0.01	0.7906	15%	0.1	0.1	-0.00	0	70%	0.1	0.9	0.0	0.1	0.9	5.6%	1955	1947	1970
Papua	F	46	0.01	0.3754	-1%	0.1	0.2	-0.03	0	70%	0.3	0.7	0.1	0.5	0.85	3.4%	1993	1987	2007
Sharjah	G	48	0.01	0.504	-8%	0.3	1.5	-0.04	0	1000%	0.2	0.7	0.1	0.3	0.8	5.6%	1998	1960	1998
Albania	B	6	0.01	0.5407	-1%	-	0.2	-0.04	0	85%	0.2	0.7	0.1	0.3	0.8	0.9%	1983	1928	1986
Chile	D	10	0.01	0.4283	9%	0.2	0.2	-0.02	0	400%	0.0	0.5	0.0	0.1	0.5	4.8%	1982	1960	1979
REGIONS																			
ME.Gulf	A	19962	7.1	245	1%	641	696	-82	0	187%	373	618	51	424	669	1.6%	1974	1948	2016
Eurasia	B	14307	5.2	191	6%	82	97	-11	-37	67%	146	337	33	179	370	2.8%	1987	1964	2003
N.America	C	4659	1.7	193	-2%	28	201	0	-184	667%	30	223	3	33	226	4.9%	1972	1930	1973
L.America	D	7672	2.8	110	-1%	88	112	-1	-43	158%	71	181	11	82	192	3.3%	1998	1977	1999
Africa	E	7438	2.7	83	2%	101	100	-5	-16	154%	65	148	15	81	163	3.0%	2004	1961	2004
Europe	F	5431	2.0	47	0%	16	16	-16	0	72%	22	68	7	29	75	6.5%	2000	1974	2000
East	G	3735	1.4	44	-1%	88	18	-3	-1	77%	23	67	7	29	74	4.4%	1999	1967	1999
ME.Other	H	2877	1.1	27	0%	40	33	-8	-25	235%	14	41	2	16	43	6.2%	1998	1965	1999
Other		647	0.2	4	7%	-	0	0	0	93%	1	5	5	6	10	3.7%	2007	1956	2009
Unforeseen											20		8	28	28				
Non-MEast		46768	16.8	682	1%	375	577	-29	-305	147%	392	1073	108	499	1181	3.3%	2004	1956	1996
WORLD		66130	24.1	944	1%	1016	1273	-111	-305	167%	764	1708	142	906	1850	2.6%	2004	1964	2003

Appendix III

PRODUCTION PROFILES

MIDDLE EAST GULF

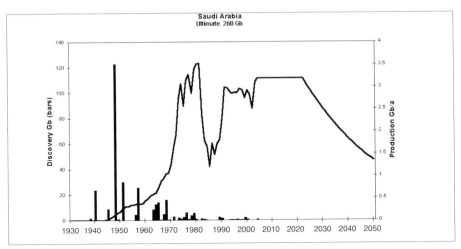

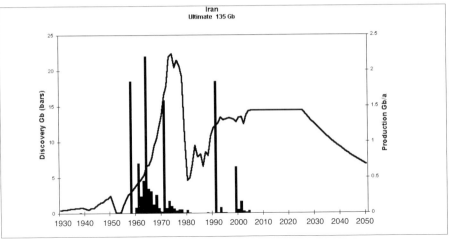

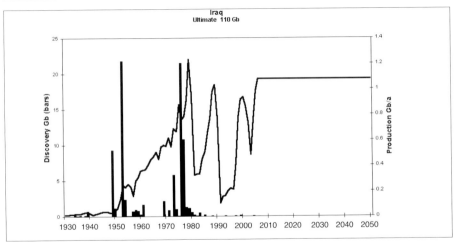

MIDDLE EAST GULF

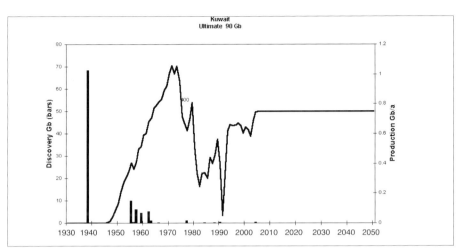

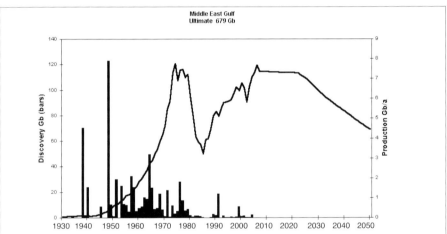

EURASIA

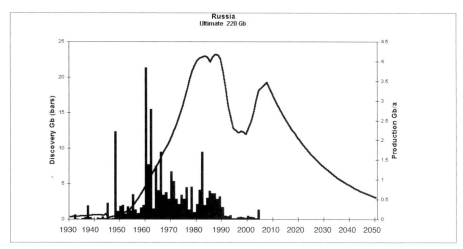

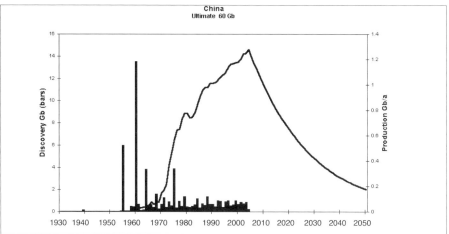

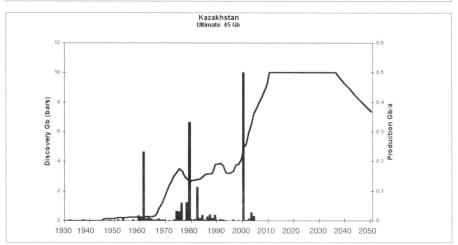

EURASIA

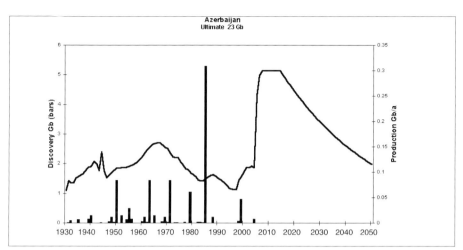

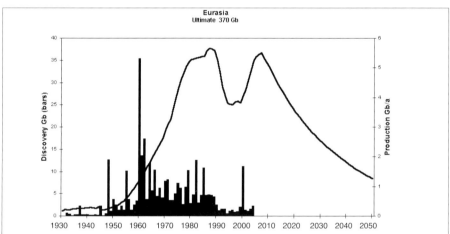

NORTH AMERICA

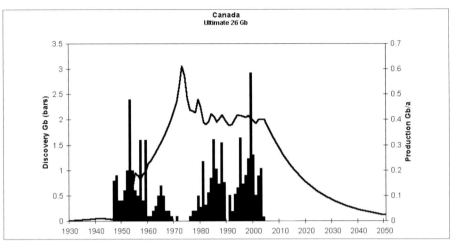

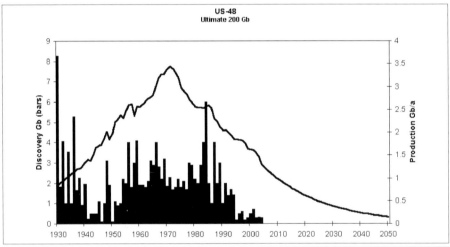

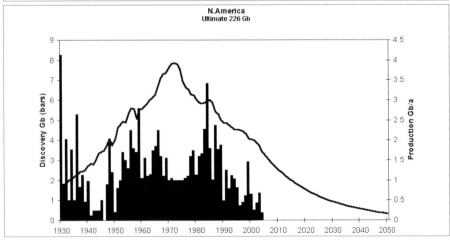

LATIN AMERICA

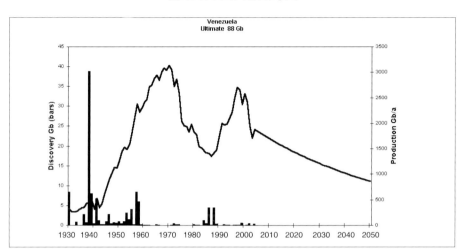

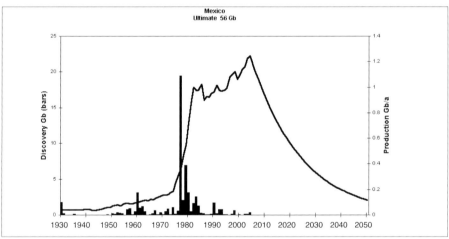

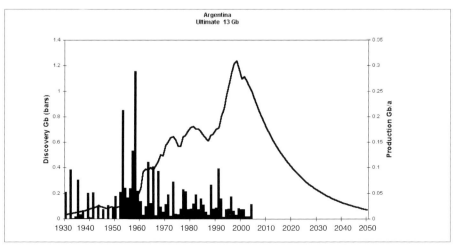

LATIN AMERICA

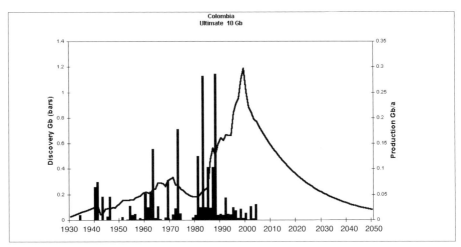

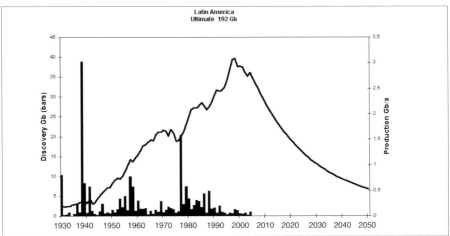

AFRICA

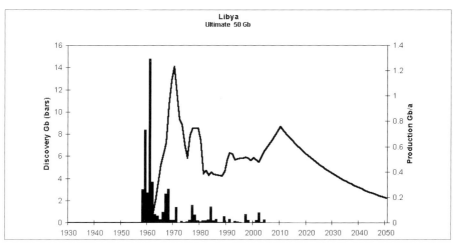

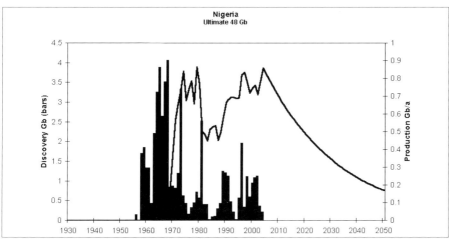

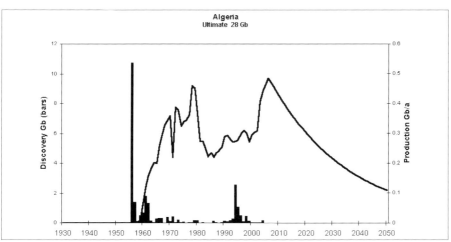

AFRICA

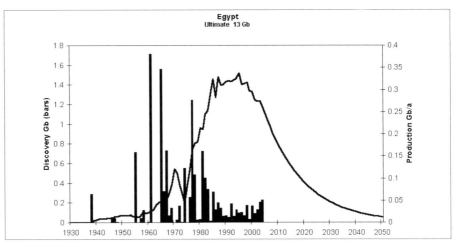

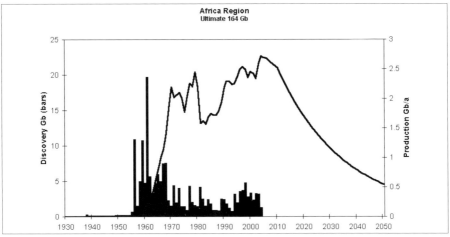

EUROPE

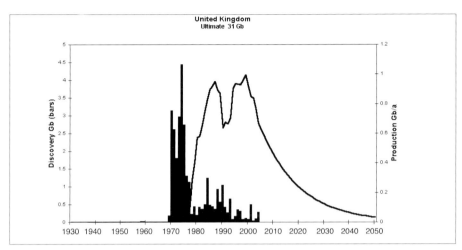

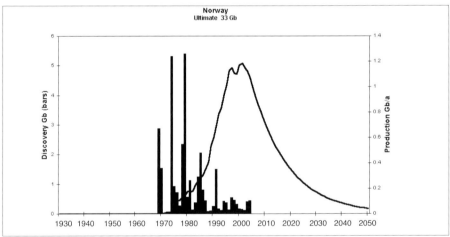

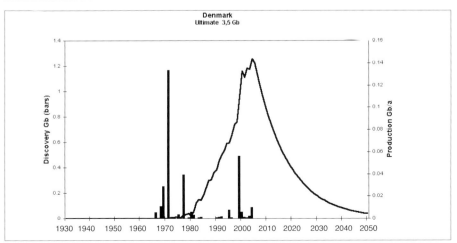

EUROPE

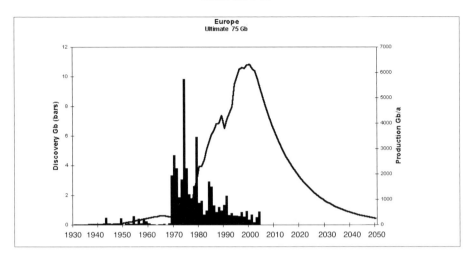

EAST

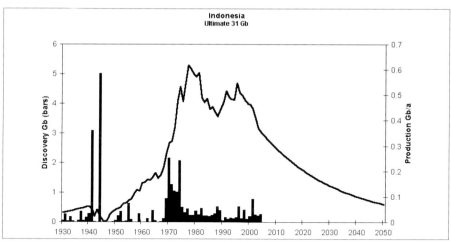

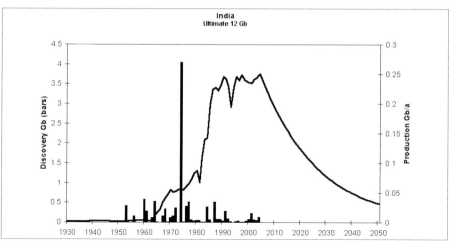

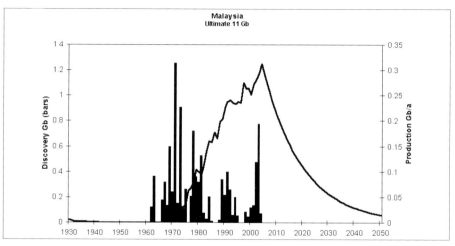

EAST

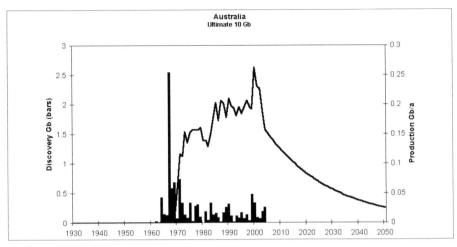

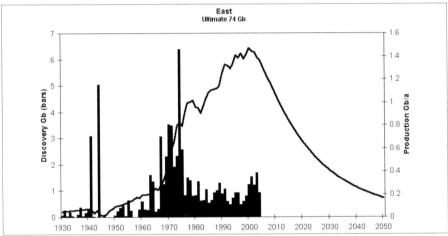

MIDDLE EAST (OTHER)

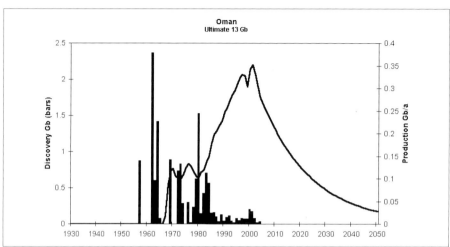

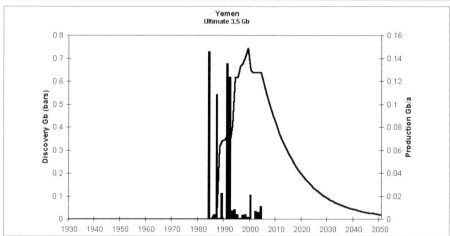

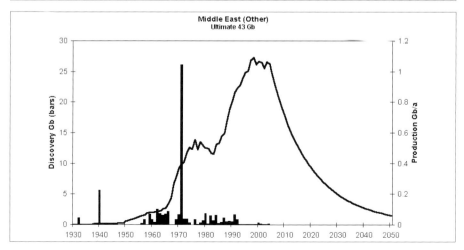

DEEPWATER

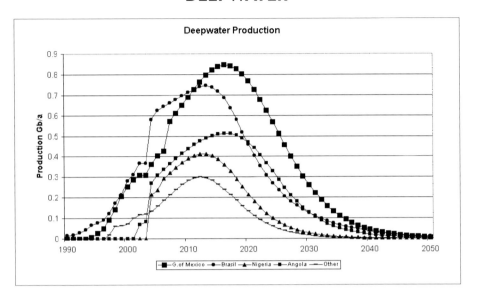

INDEX